Powering the Future

Constructing Small Modular Reactor Power Stations for Sustainable Energy

Richard Skiba

AFTER MIDNIGHT
PUBLISHING

Skiba, Richard (author)

Powering the Future: Constructing Small Modular Reactor Power Stations for Sustainable Energy

ISBN 978-1-7637696-7-0 (Paperback) 978-1-7637696-8-7 (eBook) 978-1-7637696-9-4 (Hardcover)

Non-fiction

Contents

Introduction 1

Chapter 1: Introduction to Small Modular Reactors (SMRs) 4

Understanding Small Modular Reactors 4

Evolution of Nuclear Technology 51

Benefits of SMRs in Modern Power Generation.. 72

SMRs vs. Traditional Nuclear Reactors 74

Key Players and Developers in the SMR Industry 75

Chapter 2: Feasibility and Site Selection for SMR Power Stations 90

Key Factors in SMR Site Selection 90

Environmental Impact and Regulatory Requirements 117

Assessing Geographical and Geological Considerations 137

Engaging Local Communities and Stakeholders 153

Case Studies: Site Selection in Different Regions 159

Chapter 3: Design and Engineering of SMR Power Stations 164

Core Design Principles of SMRs 164

Safety Features and Redundancies in SMR Design 180

Modular Construction Techniques for SMRs 192

Engineering Challenges and Solutions 204

Key Components and System Integrations 210

Chapter 4: Regulatory and Licensing Framework for SMRs 213

Understanding International and National Nuclear Regulations 213

Licensing Process for SMR Construction and Operation 227

Compliance with Safety and Environmental Standards 234

Case Study: Licensing an SMR in Various Jurisdictions 242

Future Trends in Nuclear Regulatory Policies 264

Chapter 5: Construction Process of SMR-Based Power Stations 265

Planning and Timeline for SMR Construction 265

Modular Assembly and On-Site Integration... 317

Quality Control in SMR Construction... 333

Workforce Training and Skill Requirements 340

Overcoming Construction Challenges 354

Chapter 6: Infrastructure and Systems for SMR Power Stations 357

Cooling Systems and Heat Management 357

Electrical Distribution and Grid Connectivity 369

Safety and Monitoring Systems 373

Waste Management Infrastructure 375

Emergency Response and Contingency Planning 380

Chapter 7: Economic and Financial Aspects of SMR Projects 383

Cost-Benefit Analysis of SMR Implementation 383

Funding and Investment Models for SMR Projects 389

Projected Return on Investment and Break-Even Points 399

Financial Risks and Mitigation Strategies 405

Long-Term Economic Impact of SMR Power Stations 406

Chapter 8: Environmental Impact and Sustainability of SMRs 409

Carbon Footprint and Emissions Reductions 409

Waste Management and Fuel Recycling Options 412

SMRs in the Context of Sustainable Energy Go 413

SMRs and Integration with Renewable Energy 415

References 423

Index 440

Introduction

Small Modular Reactors (SMRs) represent a significant advancement in nuclear technology, designed to produce less power than traditional large nuclear reactors. Conventional reactors typically generate over 1,000 megawatts of electricity (MWe), whereas SMRs usually produce less than 300 MWe per module [1, 2]. This reduced power output, combined with their modular design, enhances their versatility, allowing them to be deployed in a variety of applications and locations that may not be suitable for larger reactors [3].

The modular design of SMRs is one of their defining features, facilitating standardized manufacturing and potentially accelerating construction timelines. This modularity allows for the combination of multiple units to scale power production according to demand, thus providing flexibility in energy supply [2]. Furthermore, the compact size of SMRs necessitates less land and infrastructure, making them particularly suitable for remote or space-constrained areas [4]. Enhanced safety is another critical characteristic of SMRs; they incorporate passive safety systems that utilize natural processes such as gravity and convection to cool the reactor, which significantly reduces the risk of overheating and catastrophic failures [5]. This safety feature is particularly advantageous for deployment in isolated regions, islands, or small communities, where traditional reactors may not be feasible [6].

Economically, SMRs present several advantages. Their smaller initial investment costs make them more accessible, especially for countries or companies that may lack the capital for large-scale nuclear projects [7]. The modular construction and factory assembly processes streamline the building phase, potentially lowering overall costs and reducing construction timelines [7]. Additionally, the smaller physical footprint and enhanced safety measures of SMRs can mitigate their environmental and ecological impacts compared to conventional reactors [8].

In terms of applications, SMRs are poised to play a crucial role in the transition to clean energy. They can complement renewable energy sources in regions where renewables are limited by geographical or intermittency challenges [4]. Various countries and companies are exploring the potential of SMRs as a sustainable, low-carbon energy solution, emphasizing their capability to meet future energy demands while supporting global climate goals [8]. Their adaptability for diverse applications, including electricity generation, desalination, district heating, and industrial heat supply, further underscores their potential as a versatile energy source [9].

Interest in Small Modular Reactors (SMRs) has been expanding across various sectors, especially among those focused on energy, environmental sustainability, and economic development. SMRs represent a cleaner, low-carbon energy option, appealing to environmental advocates and policymakers who are working to reduce greenhouse gas emissions. These groups may find SMRs valuable because they offer an alternative to fossil fuels and can complement renewable sources like wind and solar. Learning about SMRs enables policymakers to design informed energy strategies and incentives, making SMRs a vital topic of interest for those shaping the future of sustainable energy.

For energy professionals and utility companies, SMRs offer flexibility and cost-effectiveness in delivering reliable power. Their ability to fill in when renewable sources fall short makes SMRs critical for maintaining grid stability and meeting fluctuating energy demands. Additionally, industries in remote locations, like mining operations and oil refineries, often require independent, dependable power sources. Understanding SMRs provides decision-makers in these sectors with potential solutions to power their operations without relying on distant grids.

Academics and researchers in fields such as nuclear physics, environmental science, and sustainable development also have an interest in SMRs. They study SMRs to explore technical capabilities, limitations, safety protocols, and their potential impact on global energy systems. This research not only advances academic knowledge but also supports the development of sustainable technologies and energy resilience solutions.

Several groups are particularly interested in the practical aspects of constructing SMRs. Nuclear engineers, mechanical and electrical engineers, and technicians in nuclear fields have a direct role in SMR design and construction. Knowledge of SMR technology allows these professionals to enhance their skills and align their expertise with emerging energy solutions. Construction and infrastructure firms, too, find SMRs appealing as they demand specialized skills in nuclear construction. For these firms, SMRs present a new growth area in clean energy infrastructure, especially as the demand for sustainable projects rises.

Government and military organizations often have strategic interests in SMR technology. Military bases in remote areas, for example, need reliable, independent power, and SMRs offer a potential solution. Defence and government agencies interested in energy security may find it worthwhile to understand SMR construction and implementation to support their missions and achieve energy independence. Similarly, energy and utility companies view SMRs as a potential long-term investment for diversifying their energy offerings, especially as public demand for clean energy options grows. SMRs could help these companies meet low-carbon energy goals and offer flexible power solutions.

Entrepreneurs and investors are also drawn to the potential of SMRs in the clean energy market. As interest in sustainable energy expands, understanding SMR construction and its regulatory landscape allows these stakeholders to make informed investment decisions in SMR technology and related projects.

This book provides content for learning about SMRs and their construction and suits those looking to contribute to the clean, reliable energy landscape. With the world increasingly focused on sustainable practices, SMRs offer a unique and scalable solution that bridges the gap between traditional energy sources and renewable options.

Chapter 1
Introduction to Small Modular Reactors (SMRs)

Understanding Small Modular Reactors

Overview

Small Modular Reactors (SMRs) represent a significant evolution in nuclear technology, characterized by their reduced size and power output compared to traditional nuclear reactors. Typically, SMRs generate up to 300 megawatts of electricity (MWe), contrasting with conventional reactors that can exceed 1,000 MWe. This smaller scale allows for modular construction, where components are manufactured in factories and then transported to the installation site, enhancing adaptability and deployment flexibility across various settings [2, 10, 11]. The modularity of SMRs not only facilitates faster construction timelines but also enables them to be deployed in locations with limited space, making them suitable for remote areas or regions with constrained infrastructure [11, 12].

The design of SMRs incorporates several features that enhance their appeal. Their modular nature allows for the assembly of smaller, standardized units, which can be combined to meet specific power generation needs. This flexibility is particularly advantageous in meeting diverse energy demands efficiently [11, 13]. Furthermore, SMRs typically require less supporting infrastructure than larger reactors, which contributes to their lower environmental impact and makes them a more sustainable option for nuclear energy [12, 13]. The compact design also allows for their integration with renewable energy sources, providing stable power in areas where grid access is limited [14].

Safety is a paramount concern in nuclear energy, and SMRs are designed with advanced safety features, including passive safety systems that utilize natural processes such as gravity and convection to maintain core cooling. This design minimizes the risk of overheating and enhances safety even in emergency situations where external power may be unavailable [11, 14]. The inherent safety characteristics of SMRs, coupled with their ability to operate independently or alongside renewable energy sources, position them as a viable solution for providing reliable, low-carbon power to small communities and industrial sites [15, 16].

The economic advantages of SMRs are also noteworthy. Their smaller size translates to lower initial capital investments, making nuclear energy more accessible to smaller markets and developing regions [17]. Additionally, the modular construction approach reduces construction times compared to traditional nuclear plants, further enhancing their economic viability [12, 13]. As countries increasingly seek clean energy solutions, SMRs are gaining traction as a complementary technology to renewable energy sources, capable of providing consistent power while supporting global sustainability goals [14, 15].

SMR Operation Principles

Small Modular Reactors (SMRs) represent a significant advancement in nuclear technology, utilizing nuclear fission to generate heat, which is subsequently converted into electricity. Similar to traditional nuclear reactors, SMRs operate by splitting uranium or other fissile materials in a controlled chain reaction. This process releases substantial amounts of heat, which is captured by a coolant, typically water, that circulates around the nuclear fuel. However, SMRs are distinct from conventional reactors in several key aspects, including their design, size, and safety features, which enhance their operational flexibility and efficiency across various applications.

The defining characteristic of SMRs is their electrical output, which is generally less than 300 MWe. This smaller scale allows for a variety of innovative designs that can be constructed more quickly and with lower financial risk compared to larger reactors. For instance, the integral pressurized water reactor (iPWR) is currently a leading design for near-term licensing and deployment, showcasing the modular construction and passive safety features that are hallmarks of SMR technology [11, 18, 19]. These reactors are particularly advantageous in locations lacking robust transmission or distribution infrastructure, as they can provide localized power generation for large population centres and specific industrial applications [3, 20, 21].

At the core of Small Modular Reactors, nuclear fuel is typically composed of uranium dioxide pellets that are encased in long fuel rods, arranged in assemblies within a reactor vessel. When fission occurs, uranium atoms split, releasing neutrons and energy. These neutrons can then collide with other uranium atoms, triggering further fission reactions, thereby establishing a self-sustaining chain reaction. This process is critical for the operation of nuclear reactors, including SMRs, as it generates the heat necessary for electricity production [5].

The heat generated from fission is absorbed by a coolant circulating around the fuel rods. This coolant, often water or a liquid metal such as sodium, plays a vital role in transferring heat away from the reactor core to prevent overheating and maintain safe operational conditions [22]. The heated coolant is subsequently directed to a steam generator or heat exchanger, where it transfers heat to water or another fluid, converting it into steam. This steam is then utilized to drive a turbine connected to an electrical generator, effectively converting thermal energy into mechanical energy, which is subsequently transformed into electricity [23].

SMRs are designed to produce approximately 300 megawatts of electricity (MWe) per module, although multiple modules can be installed together to meet larger power demands [24]. After passing through the turbine, the steam is condensed back into water and returned to the heat exchanger in a closed loop. This closed-loop system allows for the continuous recycling of water, ensuring a consistent and efficient cycle for electricity generation [25]. The integration of advanced heat exchangers is essential for optimizing this process, as they enhance heat transfer efficiency and contribute to the overall safety and reliability of the reactor system [26].

The operational principles of SMRs hinge on the effective management of nuclear fission reactions, the efficient transfer of heat through advanced cooling systems, and the conversion of thermal energy into electrical power. The design and implementation of these systems are important for maximizing the efficiency and safety of nuclear power generation [27].

Small Modular Reactors (SMRs) represent a significant advancement in nuclear reactor design, particularly due to their incorporation of passive safety systems. These systems leverage natural processes, such as gravity and convection, to maintain core cooling without the need for active mechanical controls or external power sources. For instance, the passive safety features of SMRs allow for gravity-fed coolant circulation, which can effectively cool the reactor core even during emergencies when external power or human intervention is unavailable. This design significantly mitigates the risk of overheating or meltdown, as it reduces reliance on pumps and other active cooling mechanisms that are prone to failure [5, 28, 29].

The development of passive safety systems in SMRs has its roots in the lessons learned from past nuclear accidents, such as those at Three Mile Island and Fukushima. These incidents highlighted the necessity for designs that inherently minimize risks. The concept of passive safety was formalized in the 1980s and has since evolved into the modern designs seen in SMRs today, which are engineered to be inherently safe and capable of managing decay heat without operator action [5, 29]. For example, the Westinghouse SMR incorporates an integral pressurized water reactor (iPWR) design that houses all components within a single pressure vessel, further enhancing safety and simplifying the reactor's overall design [28].

In addition to their passive safety features, SMRs are characterized by their compact design. Many models integrate the reactor core, steam generator, and containment systems into a single module, which simplifies the reactor architecture and reduces the number of components that could potentially malfunction [11, 29]. This compact structure not only enhances safety but also allows for installation in various environments, including below ground or within specially designed containment pools. Such design choices provide additional protection against external threats, including natural disasters and security breaches [30, 31]. The compact nature of SMRs also facilitates easier transportation and deployment, making them suitable for a wider range of applications, including those in developing regions with limited infrastructure [11, 29].

Furthermore, the passive cooling mechanisms employed in SMRs, such as natural circulation and convection, are designed to operate effectively without the need for electrical power. This is particularly crucial in scenarios where power loss occurs, as demonstrated by the incorporation of passive containment cooling systems in designs like the AP1000 [32, 33]. The ability of these systems to maintain cooling through natural processes not only enhances the safety profile of SMRs but also aligns with the growing emphasis on sustainability and resilience in nuclear power generation [34].

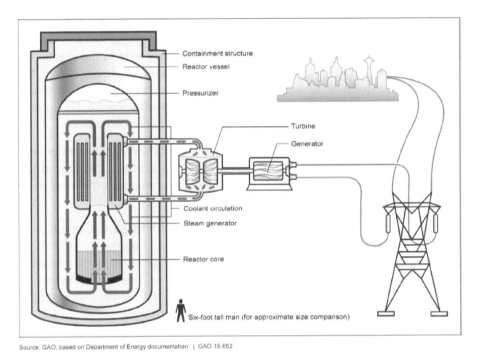

Figure 1: Illustration of a light water small modular nuclear reactor (SMR). U.S. Government Accountability Office from Washington, DC, United States, Public Domain, via Wikipedia.

Figure 1 shows a simplified diagram of a Small Modular Reactor (SMR) system, depicting its main components and the process of converting nuclear fission into electricity. The SMR is housed within a containment structure, designed to ensure safety and isolate the reactor from the external environment in case of any operational issues. Inside this containment structure lies the reactor vessel, where the nuclear fission reaction takes place. The reactor vessel contains the reactor core, made up of fuel rods with fissile material, such as uranium, which undergoes controlled fission reactions to generate heat.

Above the reactor core, the diagram shows the pressurizer. The pressurizer maintains the reactor's pressure, keeping the coolant water at high temperatures without boiling. This high-pressure environment enables efficient heat transfer from the reactor core to the coolant circulating through the system. The coolant absorbs the heat produced by the fission reactions in the core and flows upwards and out of the reactor vessel.

The heated coolant is then directed to a steam generator. In this component, the heat from the coolant is transferred to a separate loop of water, turning it into steam. This separation ensures that the water producing steam remains free of radioactive contamination. The steam generator, therefore, acts as a heat exchanger, allowing heat transfer without direct contact between the radioactive coolant and the steam that will eventually drive the turbine.

The steam produced by the steam generator flows into a turbine, where it expands and spins the turbine blades. The turbine is connected to a generator, which converts the mechanical energy from the turbine's rotation into electrical energy. This generated electricity is then transmitted through power lines to the grid, where it can supply power to homes, industries, and other infrastructure. The image illustrates this by showing the generator connected to an electrical transmission tower and a city, symbolizing the end-use of the generated power.

Once the steam has passed through the turbine and released its energy, it is condensed back into water and returned to the steam generator in a closed-loop system. This closed-loop design allows the steam to be reused, conserving water and maintaining efficiency in the power generation process.

A small figure of a six-foot-tall person is included in the diagram for size comparison, indicating that the containment structure housing the SMR is compact relative to traditional nuclear reactors. This compact design is a key characteristic of SMRs, allowing them to be deployed in a wider range of locations, including remote or space-limited sites. The compact size and modular nature of SMRs are crucial for their flexibility and scalability in different energy settings.

Nuclear Fission

Nuclear fission is a process in which the nucleus of a heavy atom, such as uranium or plutonium, splits into two smaller nuclei, releasing a significant amount of energy in the form of heat and radiation. This process occurs when a neutron collides with the nucleus of a fissile atom, causing it to become unstable and split as shown in Figure 2. Nuclear fission is the fundamental reaction used in nuclear power plants and atomic bombs, although the controlled environment in a nuclear power plant allows it to be used safely for energy production.

Nuclear Reaction

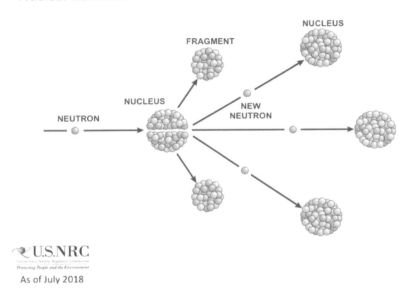

Figure 2: Nuclear reaction. Nuclear Regulatory Commission, CC BY 2.0, via Flickr.

The fission process begins when a neutron is absorbed by the nucleus of a fissile atom, like uranium-235 or plutonium-239. The absorption of this neutron destabilizes the nucleus, causing it to stretch and eventually split into two smaller nuclei, known as fission fragments. Alongside this splitting, two or three additional neutrons are usually released, along with a large amount of energy. This energy comes from the strong nuclear force, which holds the protons and neutrons together in the nucleus. When the nucleus splits, some of this binding energy is released as kinetic energy of the fission fragments and the emitted neutrons. This kinetic energy is then converted into heat as the particles collide with surrounding atoms.

The emitted neutrons from the fission event can collide with other nearby fissile nuclei, causing further fission reactions in a self-sustaining chain reaction. This chain reaction is the basis for generating energy in nuclear reactors. In power plants, materials like uranium-235 are arranged in such a way that the chain reaction can be carefully controlled. Control rods, typically made from materials that absorb neutrons like boron or cadmium, are used to regulate the reaction rate. By adjusting the position of these control rods, operators can slow down or speed up the fission process, thereby controlling the amount of heat generated.

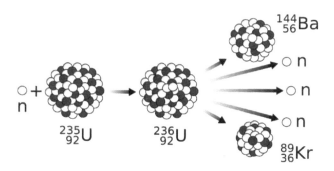

Figure 3: Illustration of a typical nuclear fission reaction. MikeRun, CC BY-SA 4.0, via Wikimedia Commons.

Figure 3 illustrates the process of nuclear fission, specifically showing how a uranium-235 nucleus undergoes fission when struck by a neutron. This chain of reactions is fundamental to the operation of nuclear reactors, as it releases a significant amount of energy. The illustration shows the sequence of steps involved in a single fission event and highlights the products of this process.

In the initial stage, a neutron (represented as "n" in the diagram) collides with a uranium-235 $\binom{235}{92}U$ nucleus. Uranium-235 is a fissile isotope, meaning it can undergo fission when struck by a neutron, especially a slow-moving, or "thermal," neutron. Upon absorbing the neutron, the uranium-235 nucleus becomes highly unstable and temporarily forms uranium-236 $\binom{236}{92}U$.

This unstable uranium-236 nucleus cannot hold together for long and soon splits into two smaller, more stable nuclei, known as fission fragments. In this example, the fission of uranium-236 produces two specific fragments: barium-144 $\binom{144}{56}Ba$ and krypton-89 $\binom{89}{36}Kr$. These nuclei are lower in atomic number and more stable than uranium, but they are still radioactive and will eventually decay to achieve stability.

In addition to the barium and krypton nuclei, the fission reaction releases a significant amount of energy, primarily in the form of kinetic energy of the fission fragments, which eventually translates into heat. This heat is what nuclear reactors capture to produce steam and generate electricity.

Moreover, the fission process releases additional neutrons—typically two or three—as shown in the image. These neutrons are crucial for sustaining the fission reaction in a chain reaction. When released, these neutrons can go on to strike other uranium-235 nuclei, causing further fission events. In a controlled environment, such as a nuclear reactor, this chain reaction is carefully managed to produce a steady output of energy. In an uncontrolled environment, such as a nuclear bomb, the chain reaction accelerates rapidly, resulting in a massive energy release.

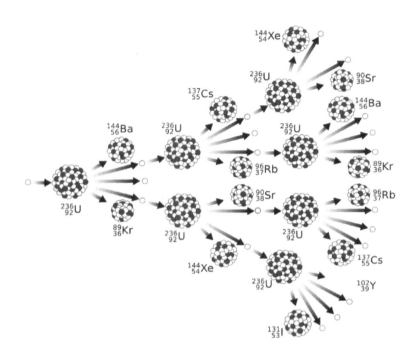

Figure 4: Illustration of a nuclear fission chain reaction. MikeRun, CC BY-SA 4.0, via Wikimedia Commons.

Figure 3 captures the essential elements of nuclear fission: the absorption of a neutron by a uranium-235 nucleus, the formation and subsequent splitting of an unstable uranium-236 nucleus, the production of fission fragments (barium and krypton in this case), the release of additional neutrons, and the substantial energy release. This process is foundational to nuclear energy, as the energy from repeated fission events is harnessed to generate electricity in nuclear reactors. The careful control of this reaction is what differentiates peaceful nuclear power generation from the explosive power of nuclear weapons.

Figure 4 illustrates a chain reaction process in nuclear fission, showing how a single fission event can initiate a series of subsequent fission reactions. This cascade effect is central to the function of nuclear reactors and atomic bombs, as it demonstrates how the energy release from fission can be sustained and amplified. The diagram specifically uses uranium-235 $\left(^{235}_{92}U\right)$ nuclei to show this process, highlighting the fission fragments and additional neutrons produced in each reaction.

At the beginning of the chain reaction, a neutron collides with a uranium-235 nucleus, transforming it into uranium-236 $\left(^{236}_{92}U\right)$, which is highly unstable. This instability causes the uranium-236 nucleus to split into two smaller nuclei, known as fission fragments, along with the release of additional neutrons. In this case, the fission fragments include elements like barium-144 $\left(^{144}_{56}Ba\right)$, krypton-89 $\left(^{89}_{36}Kr\right)$, and cesium-137 $\left(^{137}_{55}Cs\right)$, among others. These fission fragments vary depending on the specific fission event, as uranium-235 can split in multiple ways, producing different pairs of elements each time.

Each fission event also releases additional neutrons, typically two or three per reaction. These neutrons are critical for sustaining the chain reaction, as they can go on to collide with other nearby uranium-235 nuclei, inducing further fission reactions. In the image, these subsequent reactions are depicted branching out from the initial fission event, with each new uranium-235 nucleus undergoing fission after being struck by a neutron. This process creates a branching pattern, with each generation of reactions leading to more fission events, releasing more energy and more neutrons in an exponential manner.

The fission fragments shown in the diagram include a variety of elements, such as rubidium-96 $\left(^{89}_{36}Kr\right)$, strontium-90 $\left(^{90}_{38}Sr\right)$, xenon-144 $\left(^{144}_{54}Xe\right)$, and iodine-131 $\left(^{131}_{53}I\right)$. These fission products are radioactive and will eventually decay into stable isotopes over time, but their immediate byproducts contribute to the radioactivity of spent nuclear fuel, which requires careful handling and storage.

The released neutrons and the energy produced in each fission event are fundamental to both nuclear power generation and nuclear weapons. In a nuclear reactor, this chain reaction is carefully controlled to maintain a steady rate of fission, with only one of the released neutrons from each fission event continuing the reaction to the next uranium-235 nucleus. Control rods made of neutron-absorbing materials, such as boron or cadmium, are used to capture excess neutrons and regulate the reaction rate, preventing it from accelerating uncontrollably.

However, in an atomic bomb, the chain reaction is designed to proceed without control, allowing each fission event to trigger multiple subsequent reactions. This creates a rapidly accelerating chain reaction, releasing a massive amount of energy in a very short time, resulting in an explosion.

Overall, Figure 4 illustrates the self-sustaining nature of a nuclear chain reaction, showing how the fission of uranium-235 nuclei can lead to successive reactions that produce energy and additional neutrons. This process underpins the operation of nuclear reactors, where the reaction rate is controlled for safe energy production, as well as nuclear weapons, where the chain reaction is allowed to proceed uncontrolled for explosive force. The branching effect shown in Figure 4captures the exponential nature of the chain reaction, highlighting both the power and the importance of careful management in nuclear fission technology.

The heat produced by nuclear fission is harnessed to produce electricity. In a nuclear reactor, the fission reaction heats a coolant—usually water—that flows around the reactor core. This heated coolant is then used to generate steam, which drives a turbine connected to a generator, converting the thermal energy from fission into electrical energy. In contrast, in an atomic bomb, the fission reaction is uncontrolled, allowing a rapid, exponential chain reaction that releases a massive amount of energy almost instantaneously.

Fission reactions also produce radioactive waste, as the fission fragments are typically unstable and emit radiation as they decay into stable forms. The handling and disposal of these radioactive byproducts are critical considerations in nuclear power, as they require secure storage for long periods to prevent environmental contamination. The energy released by fission is substantial; just a small amount of uranium can produce more energy than several tons of coal or oil, making it an efficient but complex energy source.

One of the challenges in nuclear fission is managing the reaction rate to maintain a stable output of energy. In a power reactor, this is achieved by maintaining a critical state, where each fission event, on average, leads to exactly one more fission event, ensuring a steady release of energy. If the chain reaction accelerates uncontrollably, it could lead to overheating, which is why safety measures such as cooling systems, containment structures, and emergency shutdown procedures are integral to nuclear reactor design.

Managing the reaction rate in nuclear fission is essential for ensuring a stable and controlled energy output in nuclear reactors. Fission, the process of splitting atomic nuclei, releases energy and produces neutrons that can initiate further fission reactions in adjacent nuclei. This chain reaction must be meticulously regulated to prevent fluctuations in energy output, which could lead to hazardous conditions, such as overheating or reactor core damage [35]. The objective is to achieve a critical state where, on average, each fission event results in exactly one subsequent fission event, thereby maintaining a steady and manageable energy release [35].

If the reaction rate surpasses this critical state, where each fission event leads to more than one subsequent event, the chain reaction can escalate uncontrollably, resulting in a condition known as supercriticality. This scenario poses significant risks, including overheating of the reactor core, potential fuel melting, and damage to the reactor structure, which could lead to radiation release if containment measures fail [35]. To mitigate these risks, nuclear reactors are equipped with various systems designed to monitor and control the reaction rate with precision [35].

A primary safety feature in nuclear reactors is the implementation of control rods, which are composed of materials that absorb neutrons, such as boron or cadmium. By adjusting the position of these control rods—either inserting or withdrawing them from the reactor core—operators can regulate the number of neutrons available to sustain the chain reaction. This mechanism allows for the slowing down of the reaction rate, ensuring that it does not accelerate uncontrollably [35]. Operators must be adept at positioning these rods to maintain the desired critical state, adjusting for any changes in reactor conditions or power demand [35].

In addition to control rods, effective cooling systems are vital for managing the heat generated during fission. These systems circulate coolant, typically water, around the reactor core to absorb and dissipate heat produced by the fission process. A failure in the cooling system could lead to a dangerous rise in core temperature, risking core damage and potential meltdown [35].

Therefore, cooling systems are designed with redundancies and backup measures to ensure reliable operation, even in the event of primary system failure [35].

The containment structure surrounding the reactor is another critical safety feature. This robust, sealed building is designed to prevent the escape of radioactive materials in case of an accident. In the unlikely event of a malfunction or breach within the reactor, the containment structure serves as a final barrier, protecting the environment and public from radiation exposure [36]. Typically constructed from thick concrete and steel, the containment building is engineered to withstand extreme conditions, including potential explosions and natural disasters [36].

Finally, emergency shutdown procedures, commonly referred to as SCRAM systems, are implemented as an immediate safety response if the reactor exhibits signs of instability. A SCRAM system rapidly inserts all control rods fully into the core, effectively halting the fission process by absorbing the majority of neutrons and stopping the chain reaction [35]. This rapid shutdown mechanism is crucial for preventing overheating or other dangerous scenarios when the reactor's automatic control systems detect criticality issues or equipment malfunctions [35].

The combination of control rods, cooling systems, containment structures, and emergency shutdown procedures constitutes a comprehensive safety framework for managing the fission reaction rate in nuclear reactors. By maintaining a steady and controlled chain reaction, these systems ensure the safe and efficient operation of nuclear power plants, allowing for reliable energy generation while minimizing the risk of accidents. Properly balancing the fission reaction rate is therefore fundamental to nuclear reactor design, facilitating sustained power generation without compromising safety [35].

SMR Safety Systems

Safety is a paramount concern in nuclear energy, and SMRs are designed with enhanced safety features that differentiate them from traditional reactors. Many SMR designs incorporate passive safety systems that do not rely on active mechanical systems to maintain safety during abnormal conditions. This design philosophy stems from lessons learned in the nuclear industry, particularly after incidents such as the Three Mile Island accident, which highlighted the need for inherently safe reactor designs [29, 37]. The passive safety features of SMRs allow them to operate with a reduced risk of catastrophic failures, making them an attractive option for both developed and developing regions [3, 38].

Small Modular Reactors (SMRs) are emerging as a pivotal innovation in nuclear energy, primarily due to their advanced safety systems designed to enhance reliability and minimize the risk of accidents. Unlike traditional large-scale nuclear reactors, SMRs incorporate both active and passive safety features that contribute to their inherent safety profile. These systems are engineered to prevent overheating, manage radiation containment, and ensure safe shutdown processes, thereby reducing the necessity for human intervention during emergencies [11, 16, 29].

A key aspect of SMR safety is the implementation of passive safety systems. These systems leverage natural phenomena such as gravity, convection, and conduction to maintain reactor cooling without relying on external power or mechanical components. For instance, in the event of a power loss, passive cooling mechanisms can continue to operate autonomously, utilizing gravity-driven coolant flow to avert overheating [39, 40]. This contrasts sharply with traditional reactors, where active cooling systems are vulnerable to failure during power outages or mechanical malfunctions. The reliance on passive systems in SMRs significantly enhances their resilience during emergencies, thereby providing a robust alternative for maintaining core temperatures [16, 28].

Moreover, the modular design of SMRs contributes to their safety by confining critical components within a single, compact unit. This configuration reduces system complexity and minimizes potential failure points, which are often exacerbated by human error or equipment malfunction [11, 41]. By integrating the reactor core, cooling systems, and containment structures into a single module, SMRs facilitate easier maintenance and repair processes, as components can be isolated without impacting the entire system [42, 43]. Additionally, the capability to install SMRs below ground or within protective containment pools offers an extra layer of security against external threats such as natural disasters or security breaches [44].

Containment systems in SMRs are meticulously designed to prevent the release of radioactive materials during accidents. These systems typically feature multiple barriers, including a steel-lined containment vessel encased in thick concrete walls, which provide substantial shielding against radiation leaks [45, 46]. Furthermore, the smaller fuel quantities used in SMRs compared to traditional reactors inherently reduce the potential scale of any radioactive release. Some designs even utilize coolants like molten salt, which are less prone to vaporization under high temperatures, thereby further mitigating the risk of pressurized releases [5, 39].

Automatic shutdown capabilities are another critical safety feature of SMRs. Many designs are equipped with systems that can detect abnormal conditions, such as overheating, and initiate an immediate shutdown sequence autonomously. This "SCRAM" system rapidly inserts control rods into the reactor core, effectively halting the fission process and allowing passive cooling systems to maintain safe temperatures [16, 47]. The simplicity and fewer components in SMR designs enhance the reliability and speed of these shutdown systems, which can act within seconds during emergencies [40, 41].

Fail-safe designs are integral to SMR technology, reducing risks associated with human error and unexpected events. SMRs often utilize straightforward, intuitive control systems that minimize the likelihood of operator mistakes. Additionally, many designs operate at low pressures, which are less likely to fail catastrophically compared to high-pressure systems, further enhancing their safety profile [11, 39, 40]. This layered approach to safety, which combines both passive and active systems with redundancies, positions SMRs as exceptionally resilient against potential disruptions [16, 48].

Modular Functionality

The modular nature of SMRs means they can be built in factories under controlled conditions and then transported to their operational sites. This approach enables faster and more efficient construction than traditional, large-scale nuclear plants, which must often be built on-site. Because of their smaller size and lower power output, SMRs are suitable for diverse applications beyond traditional power generation. They can provide energy to isolated or remote communities, supplement renewable energy sources in areas with inconsistent power supply, and even be used for industrial processes, desalination, or hydrogen production.

SMRs generate electricity through nuclear fission, using a compact, modular design that incorporates passive safety features and integrated systems for heat exchange and containment. This design enhances the safety, efficiency, and adaptability of SMRs, making them a versatile solution for reliable, low-carbon energy in various settings. Their unique construction and operation make SMRs a promising component of the future energy landscape, bridging the gap between traditional nuclear power and the needs of modern, sustainable energy systems.

Small Modular Reactors (SMRs) can be combined in parallel to generate more megawatts of electricity (MWe), creating what is known as a multi-module nuclear power plant. This modular approach allows for scalability in power generation to meet specific energy needs, enabling a plant to start with a small number of SMRs and add more modules over time as demand increases. However, SMRs are not typically combined in series; rather, they are connected in parallel. Here's how it works and why this design is beneficial:

Parallel Configuration of SMRs

In a parallel configuration, each SMR operates independently and feeds its generated electricity to a shared power grid or distribution system. Each reactor module has its own reactor core, cooling system, steam generator, and turbine, functioning as a self-contained unit. By connecting multiple SMRs in parallel, a nuclear facility can increase its total power output, with each module contributing a set amount of megawatts to the overall generation capacity.

For instance, if each SMR module produces 77 MWe, then a multi-module plant with 6 SMRs operating in parallel would have a total generating capacity of around 462 MWe. This modularity provides flexibility, allowing power plant operators to add or remove modules based on demand or maintenance requirements without affecting the operation of other modules.

Parallel operation in SMRs works as follows:

1. **Independent Operation and Redundancy**: Each SMR module operates independently, with its own control and safety systems. This design allows each module to be shut down or serviced without affecting the rest of the plant. If one module needs maintenance, the other modules can continue to supply power, providing reliability and redundancy that large reactors lack.

2. **Shared Electrical Output**: The power generated by each SMR module is directed to a common electrical bus, which consolidates the power output and delivers it to the grid. By using multiple modules, the combined output of the plant can meet a wide range of

power demands. For example, an SMR facility could add modules over time to reach a desired capacity, scaling up as energy needs grow or replacing older modules without disrupting the overall power supply.

3. **Load Following and Flexibility**: SMRs configured in parallel provide load-following capabilities, meaning they can adjust output to meet varying demand. In times of lower demand, some modules can be operated at reduced power or temporarily shut down. During peak demand, additional modules can be brought online to provide extra power. This flexibility is especially beneficial for integrating with renewable energy sources like wind and solar, which are intermittent. SMRs can provide consistent power when renewable sources fluctuate, ensuring grid stability.

4. **Centralized and Efficient Cooling**: Although each module has its own cooling system, SMR plants often use a centralized cooling infrastructure that serves all modules. The cooling system can be designed to scale with the number of modules, allowing a single plant to efficiently manage heat from multiple reactors. This centralized cooling system streamlines operation and maintenance, as it consolidates cooling infrastructure for multiple reactors.

Combining reactors in series is not practical in nuclear power design. In a series configuration, each reactor would depend on the output of the previous one, creating interdependencies that could compromise safety and reliability. In nuclear reactors, each module needs to operate as an independent unit with direct access to its coolant and safety systems. A series setup would complicate the design, making it more difficult to control individual reactor operations, manage safety, and ensure that all modules can be safely isolated if necessary.

Moreover, SMRs are designed to produce electricity directly from each module's steam cycle, driving its own turbine-generator setup. In a series configuration, the heat or steam from one reactor would need to pass through another reactor, which is not feasible due to the temperature, pressure, and complexity of reactor operations.

The benefits of parallel SMR configurations include:

1. **Scalability**: Parallel configurations allow plant operators to scale up power generation incrementally. New modules can be added as demand increases, making it easier to expand capacity compared to a single large reactor.

2. **Flexibility**: Parallel operation supports load-following, allowing operators to adjust power output based on real-time demand. Some modules can be run at reduced power or shut down when less electricity is needed.

3. **Enhanced Safety and Reliability**: The independence of each module allows for easy isolation in case of issues. If one SMR needs maintenance or encounters an issue, the remaining modules can continue operating without interruption, enhancing reliability.

4. **Simplified Maintenance and Reduced Downtime**: Since each SMR is a self-contained unit, maintenance can be performed on individual modules without affecting the

operation of the entire plant. This setup minimizes downtime and enhances operational flexibility.

5. **Cost-Effective Expansion**: SMRs can be deployed in stages, spreading out costs over time rather than requiring a large upfront investment. This staged approach makes it easier for utilities and governments to finance projects.

An example of a multi-module SMR plant is the proposed NuScale Power Plant. Each NuScale Power Module produces 77 MWe, and up to 12 modules can be combined within a single plant, generating a total output of 924 MWe. This setup enables the plant to adjust its output according to grid demand and provides redundancy by ensuring that each module can be independently operated, maintained, or shut down without impacting the others.

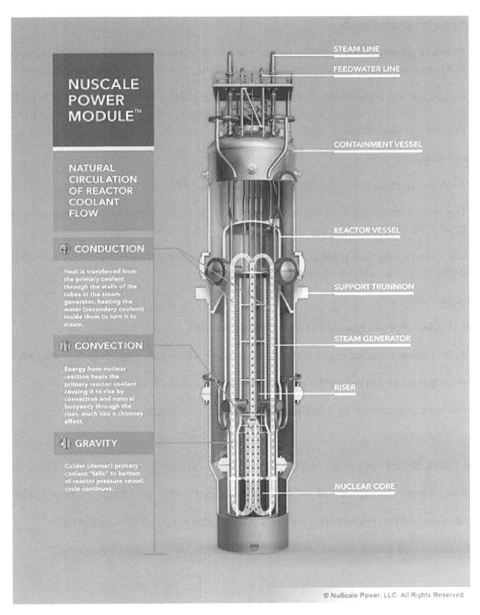

Figure 5: Diagram of a NuScale reactor. NuScale, CC BY-SA 3.0, via Wikimedia Commons.

As an example of application of multiple SMRs, to estimate the number of Small Modular Reactors required to power cities of various sizes, we need to look at average power consumption rates for cities of these populations. Power consumption can vary significantly depending on factors such as climate, industrial activity, and lifestyle, but we can use approximate per capita power consumption values to make a rough estimate.

Assumptions for Calculation:

1. **Average power consumption per person in developed countries**: Approximately 4 kW per person (combining residential, commercial, and industrial demand).

2. **Output of a typical SMR**: Modern SMRs typically produce between 50 MWe to 300 MWe per unit. For this calculation, we'll use a commonly referenced SMR output of 100 MWe per unit for simplicity.

Using these assumptions:

1. 4 kW per person is approximately equivalent to 4 MW per 1,000 people.

2. The total power requirement can be calculated by multiplying the population by 4 MW per 1,000 people.

3. We then divide this total power requirement by the output of one SMR (100 MWe) to determine the number of reactors needed.

Calculating for each city:

1. City of 100,000 People (e.g., Fairfield, California, US)

- **Total Power Requirement** = 100,000 people × 4 MW per 1,000 people = **400 MW**

- **Number of 100 MWe SMRs needed** = 400 MW ÷ 100 MWe = **4 SMRs**

So, it would take approximately **4 SMRs** to power a city of 100,000 people.

2. City of 1,000,000 People (e.g., Austin, Texas, US)

- **Total Power Requirement** = 1,000,000 people × 4 MW per 1,000 people = **4,000 MW**

- **Number of 100 MWe SMRs needed** = 4,000 MW ÷ 100 MWe = **40 SMRs**

Therefore, around **40 SMRs** would be needed to power a city of 1,000,000 people.

3. City of 5,000,000 People (e.g., Melbourne, Victoria, Australia)

- **Total Power Requirement** = 5,000,000 people × 4 MW per 1,000 people = **20,000 MW**

- **Number of 100 MWe SMRs needed** = 20,000 MW ÷ 100 MWe = **200 SMRs**

Thus, **200 SMRs** would be required to meet the power needs of a city of 5,000,000 people.

4. City of 10,000,000 People (e.g., Nagoya, Japan)

- **Total Power Requirement** = 10,000,000 people × 4 MW per 1,000 people = **40,000 MW**

- **Number of 100 MWe SMRs needed** = 40,000 MW ÷ 100 MWe = **400 SMRs**

For a large metropolis of 10,000,000 people, approximately **400 SMRs** would be needed.

Summary of Results

City	Population	Estimated Power Requirement (MW)	Number of 100 MWe SMRs Needed
Fairfield, CA, US	100,000	400	4
Austin, TX, US	1,000,000	4,000	40
Melbourne, Australia	5,000,000	20,000	200
Nagoya, Japan	10,000,000	40,000	400

Considerations

- **Redundancy and Load Management**: In practice, a few additional SMRs might be added to account for redundancy, maintenance, and peak demand periods, so the actual number could be slightly higher.

- **Variability in Power Demand**: Power demand per capita varies between cities and countries, so a city with significant industrial activity or extreme climate conditions may require more SMRs.

- **SMR Output Variability**: Some SMRs are designed to produce more (up to 300 MWe), meaning fewer reactors would be needed if using higher-output models.

These numbers provide a general estimate, but actual deployment would depend on detailed assessments of the city's specific power needs and the SMR models used.

these calculations make a compelling argument for the potential of Small Modular Reactors (SMRs) in addressing the power needs of cities of varying sizes. The results highlight several key advantages and applications of SMRs, especially in terms of scalability, flexibility, and suitability for diverse environments. Here's a deeper look at how these numbers support the case for SMRs and what they reveal about their applications:

1. Scalability and Flexibility: The calculations show that SMRs can be deployed in modular units to match the specific power demands of a city. For a smaller city of around 100,000 people, only a few SMRs would be necessary, while a metropolis of 10 million people would require many more. This modular approach allows for incremental, flexible power generation that can grow alongside a city's population and industrial demands. Unlike large traditional nuclear plants, which are often built to produce thousands of megawatts at once, SMRs can be added over time, providing a tailored solution for both urban and rural areas.

2. Suitability for Smaller or Remote Communities: The basic analysis suggests that SMRs are particularly well-suited to smaller cities and remote communities that may not need the vast output of a full-sized nuclear plant. For example, a small city or isolated area could rely on a few SMRs to provide steady, reliable power without requiring the high investment and large infrastructure footprint associated with traditional nuclear plants. This makes SMRs a practical solution for regions with limited access to large power grids or those looking to achieve energy independence, such as island communities or areas where renewable sources alone may not be reliable.

3. Complementary to Renewable Energy Sources: SMRs can act as a stable, low-carbon energy source that complements renewable sources like wind and solar. In cities with fluctuating energy demands or variable renewable output, SMRs can provide consistent "baseload" power, helping to balance the grid when renewables are not generating. Their modular design allows utilities to add SMRs as needed, potentially reducing reliance on fossil fuels and ensuring grid stability as renewables continue to expand. For larger cities looking to decarbonize, SMRs could provide the dependable power necessary to offset the intermittency of renewables.

4. Reduced Initial Investment and Lower Financial Risk: Building a large, traditional nuclear plant requires a massive upfront investment and several years (or even decades) of construction time. This high cost and long timeline are significant barriers, particularly for smaller cities or developing regions. SMRs, however, offer a more incremental approach to nuclear power. Since each SMR can be manufactured in a factory and then transported and installed on-site, the initial investment is lower, and reactors can be brought online in stages. This staged approach also lowers financial risk, as communities or utilities can invest in additional SMRs only as demand increases or as budgets allow.

5. Enhanced Safety and Reduced Environmental Impact: SMRs are designed with advanced safety features, such as passive cooling systems and containment structures, that reduce the risk of accidents and make them safer for deployment closer to populated areas. Their smaller size also means that they have a reduced environmental footprint compared to large nuclear plants. Additionally, because SMRs produce less total waste due to their modular deployment, they may be easier to manage from a waste disposal and decommissioning perspective. This makes SMRs a viable option for cities aiming to transition to clean energy without the environmental and logistical challenges posed by large nuclear facilities.

6. Applications Across Different City Sizes and Needs: The calculations show that SMRs are versatile enough to power a wide range of urban centres—from smaller cities needing just a few SMRs to large metropolises that may require dozens or even hundreds. This versatility underscores the broad applicability of SMRs across different types of communities, from small towns to large industrial hubs. SMRs could serve as the primary power source for smaller cities, a supplementary source for medium-sized cities, or a stabilizing, grid-balancing source for large cities that rely on a mix of energy sources.

7. Potential for Decentralized Power Generation: The modularity of SMRs allows for a decentralized approach to power generation. Instead of relying on a few large, centralized plants, cities and regions could deploy multiple SMRs across different locations. This distributed model enhances energy resilience, as each module operates independently, making the power supply less vulnerable to localized disruptions or grid failures. This approach could be particularly beneficial in regions prone to natural disasters, as power generation can continue in unaffected areas.

Conclusion: SMRs as a Flexible and Scalable Solution: These calculations reinforce the argument that SMRs are a flexible, scalable, and safe solution for the evolving energy needs of modern cities. Their ability to provide reliable, low-carbon power across different scales makes them a valuable addition to the global energy landscape, especially as cities and countries seek to transition away from fossil fuels and achieve energy independence. SMRs' adaptability for various city sizes, applications, and geographic settings illustrates their potential as a versatile technology, capable of supporting clean energy goals, stabilizing grids, and offering a low-risk nuclear power alternative that can grow alongside future energy demands.

SMRs offer an innovative path forward for nuclear energy, one that is adaptable to both small and large cities, capable of complementing renewables, and potentially more acceptable to the public due to enhanced safety features and a smaller environmental footprint. This versatility makes SMRs a promising option for addressing the global challenge of sustainable and resilient energy production.

Deciding between deploying Small Modular Reactors (SMRs) or building a full-scale nuclear reactor depends on multiple factors, including energy demand, cost-effectiveness, project timelines, infrastructure, and the long-term energy strategy of a region or utility. Below are the key considerations for determining when a full-scale nuclear reactor may be more effective than multiple SMRs.

In regions with consistently high energy demands, building a full-scale nuclear reactor can be more effective than deploying multiple SMRs. A full-scale reactor typically generates 1,000–1,600 MWe in a single unit, far more than an SMR's typical output of 50–300 MWe [49]. Large metropolitan areas or highly industrialized regions often require thousands of megawatts of baseload power, which a full-scale reactor can supply continuously and efficiently. Similarly, regions with energy-intensive industries, such as aluminium smelting or steel production, benefit from a large reactor's ability to provide a constant and high level of power. In these cases, building one large reactor to meet substantial energy needs can be more efficient than managing numerous SMRs, especially if the demand exceeds what SMRs can realistically supply.

Full-scale nuclear reactors benefit from economies of scale, which often translates to a lower cost per megawatt once they are constructed [50]. Although SMRs have a lower initial cost and allow for incremental capacity additions, a single large reactor can produce electricity at a lower unit cost due to operational efficiencies and reduced per-unit construction costs. A single large reactor, with centralized maintenance, reduces operational complexities compared to multiple SMRs, each requiring individual monitoring and management. Moreover, spreading fixed costs over a larger capacity can make the long-term financial savings of a full-scale reactor outweigh its higher initial investment, making it more economically viable over the reactor's lifespan in regions with high energy demands.

Large nuclear reactors need significant land and robust infrastructure, including access to ample cooling water, transportation for large reactor components, and a strong grid connection to distribute high output. If a suitable site with the necessary infrastructure is available, building a full-scale reactor may be more efficient, particularly if the location is close to major transmission infrastructure that can handle high output without costly upgrades. Additionally, if regulatory and community approval for a large site is already secured, placing a single reactor on an established site can simplify the development process.

Full-scale reactors are ideal for providing stable baseload power in regions with predictable, stable demand, where frequent load adjustments are unnecessary. While SMRs offer flexibility, making them suitable for areas with variable or growing demand, a full-scale reactor might be preferable when the region has a constant demand profile that doesn't require frequent output adjustments. In grids primarily relying on baseload power sources, with limited dependence on intermittent renewables like wind and solar, full-scale reactors operate most efficiently at full capacity and are well-suited to maintain grid stability.

Operating multiple SMRs increases operational complexity since each reactor requires its own maintenance, safety checks, and potential downtime for outages. In contrast, a single large reactor simplifies management, training, and emergency response. For utilities or operators seeking simplified management and reduced logistical challenges, a full-scale reactor presents a more streamlined approach, as it requires fewer staff and less oversight than a facility with numerous SMRs.

Building and operating a full-scale nuclear reactor requires a highly skilled workforce, advanced training, and extensive regulatory compliance. In regions with established nuclear industries, regulatory support, and an available skilled workforce, constructing a large reactor may be more feasible than deploying multiple SMRs. An established regulatory framework and licensing process for large reactors can streamline approval, while a trained workforce experienced in managing large nuclear plants can efficiently handle a full-scale facility without the operational complexities of multiple smaller units.

If the region's goal is rapid, large-scale decarbonization, a single full-scale nuclear reactor may be more effective than SMRs. A large nuclear plant can provide substantial clean energy to replace fossil fuel generation in one significant investment, rather than incrementally deploying SMRs over time. This approach is particularly relevant for regions with carbon reduction targets that require immediate, large-scale replacement of high-emission power sources with dependable baseload power. For areas prioritizing rapid clean energy deployment as part of a climate action plan, the high capacity of a full-scale reactor meets the need efficiently.

While full-scale reactors have clear advantages in high-demand, stable, and well-supported environments, SMRs offer unique benefits in other contexts. SMRs are often more suitable when:

- **Incremental or flexible capacity** is needed, such as in growing cities or regions with fluctuating demand.

- **Remote areas or smaller communities** require independent power sources.

- **Regions with limited capital or access to infrastructure** prefer modular, lower-cost initial investments.

- **Complementing renewable sources** with flexible, load-following capabilities is necessary.

- **Quick deployment** is needed, as SMRs can be manufactured in a factory and assembled on-site faster than full-scale reactors.

Building a full-scale nuclear reactor is generally more effective in areas with high, stable energy demand, available infrastructure, and the need for cost-effective, large-scale baseload power. Full-scale reactors make sense for regions that can support significant capital investment and have the regulatory and operational framework to manage a large plant. However, SMRs offer advantages in flexibility, incremental deployment, and lower upfront costs, making them ideal for areas with smaller or variable demand, limited infrastructure, or specific needs like remote power generation.

Ultimately, the choice between full-scale reactors and SMRs depends on the region's specific energy, financial, and infrastructure requirements, as well as long-term goals for flexibility, scalability, and decarbonization.

Structure of Small Modular Reactors

The structure of a Small Modular Reactor (SMR) is designed to be compact, efficient, and safe. SMRs contain similar components to traditional large nuclear reactors but are engineered with a modular approach that simplifies their design, enhances safety, and allows for factory-based manufacturing. Here is a detailed breakdown of the various components found in an SMR and their specific roles:

1. Reactor Core: The reactor core is the heart of the SMR, where nuclear fission takes place. The core consists of fuel rods filled with fissile material, usually uranium-235 or a similar isotope capable of sustaining a nuclear chain reaction. The uranium is typically in the form of uranium dioxide pellets, which are stacked in long rods made of a heat-resistant material, such as zirconium alloy. The core configuration is optimized for efficient heat production and neutron moderation, ensuring a stable chain reaction. In some advanced SMRs, alternative fuels, such as thorium or molten salts, may be used.

2. Reactor Vessel: The reactor vessel is a large, high-strength steel container that holds the reactor core, coolant, and internal structural components. This vessel is designed to withstand high temperatures and pressures generated during fission. It acts as the primary containment barrier for radioactive materials, keeping the core isolated and ensuring that, in case of an incident, radioactive materials remain confined within the vessel. In SMRs, the reactor vessel is often compact and designed to integrate closely with other components, reducing complexity and potential points of failure.

3. Coolant and Heat Transfer Systems: The coolant in an SMR is typically water, which circulates through the reactor vessel to absorb the heat generated by fission. In pressurized water SMR designs, the coolant is kept at high pressure to prevent it from boiling, allowing it to reach higher temperatures and efficiently transfer heat. Some advanced SMRs use alternative coolants, such as molten salt, liquid metal (like sodium), or gas (such as helium), which can operate at even higher temperatures and improve thermal efficiency.

The heat transfer system usually includes a primary loop and a secondary loop. In the primary loop, the coolant flows through the reactor core, absorbs heat, and then transfers this heat to a secondary loop through a heat exchanger or steam generator. The secondary loop converts water into steam, which drives a turbine to generate electricity. By separating the radioactive primary coolant from the steam-generating loop, the secondary loop ensures that steam remains uncontaminated.

4. Steam Generator: The steam generator is a critical component in SMR designs that use water as a coolant. Located next to or integrated within the reactor vessel, the steam generator transfers heat from the primary coolant loop to a secondary loop, where water is converted to steam. This steam then drives the turbine, producing electricity. In some SMR designs, such as integral pressurized water reactors (iPWRs), the steam generator is integrated within the reactor vessel, which further reduces the size and complexity of the system and minimizes the risk of leaks.

5. Control Rods and Shutdown Systems: Control rods are made of materials that absorb neutrons, such as boron or cadmium. These rods are inserted or withdrawn from the reactor core to control the rate of the nuclear chain reaction. When inserted deeper into the core, they absorb more neutrons, slowing down or halting the reaction. Control rods are crucial for adjusting the reactor's power output and ensuring safe shutdowns when needed. SMRs are equipped with automatic SCRAM systems that can rapidly insert all control rods into the core in case of an emergency, stopping the reaction almost immediately to prevent overheating.

6. Pressurizer: The pressurizer is a component specific to pressurized water reactor (PWR) designs, including many SMRs. It maintains the pressure of the primary coolant loop, preventing the coolant from boiling even at high temperatures. By keeping the water under pressure, the pressurizer allows the coolant to operate as an efficient heat transfer medium. If pressure drops or rises unexpectedly, the pressurizer can adjust accordingly to maintain safe operating conditions within the reactor vessel.

7. Containment Structure: The containment structure is a robust, reinforced enclosure surrounding the reactor vessel and primary components. Made of thick concrete and steel, this structure is designed to contain radiation and prevent the release of radioactive materials in the event of an accident. Some SMRs are placed in underground containment pools or vaults, providing additional protection from natural disasters, security threats, or potential explosions. This structural design also assists in passive cooling, as the surrounding water or earth helps dissipate heat from the reactor in case of a shutdown.

8. Passive Cooling Systems: A unique feature of SMRs is the incorporation of passive cooling systems, which rely on natural forces, such as gravity and convection, to circulate coolant without the need for pumps or external power. In an emergency, passive cooling can keep the reactor core from overheating by allowing coolant to circulate naturally. For instance, in a loss-of-power situation, these systems use gravity-driven water flow to cool the reactor core, preventing a meltdown. Passive cooling systems reduce dependency on active mechanical components, enhancing reliability and safety in the event of an unexpected shutdown.

9. Turbine and Generator: The turbine and generator are located outside the containment structure and are essential for converting thermal energy into electricity. In the secondary loop, steam generated in the steam generator flows to the turbine, where it expands and drives the turbine blades. The turbine is connected to a generator, which transforms the rotational energy into electrical energy. After passing through the turbine, the steam is condensed back into water and returned to the steam generator in a closed-loop cycle.

10. Emergency Core Cooling System (ECCS)" The Emergency Core Cooling System (ECCS) is a backup system designed to provide additional cooling in the event of an emergency. If the primary cooling system fails or the reactor overheats, the ECCS can inject coolant directly into the reactor vessel to bring temperatures down quickly. Many SMR designs have multiple layers of redundancy, including passive ECCS components, which do not require external power and can operate even in severe conditions. The ECCS is an important safety measure to prevent core damage and ensure that the reactor remains in a safe state during unexpected situations.

11. Instrumentation and Control Systems: Instrumentation and control systems monitor the reactor's operation, including temperature, pressure, power output, and radiation levels. These systems provide real-time data and control over the reactor, allowing operators to make adjustments as necessary. Many SMRs are designed with advanced, simplified control systems that can be managed remotely or require minimal human intervention, thanks to automated controls. These systems are also equipped with alarms and automatic shutdown capabilities to respond immediately to any abnormal readings, further enhancing safety.

The structure of an SMR integrates all of these components into a compact, modular design focused on safety, efficiency, and scalability. The reactor core and reactor vessel are central to the fission process, with control rods and pressurizers managing the reaction rate and pressure. Coolant systems and steam generators facilitate heat transfer, while passive cooling systems and emergency core cooling add multiple layers of safety. The containment structure and instrumentation ensure the reactor operates within safe limits, while turbines and generators convert thermal energy to electricity.

Together, these components create a streamlined, modular reactor design that can be scaled up or down to meet energy demands, making SMRs a promising solution for the future of nuclear power. The integration of passive safety features, modularity, and simplified control systems positions SMRs as an advanced and secure nuclear technology for reliable, low-carbon energy generation.

Differences between SMRs and Traditional Reactors

Understanding how nuclear reactors differ in energy output, size, and resource and land requirements is crucial for making informed choices about energy sourcing. This knowledge is increasingly important as industry, communities, and nations face rising energy demands, pushing for sustainable, reliable, and flexible power solutions. Among nuclear reactor types, microreactors, Small Modular Reactors (SMRs), and traditional large-scale nuclear plants each offer unique characteristics, making them suitable for varied applications [51].

Microreactors: Microreactors are an innovative development within nuclear energy, delivering reliable, carbon-free baseload power through a compact design that is 100 to 1,000 times smaller than conventional reactors. Definitions for microreactors vary slightly, with some setting the maximum output at under 50 megawatts of electricity (MWe) and others at less than 10 MWe, but generally, reactors generating less than 20 MWe are classified as microreactors. Due to their small size, microreactors have minimal land requirements—typically just a few acres—and need a smaller on-site workforce. This compact, low-maintenance nature makes them ideal for localized energy applications, such as powering remote industrial sites, military installations, and smaller communities [51].

An additional advantage of microreactors is modularity. Many designs are fully modular, meaning that components are manufactured in factories and then transported to the site for quick assembly. This modularity allows for faster deployment and scalability, enabling incremental expansion based on energy demand. As a result, microreactors can be quickly installed (usually within 24 months) and expanded as needed. An example is Last Energy's PWR-20, which outputs 20 MWe, occupies only 0.3 acres, and can be assembled on-site in four months. The PWR-20's air-cooled design eliminates the need for water-based cooling, allowing it to be sited virtually anywhere, which adds flexibility to its applications [51].

Small Modular Reactors (SMRs): Small Modular Reactors represent a mid-scale option in nuclear innovation, with energy outputs typically between 20 and 300 MWe. This power range makes SMRs capable of serving medium to large communities or industries while still being smaller and more flexible than traditional nuclear plants. SMRs require a larger footprint than microreactors, often around 100 acres or more, depending on the design, but still considerably less land than a traditional nuclear plant. SMRs typically need more staff than microreactors, with an estimated average of 1.5 personnel per MWe, leading to staffing requirements of 30 to 150 or more individuals, depending on the reactor's output [51].

While "modular" is part of the SMR name, the level of modularity varies across designs. A truly modular SMR would involve factory-fabricated components, minimal site preparation, and efficient on-site assembly. Modular SMRs also offer flexibility in financing and scaling, allowing utilities to add additional units incrementally to meet growing energy demand. SMRs typically take three to five years to deploy, and they require moderate water resources for cooling, depending on the specific design. The scalability of SMRs makes them suitable for larger industrial sites, remote regions, and cities that need reliable, low-carbon energy but do not require the extensive power output of a full-scale nuclear plant [51].

Traditional Nuclear Plants: Traditional nuclear power plants, or gigawatt-scale reactors, have long served as reliable power sources capable of meeting the demands of large populations. These plants typically generate around 1,000 MWe (1 GW) per reactor, making them ideal for supplying large cities or regions with substantial power needs. However, these reactors have large physical footprints, often requiring over a square mile of land, and require significant cooling resources, typically needing access to a substantial water source. The staffing ratio is lower than SMRs at around 0.7 person per MWe, but given the high output, traditional reactors still require thousands of personnel to operate and maintain the facility [51].

The complexity of building and operating a large-scale nuclear plant comes with notable challenges. Construction timelines are lengthy, averaging around 10 years, and costs often exceed the initial budget, frequently going over by as much as 200%. These plants also require rigorous regulatory assessment due to their size and environmental impact. While large reactors are efficient in terms of energy production, the significant infrastructure and regulatory requirements make these builds challenging and costly. Despite these factors, gigawatt-scale reactors remain a reliable option for regions that need substantial baseload power for large communities or highly industrialized areas [51].

Comparing Microreactors, SMRs, and Traditional Plants

The evolving nuclear landscape offers diverse options for meeting different energy needs. Microreactors, like Last Energy's PWR-20, represent one of the most promising solutions for decarbonizing industries and remote installations, with their small size, rapid deployment, and air-cooled design making them flexible and adaptable. SMRs provide a mid-scale solution, offering more substantial power output than microreactors while maintaining the modularity and scalability that appeal to cities or industrial regions with medium energy demands. Traditional nuclear plants, though complex and resource-intensive, remain unmatched in their ability to deliver large-scale power to highly populated regions.

Each type of nuclear technology plays a vital role in the transition to clean energy. Microreactors, SMRs, and traditional nuclear plants collectively offer a range of scalable, flexible, and sustainable power options, which allows governments and utilities to select the best fit based on regional needs, resources, and long-term energy goals. As the world shifts toward low-carbon energy, these nuclear technologies will be instrumental in meeting diverse power demands and driving progress toward a more sustainable future.

Nuclear reactors are diverse in their designs, fuels, coolants, and applications, each tailored to meet specific energy, safety, and efficiency requirements. Understanding these variations is crucial for selecting the most suitable type based on operational goals and resource availability. Here's an in-depth look at the main types of nuclear reactors and how they differ.

1. Pressurized Water Reactors (PWRs): Pressurized Water Reactors (PWRs) are among the most widely used nuclear reactors globally, known for their stability and safety features. They use pressurized water as both a coolant and a neutron moderator. In a PWR, water is kept under high pressure in the primary loop to prevent it from boiling. The heat from the reactor core transfers to a secondary loop via a steam generator, where it produces steam to drive a turbine. This separation of the radioactive primary loop from the turbine significantly enhances safety. PWRs typically use enriched uranium as fuel, commonly in the form of uranium dioxide. These reactors are prized for their high operational stability, ease of control, and low risk of coolant boiling. They are primarily used in electricity generation in countries like the United States, France, and Russia.

2. Boiling Water Reactors (BWRs): Boiling Water Reactors (BWRs) use a different approach: water in the reactor core is allowed to boil, generating steam directly within the reactor vessel. This steam, containing some radioactivity, is then directed to the turbine to generate electricity, eliminating the need for a secondary loop. Like PWRs, BWRs use water as both coolant and moderator and rely on enriched uranium as fuel. The simpler design of BWRs, due to the absence of a secondary loop, provides slightly higher thermal efficiency and reduced construction complexity compared to PWRs. However, additional containment measures are needed due to the radioactive steam in the turbine. BWRs are commonly found in power plants, particularly in the United States and Japan.

3. Heavy Water Reactors (CANDU Reactors): Heavy Water Reactors, such as Canada's CANDU reactors, stand out for their use of deuterium oxide (heavy water) as both a coolant and a moderator. This design allows them to use natural (unenriched) uranium, which makes these reactors efficient and cost-effective, as they do not require uranium enrichment. Heavy water reactors achieve higher neutron efficiency, enabling the use of alternative fuels like thorium. Their unique ability to use natural uranium provides fuel flexibility, reducing dependence on enrichment facilities. These reactors are widely used in Canada and India for power generation, where heavy water is relatively accessible.

4. Fast Breeder Reactors (FBRs): Fast Breeder Reactors (FBRs) differ significantly from other reactors because they do not use a moderator and operate with fast neutrons. They typically use a liquid metal, often sodium, as the coolant, allowing the reactor to operate at high temperatures without pressurization. FBRs are known for their ability to "breed" fuel, producing more fissile material (usually plutonium) than they consume. This capability makes FBRs highly fuel-efficient and capable of recycling nuclear waste by converting it into fuel. FBRs use enriched uranium or plutonium, often mixed with depleted uranium or thorium. Russia and France are leaders in using FBRs for power generation and research, as they enable efficient fuel utilization and have the potential to reduce nuclear waste.

5. Gas-Cooled Reactors (GCRs): Gas-Cooled Reactors (GCRs) use carbon dioxide or helium as a coolant and graphite as a moderator. In these reactors, high-temperature gas flows through the reactor core, transferring heat to a secondary loop to drive a turbine. GCRs can operate at higher temperatures than water-cooled reactors, which enhances thermal efficiency. Using graphite as a moderator enables slower neutron moderation, making the reactor efficient and flexible in fuel choices, such as enriched or natural uranium. GCRs were historically popular in the UK, though they are less common today. Advanced gas-cooled reactor designs continue to be used in experimental settings to study high-temperature nuclear applications.

6. Molten Salt Reactors (MSRs): Molten Salt Reactors (MSRs) offer a unique approach by using molten salt both as a coolant and a fuel carrier. In MSRs, uranium or thorium is dissolved in molten salt, which circulates through the reactor core, where it transfers heat to a secondary loop to produce electricity. This reactor type operates at low pressure but very high temperatures, enhancing thermal efficiency and reducing the risk of explosion. The molten salt can be continuously processed to remove fission products, improving fuel efficiency and safety. MSRs are considered inherently safer because they do not require pressurization. Although primarily experimental, MSRs have ongoing research in countries like the United States, China, and Canada, with potential future applications in thorium-based nuclear fuel cycles.

7. High-Temperature Gas-Cooled Reactors (HTGRs): High-Temperature Gas-Cooled Reactors (HTGRs) are a specific type of gas-cooled reactor that uses helium as a coolant and graphite as a moderator, operating at extremely high temperatures. This design enables HTGRs to generate higher-temperature heat, which makes them versatile for applications beyond electricity generation, including hydrogen production and industrial processes requiring high temperatures. They typically use enriched uranium or thorium embedded in coated fuel particles. HTGRs boast high thermal efficiency and are suitable for co-generation of electricity and industrial heat. While primarily used in research, HTGRs have significant potential for integration into hydrogen production and various industrial sectors.

Types of SMR Coolants and Fuel

Small Modular Reactors (SMRs) represent an innovative approach in nuclear technology, offering compact, efficient, and versatile designs that differ significantly based on their fuel types, coolant systems, and potential applications. Understanding the basic types of SMRs reveals how these reactors can be tailored for diverse energy needs, including electricity generation, industrial heat applications, and even hydrogen production [52].

Uranium-235 as the Common Fuel Element

Across all SMR designs, uranium-235 (U-235) is the primary nuclear fuel used for fission. This isotope is a staple in nuclear power due to its efficiency in sustaining chain reactions, and Australia, with its abundant uranium reserves—including the world's largest deposit at Olympic Dam—is well-positioned to support SMR development. In these reactors, a coolant system is critical for transferring heat from the reactor core, maintaining safe operating temperatures, and preventing overheating. SMRs are categorized based on the type of coolant used, which significantly affects their efficiency, operational temperatures, and potential applications [52].

Boiling Water Reactor (BWR) (300 Megawatts)

The Boiling Water Reactor (BWR) SMR design operates by allowing water within the reactor core to boil directly, creating steam that drives a turbine to produce electricity. This single-loop system means the reactor coolant also serves as the steam source, with the steam generated within the reactor vessel itself. BWRs use enriched uranium in the form of uranium oxide pellets as fuel. While this design simplifies the system, any fuel leak can make the water radioactive, potentially transferring radioactivity to the turbine and the rest of the coolant loop.

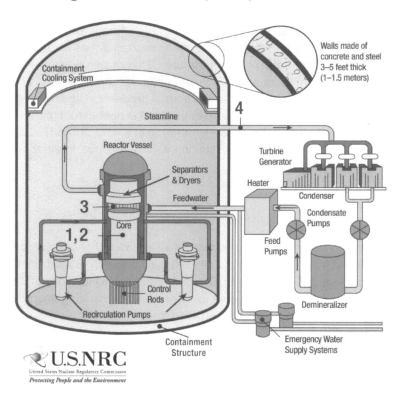

Figure 6: Typical Boiling-Water Reactor. Nuclear Regulatory Commission, CC BY 2.0, via Flickr.

With an efficiency of up to 35%, BWR SMRs are effective for electricity generation but require stringent safety measures to prevent radioactive contamination [52].

Figure 7: The La Crosse Boiling Water Reactor (LACBWR) is located on the east bank of the Mississippi River in Vernon County, WI in NRC Region III. Nuclear Regulatory Commission, CC BY 2.0, via Flickr.

The La Crosse Boiling Water Reactor (LACBWR), as shown in Figure 7, was a small, experimental nuclear power plant located near Genoa, Wisconsin, on the banks of the Mississippi River. This reactor was one of the early Boiling Water Reactors (BWRs) in the United States, initially constructed as part of a demonstration program to evaluate the potential of nuclear power in providing reliable electricity. It began operations in 1967 and was initially owned by the Dairyland Power Cooperative in partnership with the Atomic Energy Commission (AEC), which aimed to demonstrate and refine commercial nuclear technology. The reactor operated at a modest output of 50 megawatts electric (MWe), significantly smaller than most commercial reactors, making it a testing ground for the BWR technology that would later become widespread.

As a Boiling Water Reactor, the La Crosse reactor utilized a single-loop system where water in the reactor core was allowed to boil, directly generating steam that drove the turbine to produce electricity. Unlike Pressurized Water Reactors (PWRs), which keep the primary water loop under high pressure to prevent boiling, BWRs allow the water to boil naturally in the reactor vessel. This design has certain advantages, such as simplification of the plant's design, but also presents challenges, as the steam generated in the reactor core contains trace amounts of radioactivity.

This radioactive steam requires careful handling, containment, and shielding as it passes through the turbine and condenser systems before being released or recirculated.

The La Crosse Boiling Water Reactor served not only as a power generator but also as an experimental facility to advance nuclear power technology and safety protocols. The reactor operated with enriched uranium fuel, which produced a steady fission reaction and high heat levels. Dairyland Power Cooperative, with support from the AEC, used the facility to gather operational data, improve the safety and efficiency of BWRs, and contribute valuable insights that would influence the design of future BWRs. This included research into the effects of radiation on materials and fuel behaviour over time, providing essential data for the nuclear industry's growth.

By 1987, after 20 years of operation, the La Crosse reactor was shut down, as its limited power output and the challenges of maintaining an older facility rendered it economically unfeasible. The shutdown marked the beginning of the plant's decommissioning process, a complex undertaking due to the need to safely dismantle the reactor and handle radioactive materials. The U.S. Nuclear Regulatory Commission (NRC) oversaw the decommissioning, ensuring compliance with stringent safety and environmental standards. Over several decades, the site was carefully dismantled, with reactor components being removed, radioactive waste managed, and the reactor building itself eventually demolished.

The decommissioning process at La Crosse became an example for the nuclear industry, showcasing best practices in safely retiring nuclear plants. This was particularly important as many older reactors across the U.S. began reaching the end of their operational lives. La Crosse's decommissioning provided a model for waste handling, site remediation, and environmental monitoring, contributing valuable experience to the growing field of nuclear decommissioning.

Today, the site where the La Crosse Boiling Water Reactor once stood has been largely restored. Although the reactor is no longer in operation, its legacy endures as an early pioneer in the commercial use of nuclear energy. The data gathered from its operation and the lessons learned from its decommissioning have informed the development and retirement of later nuclear facilities. The La Crosse Boiling Water Reactor remains a key part of the history of nuclear energy in the United States, exemplifying the challenges and contributions of early nuclear power programs.

High Temperature Gas-cooled Reactor (HTGR) (80 Megawatts)

The High Temperature Gas-cooled Reactor (HTGR) SMR is designed to use helium as the coolant, operating at high temperatures that reach up to 850°C. This makes HTGRs not only efficient for electricity generation but also suitable for industrial process heat applications. Graphite serves as the moderator, slowing down neutrons to sustain the fission reaction, while helium is selected for its chemical inertness and lack of neutron absorption. The heated gas spins a turbine, although the design involves complex components, such as a critical magnetic bearing in the heated gas flow. Despite challenges, HTGRs reach efficiencies of up to 41%, and their high operating temperatures make them particularly useful for industries requiring intense heat [52].

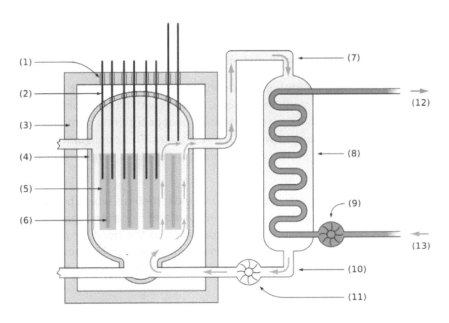

Key:
1. Charge tubes
2. Control rods
3. Radiation shielding
4. Pressure vessel
5. Graphite moderator
6. Fuel rods
7. Hot gas duct
8. Heat exchanger
9. Water circulator
10. Cool gas duct
11. Gas circulator
12. Steam
13. Water

Figure 8: Schematic diagram of a Magnox nuclear reactor. Emoscopes, CC BY-SA 3.0, via Wikimedia Commons.

A Magnox nuclear reactor. as shown in Figure 8, is one of the earliest types of gas-cooled nuclear reactors, developed in the United Kingdom during the 1950s. These reactors were primarily designed for electricity generation and producing plutonium for military purposes. The reactor's name, "Magnox," derives from the magnesium-aluminium alloy used to encase the uranium fuel, known as "magnesium non-oxidizing." This reactor type played a significant role in the UK's early nuclear power program and stands as a key historical step in nuclear technology.

Magnox reactors use natural uranium as fuel, which does not require enrichment and is relatively inexpensive. This feature made these reactors both accessible and cost-effective for early nuclear programs. The uranium fuel is encased in a Magnox alloy cladding, which prevents oxidation and protects the uranium from the carbon dioxide coolant circulating through the reactor.

The coolant in Magnox reactors is carbon dioxide (CO_2), which flows through the reactor core, absorbs the heat produced by nuclear fission, and then passes through a heat exchanger. This heat is used to generate steam that drives a turbine, producing electricity. The moderator in a Magnox reactor is graphite, which slows down the neutrons to sustain a stable fission reaction, allowing the reactor to use natural uranium fuel effectively.

The design of a Magnox reactor features a gas-cooled, graphite-moderated structure, with large cylindrical cores surrounded by graphite blocks. Channels within the graphite contain fuel rods and coolant flow, all housed within a steel pressure vessel or concrete containment structure to manage the high pressures associated with CO_2 as a coolant. This configuration allowed Magnox reactors to produce substantial power, with natural uranium as a simpler and economical fuel option.

One unique characteristic of Magnox reactors is on-load refuelling, which permits the reactor to remain operational while fuel assemblies are removed and replaced. This capability enables continuous operation and consistent power output without the need for periodic shutdowns, making these reactors more flexible for power generation. However, Magnox reactors operate at lower temperatures compared to modern reactors, resulting in lower thermal efficiency— typically around 20-30%.

Magnox reactors offered several advantages. The use of natural uranium fuel eliminated the need for costly uranium enrichment, reducing fuel processing expenses. Additionally, on-load refuelling allowed these reactors to maintain continuous operation, minimizing downtime and enhancing their utility in power generation. Another advantage was the dual-use capability of Magnox reactors; they could generate electricity while also producing plutonium, a valuable byproduct for nuclear weapons programs in the mid-20th century.

Despite their advantages, Magnox reactors had several limitations. One major drawback was their low efficiency due to the low operating temperature, making them less efficient than later nuclear reactor designs. Additionally, Magnox alloy cladding had a tendency to corrode in water, complicating fuel handling and storage and limiting the cladding's lifespan. Over time, maintaining these reactors became increasingly challenging due to the aging infrastructure, leading to decommissioning as newer, more efficient technologies emerged.

Initially, Magnox reactors served both as a source of electricity and as a means to produce plutonium for military applications. The UK operated a series of Magnox reactors for several decades, starting with Calder Hall in 1956. As nuclear technology advanced, Magnox reactors were gradually phased out, with the last operational Magnox reactor, Wylfa Unit 1 in Wales, ceasing operations in 2015.

Although no longer in active use, Magnox reactors hold significant historical value as some of the earliest power-producing nuclear reactors. Their gas-cooled, graphite-moderated design demonstrated the feasibility of nuclear energy for both power generation and plutonium production, paving the way for further developments in nuclear technology and influencing future reactor designs.

Figure 9: Hinkley Point B, Advanced Gas Cooled reactor. Reading Tom, CC BY 2.0, via Flickr.

Hinkley Point B, as shown in Figure 9, is a nuclear power station located in Somerset, England, and is home to two Advanced Gas-cooled Reactors (AGRs), a type of nuclear reactor unique to the United Kingdom. AGRs are an evolution of the earlier Magnox reactor design, intended to provide higher efficiency and power output. Construction of Hinkley Point B began in 1967, and it became fully operational in 1976. The station comprises two reactors, each originally designed to produce around 660 megawatts of electricity, enough to power approximately 1.5 million homes. Hinkley Point B marked a significant step forward in nuclear technology for the UK, improving upon earlier designs by enhancing thermal efficiency and operational performance.

The Advanced Gas-cooled Reactor design uses carbon dioxide (CO_2) as a coolant, circulating through the reactor core to absorb the heat generated by nuclear fission. This hot CO_2 gas then passes through a heat exchanger to produce steam, which drives a turbine and generates electricity. Unlike earlier gas-cooled reactors, the AGR operates at much higher temperatures—up to 650°C—improving the overall thermal efficiency of the system. The choice of graphite as a neutron moderator allows the reactor to use slightly enriched uranium as fuel, unlike the natural uranium used in the previous Magnox reactors. This combination of CO_2 coolant and graphite moderator proved effective, enabling AGRs to achieve efficiencies close to 41%, a notable improvement over previous designs.

A unique feature of the Hinkley Point B AGRs, and AGRs in general, is their fuel design and configuration. The reactor core comprises thousands of graphite blocks with channels for uranium fuel rods and CO_2 coolant flow. The uranium fuel is housed in stainless steel cladding, which offers higher resistance to corrosion than the magnesium alloy used in earlier designs. This stainless steel cladding allows for higher temperature operations, contributing to the AGR's improved efficiency and extended operational lifespan. Additionally, AGRs are designed for on-load refuelling, meaning they can replace spent fuel while the reactor continues to operate, reducing downtime and maximizing energy output.

Safety and structural features of Hinkley Point B reflect the focus on improved containment and durability. Each reactor is housed in a concrete pressure vessel, capable of withstanding high-pressure CO_2 circulation. In addition, Hinkley Point B includes multiple safety systems to ensure reactor shutdown in emergencies, including control rods that can be rapidly inserted into the core to halt the fission reaction. Backup cooling systems are also in place to prevent overheating in case of an interruption in CO_2 flow.

Despite its success, Hinkley Point B faced challenges typical of early AGRs, including issues with fuel cladding corrosion and graphite moderation over time. As the graphite in the core ages, it can lose structural integrity, affecting the core's ability to moderate neutrons effectively. These issues have led to increased maintenance demands and operational adjustments over the years to ensure continued safety. Originally designed to operate for around 25 years, Hinkley Point B exceeded its initial lifespan thanks to significant upgrades and regular maintenance. However, as the reactors aged, the safety and economic feasibility of continued operation came into question. As a result, Hinkley Point B was finally shut down in August 2022 after 46 years of service.

Hinkley Point B's legacy is notable in the UK nuclear industry, showcasing the potential of Advanced Gas-cooled Reactors to provide reliable, efficient energy over an extended period. The station played a key role in reducing the UK's reliance on fossil fuels, contributing significantly to the nation's low-carbon energy supply. The success and longevity of Hinkley Point B also informed future AGR developments and influenced the design and operation of subsequent UK reactors. As Hinkley Point B enters the decommissioning phase, its contribution to nuclear power and the insights gained from its operation continue to shape the UK's nuclear strategy and energy policies.

Molten Salt Reactors (MSRs) (35-300 Megawatts)

Molten Salt Reactors (MSRs) are unique in that they can dissolve reactor fuel in a molten salt mixture, using this molten salt as both fuel and coolant. Operating at high temperatures and low pressures, MSRs are highly efficient for thermal energy conversion and can also provide process heat. These reactors are well-suited for electricity generation, though challenges include corrosion and the need to shut down and drain the fuel/coolant mix for refuelling. Operating temperatures of 850°C make MSRs compatible with high-temperature applications. MSRs are among the most versatile SMR designs, balancing efficient energy production with potential industrial applications [52].

The Integral Molten Salt Reactor (IMSR), as shown as a schematic in Figure 10, developed by Canadian company Terrestrial Energy, is an advanced type of Molten Salt Reactor (MSR) designed for enhanced safety, simplified operations, and improved efficiency in commercial power generation. This reactor represents a significant leap in MSR technology, targeting reliable, low-carbon energy production with applications in electricity generation, industrial heat, and even hydrogen production.

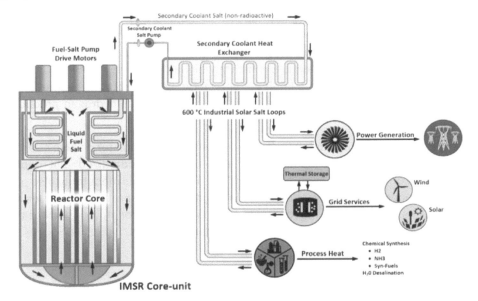

Figure 10: Schematic describing the different possible heat applications for the IMSR. Terrestrial Energy, CC BY-SA 4.0, via Wikimedia Commons.

One of the core innovations of the IMSR is its integral reactor design. The IMSR integrates all primary reactor components, including the reactor core, heat exchangers, and pumps, into a single sealed vessel known as the Core-unit. This modular approach reduces the number of moving parts exposed to the corrosive molten salt, enhancing both safety and operational reliability while lowering maintenance requirements. Each Core-unit is designed for a fixed operational period—typically around seven years—after which it is replaced, allowing for continuous reactor operation with minimal interruptions for refuelling.

The IMSR uses molten salt as both fuel and coolant, a unique feature of MSRs that eliminates the need for traditional solid fuel rods. In this design, the nuclear fuel is dissolved in the molten salt, which flows through the reactor core and absorbs the heat produced by nuclear fission. The molten salt then transfers this heat to a secondary salt loop, which can be used to produce steam or deliver high-temperature heat for industrial applications. Operating at high temperatures (up to 700°C) and low pressure, the IMSR significantly reduces the risk of high-pressure failures and is more efficient than traditional reactors operating at lower temperatures.

The IMSR also incorporates passive safety features that rely on natural forces like gravity and convection to manage reactor conditions. For instance, if the reactor overheats, the molten salt expands naturally, slowing down the fission reaction and cooling the system without active intervention. Additionally, the reactor is equipped with a drain tank that can automatically empty the molten salt fuel into a separate container if an emergency shutdown is needed. In this tank, the fuel cools and solidifies safely, providing an additional layer of passive safety.

Fuel flexibility and efficiency are other key aspects of the IMSR. The reactor can use different types of nuclear fuel, including low-enriched uranium and potentially thorium, making it adaptable to various fuel cycles. Operating at high temperatures enables the IMSR to achieve around 45% thermal efficiency, which is significantly higher than many conventional reactors. This high-efficiency output makes the IMSR suitable not only for electricity generation but also for industrial applications requiring high-temperature heat, such as hydrogen production and chemical processing.

The IMSR's modular design and scalability make it an attractive option for diverse energy needs. Its modular Core-unit simplifies construction and enables standardized unit deployment, reducing construction times compared to conventional reactors. This modularity also allows the IMSR to be scaled according to power demand, making it flexible for both small utilities and large industrial operators. Additionally, its compact size and modular construction lower the initial investment required, broadening its appeal to a wider range of users.

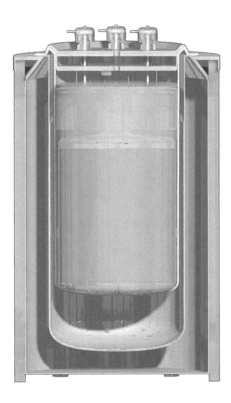

Figure 11: Reactor vessel (Core-unit) and guard vessel of the Integral Molten Salt Reactor.
Terrestrial Energy, CC BY-SA 4.0, via Wikimedia Commons.

The applications and potential impact of the IMSR extend beyond electricity generation due to its high-temperature capabilities. The heat generated by the IMSR can be utilized in various industrial applications, including hydrogen production through thermochemical processes, process heat for chemical manufacturing, and even synthetic fuel production. By supporting both power and industrial applications, IMSRs offer a multi-functional solution for energy-intensive sectors seeking to reduce their carbon footprint.

In summary, the IMSR is a next-generation molten salt reactor that emphasizes safety, efficiency, and modularity. Its innovative design positions it as a promising player in advanced nuclear technology, with the potential to serve as a low-carbon, versatile energy source for power generation and industrial applications alike. If successfully commercialized, the IMSR could make substantial contributions to decarbonizing traditionally hard-to-electrify sectors, advancing the global transition to clean energy.

Pressurized Water Reactors (PWRs) (60-300 Megawatts)

Pressurized Water Reactor (PWR) SMRs, see diagram in Figure 12, use pressurized water both as a coolant and a neutron moderator. In this design, a two-loop system separates the primary coolant from the secondary steam generator loop, which drives turbines to generate electricity. This setup prevents water in the core from boiling, maintaining stability.

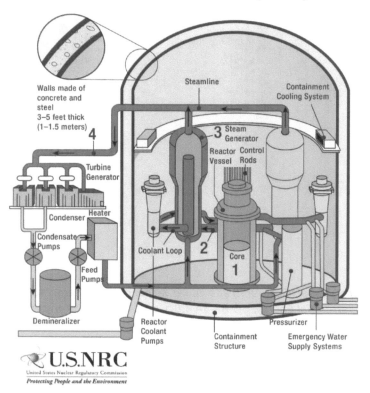

Figure 12: Typical Pressurized-Water Reactor. Nuclear Regulatory Commission, CC BY 2.0, via Flickr.

The use of pressurized water allows PWRs to reach efficiencies of up to 35%, though the high-pressure environment demands durable, high-cost components. Safety measures in PWR SMRs are similar to those of large water-cooled reactors. This design is reliable and widely used, with the two-loop system providing enhanced containment of radioactive materials [52].

Figure 13: Two Pressurized Water Reactors (PWRs) under construction at the Kudankulam nuclear power plant, India. IAEA Imagebank, CC BY-SA 2.0, via Flickr.

The Kudankulam Nuclear Power Plant (KKNPP), as shown in Figure 13 whilst under construction, is India's largest nuclear power facility, located in the Tirunelveli district of Tamil Nadu, near the southern tip of the country. Developed through a collaboration between India and Russia, the plant is operated by the Nuclear Power Corporation of India Limited (NPCIL) and uses Russian-built VVER (Water-Water Energetic Reactor) reactors. KKNPP represents a significant milestone in India's nuclear energy program, not only because of its scale but also as a model for future collaboration and advancements in nuclear technology.

Construction of the Kudankulam plant began in 2002, following years of planning and negotiation between India and Russia. The facility currently hosts two operational reactors, each with a capacity of 1,000 megawatts electric (MWe), which were connected to the grid in 2013 and 2016, respectively. These VVER reactors use pressurized water as both a coolant and moderator, a design that incorporates several modern safety features, including passive cooling systems that activate in emergencies. When completed, Kudankulam will host a total of six reactors, with the four additional units currently under construction and expected to be operational in phases over the coming years. This would ultimately increase the plant's capacity to 6,000 MWe, making it one of the largest nuclear power installations in the world.

Kudankulam's VVER reactors are known for their advanced safety features. Each reactor is equipped with a passive heat removal system that can maintain core cooling without external power, an essential feature in case of a power outage. The reactors are also enclosed in a double containment structure to prevent radioactive leakage in the event of an accident. The plant includes additional safety measures, such as hydrogen recombiners to prevent hydrogen buildup, a core catcher to contain melted fuel, and advanced seismic protection systems. These features align the Kudankulam reactors with international standards, reflecting India's commitment to safe and reliable nuclear power.

The economic and strategic importance of Kudankulam is considerable for India. Nuclear power is a key component of India's strategy to meet its growing energy demands while reducing reliance on fossil fuels. Kudankulam plays a pivotal role in this vision, as it provides a stable, low-carbon energy source that helps offset coal-fired power. Moreover, the plant is part of India's long-term nuclear expansion goals, aiming to increase nuclear capacity to 22,480 MWe by 2031. Kudankulam is a prime example of India's ambition to scale up nuclear energy, contributing not only to energy security but also to environmental sustainability.

Despite its benefits, the Kudankulam project has faced controversy and local opposition. Concerns about radiation risks, potential environmental impacts, and the plant's location in a seismic zone have led to protests from local communities and environmental groups. In response, the Indian government and NPCIL have emphasized the rigorous safety measures in place and provided transparency regarding safety protocols. The protests brought attention to the challenges of nuclear energy in densely populated areas, prompting further discussion on balancing energy needs with community concerns.

Kudankulam Nuclear Power Plant represents a landmark in India's energy landscape, showcasing the potential of nuclear energy to support economic growth while addressing environmental challenges. As construction on additional reactors progresses, Kudankulam is expected to become a cornerstone of India's nuclear infrastructure, fostering energy independence and advancing India's commitment to clean energy. The plant stands as a testament to India-Russia cooperation in nuclear technology and exemplifies India's strategy to integrate nuclear power into its diverse energy mix, aiming for a sustainable and secure energy future.

As another example of Pressurized Water Reactors, The Dukovany Nuclear Power Plant, as per Figure 14, is one of the Czech Republic's key sources of electricity, located in the Vysočina Region near the village of Dukovany, roughly 30 kilometres from the Austrian border. Operated by ČEZ Group, the largest electricity producer in the Czech Republic, Dukovany has played a crucial role in the country's energy landscape since it began operations in 1985. This plant was the Czech Republic's first commercial nuclear power plant and remains a vital contributor to the national grid, producing around 20% of the country's electricity.

Dukovany houses four pressurized water reactors (PWRs), specifically the VVER-440/V-213 models, each with a capacity of around 510 megawatts electric (MWe), giving the plant a total generating capacity of approximately 2,040 MWe. These reactors were designed by Soviet engineers and incorporate various safety improvements based on lessons learned from other early nuclear reactors. The VVER design, short for "water-water energy reactor" in Russian, uses light water as both a coolant and moderator, which helps maintain a stable and controlled fission reaction within the core. Over the years, Dukovany's reactors have undergone multiple upgrades to enhance their safety, performance, and efficiency, including systems for improved control, diagnostics, and digital monitoring.

Figure 14: Inside the reactor hall at Dukovany Nuclear Power Plant. Dukovany is one of two nuclear power plants in the Czech Republic. It has two main production units, each of them contains two pressurized water reactors. Imagebank, CC BY-SA 2.0, via Flickr.

The safety systems at Dukovany are a critical aspect of the plant's operations, especially given its age. Like many older nuclear plants, Dukovany has been subject to rigorous safety assessments and upgrades to align with modern safety standards. These include seismic reinforcements, improved reactor vessel inspections, and redundant emergency cooling systems to ensure the plant's resilience in case of an accident. The plant also has comprehensive radiation monitoring systems to track emissions and safeguard the surrounding environment. The Czech State Office for Nuclear Safety (SÚJB) oversees the regulatory framework and safety inspections at Dukovany, ensuring compliance with national and international nuclear safety protocols.

Dukovany's economic significance is considerable, as it contributes significantly to the Czech Republic's energy security. Nuclear power is a cornerstone of the Czech energy mix, and Dukovany, together with the Temelín Nuclear Power Plant, accounts for more than one-third of the country's electricity generation. This reliance on nuclear energy has allowed the Czech Republic to reduce its dependence on coal, which has historically been a primary energy source.

The Czech government views nuclear power as essential for achieving its energy policy goals, particularly in reducing carbon emissions and ensuring a stable supply of low-cost, low-carbon electricity. Plans are underway to construct additional units at Dukovany, with bids from several international reactor vendors being evaluated. These expansions aim to replace aging reactors and support the nation's energy transition.

Over the years, the lifespan of Dukovany's reactors has been extended multiple times, thanks to continuous upgrades and maintenance. The reactors, initially intended to operate for 30 years, are now expected to remain in service until at least 2035, with potential extensions further based on their condition and regulatory approval. ČEZ Group has invested significantly in modernization projects at Dukovany to ensure safe and efficient operation, including replacing aging equipment, upgrading control systems, and improving waste management practices.

However, the plant has faced opposition and concerns from neighbouring Austria, where there is significant public resistance to nuclear energy. Austrian environmental groups and some political leaders have raised concerns about the plant's proximity to their border and its operational safety, especially given its Soviet-era design. While the Czech government has reassured its neighbours about the plant's safety measures, the debates around Dukovany highlight the broader tensions in Europe regarding nuclear energy, where attitudes vary widely from one country to another.

Sodium or Lead-Bismuth-cooled Fast Nuclear Reactor (Liquid Metal FNR) (100-345 Megawatts)

Fast Nuclear Reactors (FNRs) in SMR form use liquid metals like sodium, lead, or lead-bismuth as coolants. These reactors operate with fast neutrons, which enable efficient fuel utilization and the ability to "burn" certain long-lived radioactive isotopes, reducing nuclear waste. With operating temperatures between 550°C and 800°C, Liquid Metal FNRs are suitable for both electricity and industrial heat applications. Using sodium as a coolant provides excellent heat transfer but reacts violently with water, posing a safety risk. FNRs also use advanced fuel types, such as metallic UZr with minor actinides or mononitride-mixed fuel. This design operates at low pressure and relies on natural convection for coolant circulation, minimizing mechanical components [52].

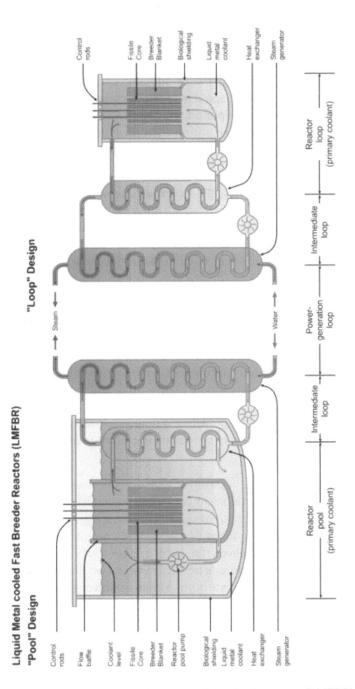

Figure 15: Summary schematic of Liquid Metal Fast Breeder Reactors (LMFBR). . Emoscopes, CC BY-SA 2.5, via Wikimedia Commons.

Liquid Metal Fast Breeder Reactors (LMFBRs), as shown as a schematic in Figure 15, are a type of nuclear reactor that use liquid metal, typically sodium or a sodium-potassium alloy, as a coolant and are designed to generate more fissile material than they consume. This reactor type is unique in its use of fast neutrons rather than moderated (slowed-down) neutrons to sustain the fission chain reaction. The "breeder" aspect of LMFBRs comes from their ability to convert non-fissile uranium-238 (U-238) into fissile plutonium-239 (Pu-239), creating new fuel as the reactor operates. LMFBRs were first developed in the mid-20th century as part of an effort to improve the efficiency of nuclear fuel use and address concerns about uranium scarcity by "breeding" additional fuel.

One of the defining characteristics of LMFBRs is their use of liquid metal coolant, with sodium being the most common choice. Sodium's high thermal conductivity allows it to effectively transfer heat away from the reactor core, maintaining the necessary temperature while operating at low pressure. This low-pressure operation reduces the risk of high-pressure failures that can occur in water-cooled reactors. However, sodium has some drawbacks; it reacts vigorously with water and can ignite if exposed to air, requiring specialized handling and containment systems to mitigate these risks. Despite these challenges, sodium's superior heat transfer properties make it an ideal coolant for the high-energy environment within an LMFBR.

In contrast to traditional reactors that use slow neutrons, LMFBRs rely on a fast neutron spectrum. Fast neutrons are not slowed by a moderator and can induce fission in a broader range of nuclear materials, including the U-238 present in most reactor fuel. This allows the reactor to "breed" new fissile material in a surrounding blanket of U-238, converting it into Pu-239, which can then be used as fuel. By producing more fissile material than they consume, LMFBRs are extremely fuel-efficient, theoretically providing a nearly self-sustaining fuel cycle. This design addresses the issue of uranium scarcity by making use of more abundant U-238, extending the life of nuclear fuel resources.

Safety features in LMFBRs are primarily passive, relying on the inherent properties of the reactor's materials and design. For example, liquid sodium's thermal inertia and high boiling point provide a buffer against temperature spikes, helping prevent overheating. Additionally, the design of most LMFBRs includes redundant cooling loops and advanced containment structures to control the highly reactive sodium coolant. However, the risk of sodium leaks and fires remains a challenge. In the event of a malfunction, the reactor's passive systems would allow the reactor to gradually shut down without the need for immediate human intervention or external power sources, adding a layer of safety to the reactor's operations.

LMFBRs offer substantial environmental benefits, especially regarding nuclear waste management. Unlike conventional reactors, which generate a significant amount of high-level radioactive waste, LMFBRs can recycle certain waste materials as fuel, particularly the long-lived actinides found in spent nuclear fuel. This capability reduces the volume and longevity of radioactive waste, presenting a more sustainable option for nuclear energy. In some cases, LMFBRs can also help "burn" plutonium from dismantled nuclear weapons, providing a productive use for otherwise hazardous materials.

Despite their advantages, LMFBRs face significant economic and engineering challenges. The reactors are expensive to build and operate due to the complex systems needed to manage liquid metal coolant safely. Sodium-cooled reactors have been especially costly, as they require specialized containment and maintenance to handle the chemical reactivity of sodium. These challenges have slowed the widespread adoption of LMFBR technology, with only a few operational examples worldwide, such as the BN-600 and BN-800 reactors in Russia and the Monju reactor (Figure 17) in Japan (now decommissioned).

Figure 16: Reactor BN-800. Rosatom. Empresa Estatal de Energía Atómica Rusa, CC0, via Wikimedia Commons.

Figure 17: Fast Breeder Reactor Monju in Japan Fuku. Nife, CC BY-SA 3.0, via Wikimedia Commons.

Nuclear Hydrogen Production with SMRs

The high temperatures generated by gas-cooled and molten salt-cooled SMRs make them suitable for hydrogen production through thermochemical processes, often termed "red hydrogen." Using nuclear heat, high-temperature electrolysis or thermochemical water-splitting cycles—such as Cu-Cl, V-Cl, Mg-Cl, and S-I cycles—can produce hydrogen from water. These cycles reuse chemicals in a closed loop, consuming only water and producing hydrogen and oxygen. Efficiency varies by cycle, ranging from 46.9% to 69%, and while nuclear-based hydrogen production has the advantage of using waste heat, renewable electrolysis methods currently achieve comparable efficiency at around 70%. Although promising, the high temperature requirements (500°–2,000°C) limit nuclear hydrogen production to specific reactor designs, like HTGRs and MSRs, which require extensive testing before commercial use [52].

Heavy Industry Use and SMR Waste Heat

The waste heat from gas-cooled, molten salt-cooled, and molten metal-cooled SMRs can be repurposed for heavy industries that require high temperatures. Industries such as chemical manufacturing, cement production, and primary metals manufacturing could benefit from waste heat in the range of 800-850°C. Specific processes, like lead and zinc smelting or ammonia synthesis via the Haber-Bosch process, align well with the heat output of SMRs, creating a potential win-win for industrial efficiency and energy utilization. However, some industries, like cement and iron production, require temperatures that exceed SMR waste heat output, necessitating additional energy input, often through electric arc furnaces, which diminishes the economic appeal [52].

SMRs offer varied designs with distinct advantages for power generation, industrial applications, and hydrogen production. BWRs provide direct steam generation, HTGRs deliver high-temperature heat for specialized applications, MSRs are flexible and efficient for process heat, PWRs offer reliable containment, and Liquid Metal FNRs excel in fuel efficiency and waste reduction. Each type of SMR has unique operational advantages, with potential for further advancements in energy and industrial applications as the technology matures [52].

Evolution of Nuclear Technology

The evolution of nuclear technology spans over a century, marked by major advancements, diverse applications, and significant changes in public perception. Beginning with early theoretical discoveries, nuclear technology progressed through multiple phases, from weapons development and energy generation to advanced reactors and emerging applications in medicine, industry, and space exploration. Each phase has built upon previous knowledge, continuously expanding nuclear technology's potential while striving to enhance safety, efficiency, and environmental sustainability.

The roots of nuclear technology can be traced back to the late 19th and early 20th centuries, with scientific breakthroughs that laid the foundation for nuclear physics. Key figures like Henri Becquerel, Marie Curie, and Ernest Rutherford explored radioactivity, identifying the properties of radioactive elements. In 1938, Otto Hahn and Fritz Strassmann discovered nuclear fission, revealing that splitting an atom could release vast amounts of energy. Shortly thereafter, Lise Meitner and Otto Frisch provided the theoretical explanation for fission, marking a turning point in the scientific understanding of atomic energy.

The discovery of nuclear fission during World War II led to the development of nuclear weapons through the Manhattan Project, a U.S.-led research initiative. In 1945, the project culminated in the creation of the atomic bomb, which was subsequently used in Hiroshima and Nagasaki, demonstrating the unprecedented destructive power of nuclear energy. This event fundamentally changed global geopolitics, highlighting the dual-use nature of nuclear technology—capable of both immense destruction and immense potential for civilian applications. The post-war era saw a nuclear arms race between superpowers, leading to significant advancements in weaponry and nuclear physics.

In the 1950s, the focus began shifting towards the peaceful use of nuclear energy, specifically for electricity generation. The first civilian nuclear reactor, Obninsk Nuclear Power Plant in the Soviet Union, began operations in 1954, becoming the world's first grid-connected nuclear power plant. Shortly after, the United States launched the Shippingport Atomic Power Station in 1957, marking the first civilian nuclear power plant in the U.S. These early reactors, often referred to as Generation I reactors, were based on rudimentary technology with basic safety features. Although they were effective in producing power, they had limitations in efficiency, reliability, and safety.

With rising energy demands, the 1960s and 1970s saw a global expansion of nuclear power, with the introduction of Generation II reactors. These reactors, including Pressurized Water Reactors (PWRs) and Boiling Water Reactors (BWRs), became the industry standard and remain among the most widely used reactor types today. Generation II reactors offered improvements in efficiency, safety, and standardization. However, this period was also marked by notable accidents, such as the Three Mile Island incident in 1979 and the Chernobyl disaster in 1986, which led to increased scrutiny, stricter regulations, and a slowdown in new reactor construction in many countries. Public concerns about safety and environmental risks prompted a shift in focus toward designing safer, more reliable reactors.

Responding to the challenges and lessons learned from earlier generations, Generation III and III+ reactors were developed in the 1990s and early 2000s. These reactors incorporated advanced safety features, such as passive safety systems that operate without human intervention or external power, enhanced fuel efficiency, and extended operational lifespans. Examples include the AP1000, EPR, and ABWR reactors (see Figure 18), which employ sophisticated containment structures, digital monitoring, and emergency cooling systems. Generation III+ reactors aim to prevent severe accidents through multiple layers of defence-in-depth and to withstand natural disasters and power outages. These designs have revitalized interest in nuclear energy as a safe, reliable, and low-carbon power source, particularly in the face of climate change concerns.

Figure 18: Kashiwazaki-Kariwa Nuclear Power Station, ABWR. (Kashiwazaki-Kariwa, Japan). IAEA Imagebank, CC BY-SA 2.0, via Wikimedia Commons.

Building upon previous designs, Generation IV reactors represent a future vision for nuclear technology, focusing on sustainability, safety, and waste reduction. Generation IV designs include Fast Breeder Reactors (FBRs), Molten Salt Reactors (MSRs), High-Temperature Gas-Cooled Reactors (HTGRs), and Lead-Cooled Fast Reactors (LFRs), among others. These reactors are designed to be more fuel-efficient, capable of recycling nuclear waste, and inherently safe, often using coolants like liquid metal, gas, or molten salt. Some designs aim to close the nuclear fuel cycle by "breeding" new fuel, making nuclear energy more sustainable. Additionally, Generation IV reactors can provide high-temperature heat for industrial processes, hydrogen production, and other non-electric applications, extending the versatility of nuclear energy.

Recent years have seen the emergence of Small Modular Reactors (SMRs) and microreactors, which promise a more flexible and scalable approach to nuclear power. SMRs are compact reactors with outputs ranging from 50 to 300 MWe, designed for factory assembly and modular deployment. Their smaller size and modular nature make them suitable for remote locations, industrial sites, and smaller grids, allowing for incremental capacity additions as needed.

Microreactors, with even smaller outputs of around 1 to 20 MWe, are envisioned as highly portable, providing localized power in remote or critical areas. Both SMRs and microreactors incorporate passive safety features, making them safer and more adaptable to diverse energy needs.

Exploring the Nature of the Atom

The journey to understanding the atom began with early scientific discoveries about elements and radiation. Uranium, a crucial element in nuclear science, was discovered in 1789 by the German chemist Martin Klaproth and named after the planet Uranus. The concept of ionizing radiation emerged in 1895 when Wilhelm Röntgen passed an electric current through an evacuated glass tube, producing continuous X-rays. A year later, Henri Becquerel found that pitchblende, a uranium-rich ore, darkened photographic plates due to emissions of beta radiation (electrons) and alpha particles (helium nuclei). Paul Villard later identified gamma rays, adding a third type of radiation similar to X-rays. In 1896, Pierre and Marie Curie coined the term "radioactivity" for these emissions and successfully isolated radium and polonium from pitchblende. These elements soon found applications, with radium, for instance, used in medical treatments. In 1898, Samuel Prescott demonstrated that radiation could sterilize food by killing bacteria [53].

Figure 19: Ernest Rutherford is considered the founder of experimental nuclear physics. Av George Grantham Bain Collection, Public Domain, via Store norske leksikon.

In 1902, Ernest Rutherford, see Figure 19, revealed that radioactivity involved the spontaneous emission of alpha or beta particles, transforming one element into another. He advanced atomic theory, showing in 1919 that firing alpha particles into nitrogen caused nuclear rearrangement, creating oxygen. Niels Bohr furthered this understanding, focusing on the electron's arrangement around the nucleus into the 1940s. By 1911, Frederick Soddy discovered that radioactive elements had isotopes with identical chemistry but different nuclear compositions, and George de Hevesy proved that these isotopes served as effective tracers due to their detectability in minute amounts [53].

The neutron's discovery by James Chadwick in 1932, followed by Cockcroft and Walton's successful nuclear transformations using proton bombardment, set the stage for artificial radioisotope production. In 1934, Irène Curie and Frédéric Joliot found that bombarding atoms could create synthetic radionuclides. Enrico Fermi expanded on this, showing that a greater variety of radionuclides could be produced using neutrons instead of protons. Fermi's experiments with uranium produced not only heavier elements but also lighter fragments, hinting at the process of nuclear fission. In 1938, Otto Hahn and Fritz Strassmann demonstrated that atomic fission had occurred by identifying the byproduct, barium. Working under Niels Bohr, Lise Meitner and Otto Frisch explained fission as the splitting of the nucleus into two parts, releasing approximately 200 million electron volts of energy. Frisch's experimental confirmation in January 1939 validated Albert Einstein's 1905 paper on mass-energy equivalence [53].

Harnessing Nuclear Fission

The discovery of fission ignited a flurry of research in laboratories worldwide. Hahn and Strassmann found that fission not only released energy but also emitted additional neutrons, which could trigger further fission in nearby uranium nuclei, creating a potential self-sustaining chain reaction. This was experimentally confirmed by Joliot and colleagues in Paris and by Leo Szilard and Enrico Fermi in New York.

Ernest Lawrence, Enrico Fermi and I. I. Rabi at Bandelier, ca 1945.

Figure 20: Ernest Lawrence, Enrico Fermi and Isidor Isaac Rabi at Bandelier. LA-Lawrence, Fermi, Rabbi, Public Domian, via Picryl.

Bohr proposed that the uranium-235 isotope, rather than the more abundant uranium-238, was more likely to undergo fission, especially with slow-moving neutrons. Szilard and Fermi confirmed this, suggesting the use of a "moderator" to slow down neutrons, enhancing the likelihood of fission in U-235. Bohr and Wheeler's analysis of fission, published in 1939, laid the theoretical foundation for nuclear chain reactions [53].

A practical challenge arose from the fact that uranium-235 made up only 0.7% of natural uranium, with uranium-238 accounting for the remaining 99.3%. This difference in isotopic composition required scientists to separate U-235 through enrichment, exploiting its slight physical differences from U-238. In 1939, Francis Perrin introduced the concept of a "critical mass," the minimum amount of uranium needed to sustain a chain reaction. Rudolph Peierls later expanded on these ideas, and his calculations were crucial for atomic bomb development. In Paris, Perrin's group demonstrated that a chain reaction could be sustained in a uranium-water mixture with neutron injection, paving the way for controlled nuclear reactions, the basis of nuclear reactors. They also introduced neutron-absorbing materials to regulate neutron multiplication, a principle still used in modern nuclear power plants [53].

The Race to Build Nuclear Weapons

With rising tensions during World War II, nuclear fission research took on military significance. Peierls, a student of Werner Heisenberg, contributed to the German nuclear project under the German Ordnance Office. By late 1939, Heisenberg theorized that a controlled chain reaction in a "uranium machine" (nuclear reactor) could generate energy, while an uncontrolled reaction could produce an explosive yield far surpassing conventional bombs. Heisenberg also proposed that using pure U-235 could create a powerful explosive, though he initially overestimated the critical mass needed for such a device [53].

In 1940, Carl Friedrich von Weizsäcker, Heisenberg's colleague, suggested that a chain reaction would convert some U-238 into element 94, later named plutonium, which could serve as an explosive material. He even applied for a patent to produce this element in a uranium machine. However, by 1942, the German project faced resource limitations and shifted focus to rocket development. Yet, the prospect of a German atomic bomb remained a motivating factor for the United States and Britain, leading to the Manhattan Project and the eventual development of the first atomic bombs [53].

The pursuit of nuclear technology spurred scientific breakthroughs, military applications, and the establishment of nuclear power as a lasting energy source. The early exploration of the atom and the harnessing of nuclear fission have left an indelible impact on science, energy, and global history [53].

Nuclear Physics in Russia

The origins of nuclear physics in Russia trace back more than a decade before the Bolshevik Revolution, with research into radioactive minerals starting around 1900. These minerals were primarily sourced from Central Asia, a region rich in naturally occurring radioactive materials. By 1909, the St. Petersburg Academy of Sciences initiated a large-scale investigation into radioactivity, marking the beginning of Russia's formal nuclear science research. This foundation laid by early Russian scientists gained momentum following the 1917 Revolution, which led to a strong governmental push for scientific research. In the years following the revolution, over ten physics institutes were established across Russia, especially in St. Petersburg, making scientific advancement a national priority.

In the 1920s and early 1930s, many Russian physicists took the opportunity to work abroad, with the regime's encouragement, to gain expertise and further elevate Russia's scientific knowledge base. The new government recognized that quickly advancing their knowledge of nuclear physics was essential, and sending scientists overseas was seen as the best way to achieve this. Kirill Sinelnikov, Pyotr Kapitsa, and Vladimir Vernadsky were among the notable Russian scientists who worked abroad during this period. Their time spent in prestigious research institutions outside Russia allowed them to absorb and bring back cutting-edge nuclear science techniques and ideas, which they later applied within Russia's burgeoning scientific institutes [53].

By the early 1930s, Russia had established multiple research centres dedicated to nuclear physics. Sinelnikov returned from Cambridge in 1931 to set up a department at the Ukrainian Institute of Physics and Technology (later renamed the Kharkov Institute of Physics and Technology, or KIPT) in Kharkov, which had been founded in 1928. Around the same time, Abram Ioffe established a group specializing in nuclear physics at the Leningrad Physics and Technical Institute (FTI). This group later became independent as the Ioffe Institute, with Ioffe himself serving as its director until 1950. The team at Leningrad included Igor Kurchatov, a young scientist who would later play a leading role in Russia's nuclear research. These centres became pivotal to advancing Russia's understanding of nuclear physics.

Figure 21: Main building of the National Science Center "Kharkiv Institute of Physics and Technology" (NSC KIPT), Pyatykhatky settlement, Kharkiv (2021). Photo: Sergiy Bobok. From Wikimedia Commons. License CC BY-SA 4.0.Фото: Serhii Bobok. 3 Wikimedia Commons. Ліцензія CC BY-SA 4.0.

By the late 1930s, cyclotrons—devices capable of accelerating atomic particles—were installed at both the Radium Institute and the Leningrad FTI. These were some of the largest cyclotrons in Europe at the time, underscoring Russia's commitment to staying at the forefront of nuclear research. However, this progress coincided with a period of intense political repression under Stalin's purges, during which many scientists suffered arrest, imprisonment, or worse. At the Kharkov Institute alone, half of the staff were arrested in 1939. Despite these setbacks, Russian nuclear research continued, and by 1940, substantial advances were made in the understanding of nuclear fission and the potential for a chain reaction, laying the groundwork for further nuclear advancements [53].

At the urging of Igor Kurchatov and his colleagues, the Academy of Sciences established a "Committee for the Problem of Uranium" in June 1940, chaired by Vitaly Khlopin. This committee aimed to explore uranium's potential as a fuel source and to investigate Central Asia's uranium deposits for scientific and possible military use. The Radium Institute had a facility in Tartarstan where Khlopin succeeded in producing Russia's first high-purity radium, a significant achievement in the country's nuclear history. However, the course of nuclear research shifted in 1941, when Germany's invasion of Russia redirected many resources and scientific efforts toward military applications. This period marked the beginning of nuclear science's intersection with national defence in Russia, setting the stage for later developments in atomic weaponry and energy.

Conception of the Atomic Bomb

The early concept of the atomic bomb was profoundly influenced by British scientists and refugee physicists who were driven by the urgency of the war. Among these physicists, Rudolf Peierls and Otto Frisch were particularly instrumental. They, both German-born physicists, had fled Nazi Germany and taken refuge in Britain, where they continued their work despite the upheaval of war. In 1940, they created a concise but groundbreaking document known as the Frisch-Peierls Memorandum. This three-page memorandum laid out the scientific foundation for an atomic bomb, predicting that a relatively small amount of uranium-235—around 5 kilograms—could produce an explosion equivalent to thousands of tons of dynamite. In their memorandum, they detailed how the bomb might be detonated, suggested possible methods for isolating U-235 from natural uranium, and explained the potential radiation effects that would follow the blast. One of their key proposals was using thermal diffusion to separate U-235 from uranium's more abundant isotope, U-238. At the time, this memorandum generated significant interest in Britain, in contrast to the more hesitant approach in the United States [53].

In response to the memorandum, the British government established a group of prominent scientists known as the MAUD Committee. This committee supervised atomic research across several prestigious institutions, including the Universities of Birmingham, Bristol, Cambridge, Liverpool, and Oxford. Their mission was to investigate the feasibility of developing an atomic bomb. Birmingham University and Imperial Chemical Industries (ICI) undertook the challenging chemical work required to produce gaseous compounds of uranium and pure uranium metal. In 1940, Dr. Philip Baxter at ICI produced the first small batch of uranium hexafluoride gas, a vital compound for uranium enrichment, for Professor James Chadwick. Later that year, ICI received an official contract to produce 3 kilograms of uranium hexafluoride, marking an essential step forward in the project. Other aspects of the research were funded directly by the universities, underscoring the shared commitment of academic institutions to the war effort.

Research at Cambridge University yielded two critical breakthroughs. First, experimental work demonstrated that a sustained chain reaction could occur when slow neutrons interacted with a mixture of uranium oxide and heavy water. This confirmed that the number of neutrons produced was greater than those consumed, proving the feasibility of a self-sustaining reaction. The second breakthrough came from Otto Frisch and Egerton Bretscher, who expanded upon earlier work by Hans Halban and Lew Kowarski, French scientists who had fled to Britain from Paris. Their research focused on the differing reactions of U-235 and U-238 when absorbing neutrons. They observed that while U-235 readily underwent fission when it absorbed slow neutrons, U-238 was more likely to transform into a new isotope, U-239. This new isotope would then undergo beta decay, creating element 93 (later named neptunium), and subsequently decay into element 94 (later named plutonium). Bretscher and Feather theorized that this element 94 would be highly fissionable by both slow and fast neutrons, and that it could be easily separated from uranium due to its distinct chemical properties. This discovery provided the theoretical groundwork for producing plutonium, which would later become a critical component in the atomic bomb [53].

This theoretical work on plutonium was independently corroborated by Edwin McMillan and Philip Abelson in the United States in 1940. At Cambridge, Dr. Kemmer suggested naming the new elements by following the pattern of the periodic table's outermost elements: neptunium for element 93, after Neptune, and plutonium for element 94, after Pluto. Coincidentally, American scientists proposed the same names, establishing an international consensus. The official discovery of plutonium in 1941 is widely credited to Glenn Seaborg and his team in the U.S., marking an important milestone in nuclear science and the development of the atomic bomb.

These breakthroughs not only expanded the theoretical understanding of nuclear fission but also set the stage for the practical development of atomic weapons. British scientists, supported by refugee researchers, established foundational knowledge and techniques that would later be pivotal in the U.S.-led Manhattan Project. Their work underscored the critical importance of collaboration and shared scientific inquiry during a period of intense global conflict. The insights gained from this research led to the realization that uranium and plutonium isotopes could produce unprecedented amounts of energy, fundamentally transforming military strategy and shaping the course of modern history [53].

By the close of 1940, significant advancements in atomic bomb research had been achieved by British scientists under the MAUD Committee. Despite the limited funding allocated to the project, the coordinated efforts among research groups yielded remarkable insights into nuclear fission and the feasibility of building an atomic bomb. While this work was carried out in utmost secrecy in Britain, scientists in the United States continued to publish on nuclear research, showing little urgency or immediate focus on weaponization. The British, however, were increasingly motivated by the ongoing war in Europe and the possibility that Germany might also be pursuing nuclear weapons.

One major breakthrough in the spring of 1941 involved confirming the fission cross-section of uranium-235 (U-235)—a critical factor in understanding how likely a U-235 nucleus was to split upon collision with a neutron. Initially, physicists Peierls and Frisch had theorized that almost every neutron collision with U-235 would cause fission, regardless of the neutron's speed. However, it was later determined that slow neutrons were far more effective at inducing fission in U-235, a finding particularly relevant to the development of nuclear reactors, though it had less impact on bomb design, where fast neutrons would still suffice. With these findings, Peierls declared that an atomic bomb was indeed possible, provided that highly enriched U-235 could be secured. They estimated the critical mass needed for a U-235 bomb at approximately 8 kilograms, a figure that could be reduced with the right neutron-reflecting material, but direct measurements remained necessary. Recognizing this, the British accelerated efforts to produce even a few micrograms of pure U-235 for testing.

In July 1941, the MAUD Committee issued two key reports. The first report, titled "Use of Uranium for a Bomb," concluded that constructing a nuclear bomb was technically feasible. The report estimated that a bomb containing around 12 kilograms of U-235 would produce an explosion equivalent to 1,800 tons of TNT and would emit significant radiation, rendering areas around the blast site hazardous for extended periods. To produce 1 kilogram of U-235 daily, the committee estimated that a facility costing £5 million would be necessary, requiring a large, skilled workforce that was also in demand for other wartime projects. Concerned that Germany might also be working on nuclear weapons, the report urged that Britain pursue the bomb project urgently and in partnership with the United States, despite America's primary focus at the time being the peaceful use of uranium for power and naval propulsion.

The second report, "Use of Uranium as a Source of Power," focused on the potential for controlled nuclear fission to generate heat for mechanical applications and produce radioisotopes as alternatives to radium. It discussed possible moderators for the fission process, such as heavy water, graphite, and even regular water if the uranium were enriched in U-235. The report anticipated that a "uranium boiler," essentially a nuclear reactor, held considerable promise for peaceful energy production in the postwar era, although it was deemed not worth pursuing during wartime. The committee recommended that scientists Hans Halban and Lew Kowarski relocate to the United States, where large-scale production of heavy water was planned. The possibility of using plutonium as a fuel instead of U-235 was also noted, prompting continued research by British scientists Bretscher and Feather into the fissionable properties of plutonium.

The MAUD reports prompted a complete restructuring of British efforts in atomic bomb development and reactor research, or the "boiler" project. The committee's work was regarded as groundbreaking; within fifteen months, it had established Britain as a leader in nuclear research and demonstrated the effectiveness of collaborative scientific oversight. Prime Minister Winston Churchill, with the backing of the Chiefs of Staff, made the decisive commitment to prioritize the bomb project. The implications of the reports extended beyond Britain, prompting a high-level review in the United States led by the National Academy of Sciences. Initially, American interest centred on nuclear power, rather than weaponization. This stance changed dramatically following the attack on Pearl Harbor on December 7, 1941, as the United States entered World War II. With this shift, the United States committed its extensive resources to developing atomic bombs, leading to the establishment of the Manhattan Project, an unprecedented collaboration that would ultimately bring the atomic bomb to fruition [53].

The Manhattan Project

The Manhattan Project marked a critical turning point in the development of nuclear weapons during World War II, rapidly surpassing British efforts and establishing the United States as the leader in atomic research. By early 1942, American scientists and engineers were actively engaged in several competing methods for enriching uranium to create a bomb, while maintaining some information exchange with British scientists. In 1942, several key British researchers travelled to the United States, where they gained full access to American atomic research. The Americans pursued three primary enrichment processes in parallel: electromagnetic separation led by Professor Ernest Lawrence at Berkeley, centrifuge separation by E. V. Murphree of Standard Oil, and gaseous diffusion coordinated by Professor Harold Urey at Columbia University. Simultaneously, Arthur Compton at the University of Chicago was tasked with building a reactor to produce plutonium, another fissile material with potential for bomb construction. At this stage, the British focused solely on gaseous diffusion, leaving the Americans to explore a broader range of methods [53].

In June 1942, the U.S. Army assumed full control over the project, overseeing process development, engineering design, material procurement, and site selection for pilot plants dedicated to all four fissionable material production methods. This massive mobilization shifted the project's focus toward an industrial scale, and at this point, communication with Britain diminished. This breakdown in information exchange posed a setback for both British and Canadian scientists, who had been collaborating with the Americans, particularly in the area of heavy water production and reactor research. In response, Winston Churchill began exploring the feasibility of constructing a diffusion plant, heavy water facility, and nuclear reactor within Britain, aiming to bolster British nuclear research independently.

Negotiations between the British and American governments continued for several months, culminating in an agreement signed by Churchill and President Franklin D. Roosevelt in Quebec in August 1943. Under this agreement, Britain handed over all of its research reports to the United States, while receiving in exchange copies of General Leslie Groves' progress reports to the President. These reports revealed the extraordinary scope and cost of the Manhattan Project, which was expected to exceed $1 billion—an unprecedented sum, all directed toward developing the atomic bomb. Unlike in Britain, the American program focused exclusively on bomb production, without diverting resources to explore other peaceful applications of nuclear energy [53].

By 1943, construction of production facilities was well underway across multiple locations. The electromagnetic separation plant, utilizing "calutrons," and the gaseous diffusion plant were progressing swiftly. Meanwhile, the first controlled nuclear chain reaction had been achieved in December 1942 at the University of Chicago, where Enrico Fermi and his team successfully

Figure 22: Dr. J. Robert Oppenheimer, atomic physicist and head of the Manhattan Project. U.S. National Archives and Records Administration, Public Domain, via Picryl.

operated an experimental graphite reactor, known as the Chicago Pile-1. To produce plutonium, a large-scale production reactor was under construction at Argonne, with additional reactors later built at Oak Ridge and Hanford. These facilities were accompanied by reprocessing plants to extract plutonium from irradiated uranium. Furthermore, four plants were established for heavy water production—three in the United States and one in Canada—emphasizing the sheer scale of the American nuclear project.

As research continued, Robert Oppenheimer (see Figure 22) and his team at Los Alamos, New Mexico, took on the critical task of designing and constructing the bombs. They worked simultaneously on two types of devices: one using highly enriched U-235 and the other using Pu-239. Both were complex undertakings, but by mid-1945, sufficient quantities of Pu-239 and U-235 had been amassed, primarily due to enrichment efforts at Oak Ridge. The uranium needed for the bombs mostly came from the Belgian Congo, a region with significant uranium deposits.

The project's culmination was the Trinity Test on July 16, 1945, at Alamogordo, New Mexico, where the first atomic device, fuelled by plutonium, was detonated successfully. The test's success demonstrated the devastating power of nuclear fission and solidified the effectiveness of the plutonium bomb design. However, the simpler U-235 bomb design was considered reliable enough that it did not require testing. On August 6, 1945, the first atomic bomb, a U-235 device, was dropped on Hiroshima, Japan, inflicting massive destruction. Just three days later, on August 9, a second bomb, containing plutonium (Pu-239), was dropped on Nagasaki. That same day, the Soviet Union declared war on Japan, intensifying pressure on the Japanese government [53].

On August 10, 1945, Japan surrendered, ending the war. The Manhattan Project had not only transformed the course of World War II but also ushered in the nuclear age, fundamentally altering global military strategy, geopolitics, and the future of energy development. The project exemplified unprecedented scientific collaboration, engineering ingenuity, and governmental support, but it also raised profound ethical and existential questions about humanity's capacity to wield such destructive power. The legacy of the Manhattan Project lives on, influencing nuclear policy, arms control, and the development of nuclear energy for generations to come [53].

The Soviet Atomic Bomb Program

Initially, Soviet leader Joseph Stalin was hesitant to allocate resources toward developing an atomic bomb, as he was focused on the ongoing war and was cautious about diverting efforts from conventional military priorities. However, intelligence reports suggesting atomic research efforts in Germany, Britain, and the United States shifted his perspective, leading him to consider the strategic implications of nuclear weapons. Consultations with prominent Soviet scientists, including Abram Ioffe, Pyotr Kapitsa, Vitaly Khlopin, and Vladimir Vernadsky, convinced Stalin that a bomb could be developed relatively quickly. Based on their advice, he initiated a modest atomic research program in 1942. Igor Kurchatov, a relatively young and emerging scientist at the time, was appointed to lead this initiative. In 1943, Kurchatov became the Director of Laboratory No. 2, a newly established research facility on the outskirts of Moscow, later renamed LIPAN and eventually known as the Kurchatov Institute of Atomic Energy. The overall responsibility for the bomb program was placed under the control of Soviet Security Chief Lavrenti Beria, with administrative oversight managed by the First Main Directorate, which later became the Ministry of Medium Machine Building [53].

The Soviet nuclear research effort focused on three primary objectives: achieving a controlled chain reaction, investigating methods for isotope separation, and exploring designs for both enriched uranium and plutonium bombs. To initiate a controlled chain reaction, scientists experimented with two types of atomic piles (nuclear reactors): one using graphite as a moderator and another employing heavy water. Additionally, they examined three methods for separating uranium isotopes to produce the necessary enrichment levels: counter-current thermal diffusion, gaseous diffusion, and electromagnetic separation. These early experiments laid the groundwork for the eventual production of fissionable materials critical to bomb construction [53].

After Nazi Germany's defeat in May 1945, Soviet authorities "recruited" German scientists to work on the bomb program, particularly in the area of isotope separation to produce enriched uranium. This recruitment expanded the Soviet research capabilities, introducing gas centrifuge technology as a new enrichment method alongside the other techniques. Despite the success of the first U.S. atomic bomb test in July 1945, the Soviet program remained largely unaffected. By this time, Kurchatov and his team were making notable progress in developing both uranium and plutonium bombs. Kurchatov had initiated designs for an industrial-scale reactor to produce plutonium, while advancements in gaseous diffusion were being made by scientists working on uranium isotope separation [53].

The atomic bombings of Hiroshima and Nagasaki in August 1945 elevated the profile of the Soviet nuclear program and underscored the importance of matching U.S. nuclear capabilities. In November 1945, construction began on a new secret city in the Ural Mountains called Chelyabinsk-40 (later known as Chelyabinsk-65 or the Mayak production association), which would house the Soviet Union's first plutonium production reactors. This facility marked the first of ten secret "nuclear cities" established in the Soviet Union, and it included a processing plant to extract plutonium from irradiated uranium. The first of five reactors at Chelyabinsk-65 became operational in 1948, signifying a major milestone in Soviet atomic production [53].

Regarding uranium enrichment, Soviet authorities decided in late 1945 to begin constructing the first gaseous diffusion plant at Verkh-Neyvinsk (later the closed city of Sverdlovsk-44), near Yekaterinburg. Specialized design bureaus were established at the Leningrad Kirov Metallurgical and Machine-Building Plant and the Gorky Machine Building Plant to support these efforts. The program also benefited from the expertise of a group of German scientists who worked at the Sukhumi Physical Technical Institute. Together, these groups contributed to the Soviet mastery of uranium enrichment techniques essential for nuclear weapons development.

In April 1946, bomb design work shifted to a newly established centre called Design Bureau-11 in Sarova, a location that later became the closed city of Arzamas-16. This move marked the intensification of efforts to design an actual atomic device, bringing in additional experts, including metallurgist Yefim Slavsky, who was tasked with producing the high-purity graphite needed for Kurchatov's plutonium production pile, known as F-1. The F-1 pile, housed at Laboratory No. 2, became operational in December 1946. Additional support came from Laboratory No. 3 in Moscow, now known as the Institute of Theoretical and Experimental Physics, which contributed reactor research critical to the program's progress [53].

Intelligence-gathering activities influenced the bomb's design, leading Soviet scientists to model their first device closely on the Nagasaki bomb (a plutonium-based weapon). In August 1947, the Soviets established a test site near Semipalatinsk in Kazakhstan, where preparations began for the first bomb test. This site became central to Soviet nuclear testing. In August 1949, just two years after the test site was completed, the Soviet Union successfully detonated its first atomic bomb, RDS-1. Even before this test, another group of Soviet scientists led by Igor Tamm and Andrei Sakharov had begun preliminary work on a hydrogen bomb, marking the start of the next phase in nuclear weapons development [53].

Revival of the 'Nuclear Boiler'

By the end of World War II, the devastating power of nuclear technology, as predicted by the Frisch-Peierls Memorandum only five and a half years earlier, had been demonstrated in the atomic bomb. As the war ended, scientific attention began to shift towards the peaceful use of nuclear energy. Although weapons development continued in both the Soviet Union and the West amid rising Cold War tensions, a new focus emerged: harnessing nuclear energy to produce steam and electricity for civilian purposes. Scientists recognized that the immense heat generated from nuclear fission, initially used for weaponry, could instead be utilized to power industries and homes. This discovery opened the door to developing compact, long-lasting power sources, potentially transforming industries like maritime shipping and submarine operations.

The first step in this direction was made by Argonne National Laboratory in Idaho, USA, with the Experimental Breeder Reactor (EBR-1). This small reactor, which started up in December 1951, became the first to generate a small amount of electricity. Although its output was minimal, it demonstrated the feasibility of nuclear reactors for power generation and laid the groundwork for future advancements in nuclear energy. Then, in 1953, President Dwight D. Eisenhower announced the Atoms for Peace program, shifting significant research efforts towards electricity generation from nuclear energy and establishing a path for civil nuclear energy development in the United States.

In the Soviet Union, nuclear energy research also accelerated post-war, focusing on refining existing reactors and creating new designs. The Institute of Physics and Power Engineering (FEI) was established in 1946 in the closed city of Obninsk, about 100 kilometres southwest of Moscow, to spearhead nuclear power technology development. In June 1954, FEI achieved a major milestone by operating the world's first nuclear-powered electricity generator: the AM-1 reactor, also known as Atom Mirny (Peaceful Atom). This graphite-moderated, water-cooled reactor had a design capacity of 30 MWt (5 MWe) and marked a significant achievement, demonstrating that nuclear technology could be used for peaceful applications. The AM-1 reactor remained in operation until 1959 for electricity production and continued as a research facility until 2000, serving as a prototype for later Soviet reactors, including the RBMK (Reaktor Bolshoi Moshchnosty Kanalny or high-power channel reactor) model used in the Chernobyl plant [53].

During the 1950s, the FEI at Obninsk also explored fast breeder reactors (FBRs) and lead-bismuth reactors for the Soviet navy. The BR-1 (Bystry Reaktor, or Fast Reactor), a fast-neutron reactor with no power output, began operation in April 1955. It led to the BR-5, which started up in 1959 with a capacity of 5 MWt, providing a platform for the Soviet Union to conduct foundational research for sodium-cooled FBRs. The BR-5 was upgraded in 1973 and underwent major reconstruction in 1983, becoming the BR-10 with an 8 MWt capacity. This reactor continues to support fuel endurance studies, materials research, and isotope production, playing a significant role in advancing nuclear technology.

In the United States, Admiral Hyman Rickover led the development of the pressurized water reactor (PWR), primarily for naval use, especially in submarines. The PWR design used enriched uranium oxide fuel and was cooled and moderated by ordinary (light) water. The Mark 1 prototype naval reactor, started in March 1953 in Idaho, set the stage for the USS Nautilus, the world's first nuclear-powered submarine, launched in 1954. By 1959, both the United States and the Soviet Union had launched their first nuclear-powered surface vessels, marking the beginning of nuclear propulsion in naval fleets. The success of the Mark 1 reactor prompted the US Atomic Energy Commission to construct the Shippingport demonstration PWR reactor in Pennsylvania, with a capacity of 60 MWe. Shippingport started up in 1957 and remained operational until 1982, showcasing the potential of PWR technology for civilian energy generation [53].

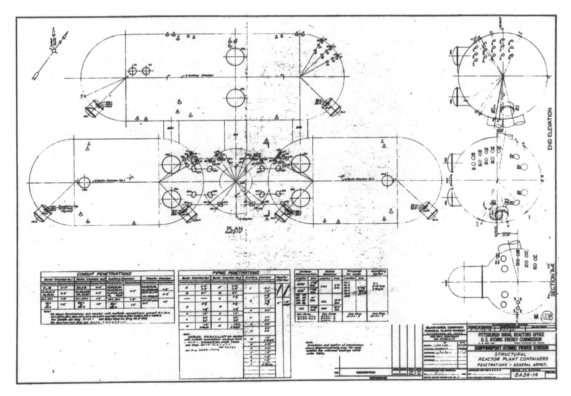

Figure 23: Plans for the Shippingport Atomic Power Station, On Ohio River, 25 miles Northwest of Pittsburgh, Shippingport, Beaver County, PA. Library of Congress, Public Domain, via Getarchive.

Figure 24: The Shippingport Atomic Power Station, the world's first full-scale atomic electric power plant devoted exclusively to peacetime production of electricity, is located near the present-day Beaver Valley Nuclear Generating Station. Photo from 1956. Library of Congress, CC BY 2.0, via Flickr.

In the United Kingdom, the post-war nuclear development took a different direction due to limited access to uranium enrichment technology, which the United States controlled. Instead, British scientists developed reactors fuelled by natural uranium metal, moderated by graphite, and cooled by gas. The Magnox reactors, with the first unit at Calder Hall starting up in 1956, represented this unique approach. These reactors, with an output of 50 MWe, were fuelled by natural uranium and ran until 2003. After building 26 units, Britain halted Magnox reactor development in 1963, shifting towards advanced gas-cooled reactors (AGRs) that used enriched uranium oxide fuel. Eventually, the UK adopted the more globally prevalent PWR design, acknowledging its efficiency and cost-effectiveness [53].

The revival of the "nuclear boiler" concept through these advancements demonstrated the potential for nuclear technology to revolutionize energy production. From the Atoms for Peace initiative to pioneering reactors in the Soviet Union and naval applications in the United States, the post-war years marked a transformative era in nuclear development. The groundwork laid in these early projects fuelled the ongoing growth of nuclear power as a viable, large-scale source of electricity and demonstrated its versatility in various sectors, including national defence and civilian energy applications [53].

The Nuclear Power Brown-Out and Revival

From the late 1970s until the early 2000s, the nuclear power industry experienced a period of decline and stagnation. The momentum that nuclear energy had gained in the mid-20th century slowed considerably, with few new reactor orders and a gradual shift in the industry's growth trajectory. By the mid-1980s, the number of new reactors coming online barely kept pace with the retirement of older ones. Although nuclear capacity grew by about one-third, and overall output increased by 60% due to improved load factors and increased operational efficiency, the industry faced challenges that prevented significant expansion. During this time, nuclear power's share in global electricity production remained steady at around 16-17%. Many reactor projects initiated in the 1970s were cancelled, resulting in a drop in uranium prices and an increase in secondary supplies. Oil companies that had invested in uranium withdrew, leading to a consolidation of uranium producers and further limiting industry growth.

Despite these setbacks, there were signs of a nuclear power revival by the late 1990s. In Japan, the commissioning of Kashiwazaki-Kariwa 6, a 1350 MWe Advanced Boiling Water Reactor (ABWR), symbolized a shift in the industry's outlook. This third-generation reactor was a major advancement, showcasing improved safety features and operational efficiency, indicating the potential for a nuclear renaissance [53].

Several factors in the early 21st century helped rekindle interest in nuclear power. The first was the projected global increase in electricity demand, particularly in rapidly developing countries. As these economies grew, the demand for reliable, large-scale energy sources became essential. Secondly, energy security emerged as a critical concern. Countries recognized the importance of having reliable, affordable energy supplies, with nuclear power providing dispatchable electricity that could meet fluctuating demand without interruption. Lastly, there was a growing need to reduce carbon emissions in response to concerns about climate change. Nuclear power, as a low-carbon energy source, offered a way to meet these demands while minimizing environmental impact.

These factors aligned with the emergence of new third-generation nuclear reactors that were safer, more efficient, and more adaptable to the needs of modern energy grids. In 2004, the first late third-generation reactor, a 1600 MWe European Pressurized Reactor (EPR), was ordered for Finland, signaling Europe's renewed commitment to nuclear power. France soon followed with plans to construct a similar unit, and the United States began building two new Westinghouse AP1000 reactors. This renewed interest in nuclear energy illustrated a global shift in energy strategy, with nuclear power viewed once again as a viable option for sustainable and secure electricity generation.

While Europe and North America pursued limited nuclear expansion, Asia emerged as the epicenter of nuclear growth. China and India embarked on ambitious nuclear programs, driven by credible political will and strong public support. China, in particular, set plans for a substantial increase in nuclear power capacity by 2030, with more than 100 new large reactors proposed. China's approach included adopting Western reactor designs and developing local adaptations that aligned with its specific energy needs. This ambitious expansion reflects Asia's central role in the future of nuclear power, as these countries work to address growing energy demands and environmental goals.

The trajectory of nuclear power has thus evolved over the decades. Starting with scientific research in Europe and initial growth in the United Kingdom and United States, nuclear energy experienced a lull but found new life in East Asia. This growth has led to an accumulation of over 17,000 reactor-years of operational experience, demonstrating nuclear energy's potential to provide a reliable and significant share of the world's electricity. As nuclear technology continues to advance, it remains a key player in the global pursuit of sustainable and low-carbon energy solutions, reflecting its adaptability and enduring relevance in meeting humanity's energy needs [53].

Nuclear Power in the World Now

Nuclear technology, which originated in the 1940s with a focus on weapon development during World War II, transitioned in the 1950s towards peaceful applications, particularly in energy generation. By harnessing nuclear fission to produce electricity, the potential of nuclear power expanded dramatically. Since the commercial nuclear industry took off in the 1960s, nuclear power plants have now accumulated around 20,000 reactor years of operational experience and are active in 31 countries (plus Taiwan). Regional transmission grids further extend the influence of nuclear power, particularly in Europe, where many countries rely partly on nuclear-generated electricity from neighbouring states [54].

Initially, the nuclear industry operated with distinct boundaries between the East and West during the Cold War. Today, however, it is defined by international collaboration and commerce. A modern nuclear reactor under construction in Asia might use components from South Korea, Canada, Japan, France, Germany, Russia, and other nations, highlighting the truly global nature of the industry. Uranium mined in Australia or Namibia could fuel reactors in the UAE, after processing stages across multiple countries—conversion in France, enrichment in the Netherlands, deconversion in the UK, and fabrication in South Korea. This global interdependence underscores the role of nuclear technology as a shared solution to energy challenges [54].

Beyond electricity generation, nuclear technology plays a crucial role in several fields. It aids in disease control, supports medical diagnostics and treatment, and powers space exploration missions. These applications reinforce nuclear technology's importance in advancing sustainable development goals worldwide. In 2023, nuclear plants collectively supplied 2602

TWh of electricity, an increase from 2545 TWh in 2022, illustrating the industry's steady growth and resilience [54].

Globally, fourteen countries in 2023 generated at least 25% of their electricity from nuclear power. France relies on nuclear energy for up to 70% of its electricity, while Ukraine, Slovakia, and Hungary produce around 50% of their power from nuclear sources. Japan, which historically depended on nuclear energy for over a quarter of its electricity, is working to restore its nuclear capacity to previous levels [54].

In 2024, several new reactors were connected to the grid. Notable examples include Kakrapar 4 in India (630 MWe), Vogtle 4 in the USA (1117 MWe), Barakah 4 in the UAE (1310 MWe), Fangchenggang 4 in China (1105 MWe), and Shidaowan Guohe One 1 in China (1400 MWe). Additionally, new reactor projects began, such as El Dabaa 4 in Egypt, Zhangzhou 3 in China, Leningrad 2-3 in Russia, and others. The steady construction and commissioning of reactors underscore nuclear energy's critical role in meeting future energy demands and reducing carbon emissions. Meanwhile, shutdowns also occurred, such as Kursk 2 in Russia, Maanshan 1 in Taiwan, and Pickering 1 in Canada, as older reactors reached the end of their operational lives [54].

In 2023, 439 reactors operated worldwide, with 66 under construction, 87 planned, and 344 proposed. Each region shows varying levels of nuclear development, reflecting both the operational and strategic diversity of nuclear power use. For instance, China leads with the most reactors in operation and under construction, with substantial uranium demand expected for its growing fleet [54].

Nuclear reactors have demonstrated significant improvements in performance over time, particularly in terms of capacity factors—the ratio of actual to potential energy output. The capacity factor for reactors has improved markedly over the last 40 years, showing that existing reactors can operate more efficiently and consistently over extended lifespans. Notably, there has been no significant decrease in average capacity factor with reactor age in recent years, suggesting that reactors continue to deliver high performance even as they age [54].

The demand for electricity continues to grow globally, necessitating new power generation capacity to replace aging fossil fuel plants (especially coal) that emit large amounts of carbon dioxide. While renewable energy sources have expanded, fossil fuels still account for 61% of electricity generation, only slightly down from 66.5% in 2005. Nuclear power offers a dependable, low-carbon solution to meet this demand. The OECD International Energy Agency (IEA) has projected that nuclear capacity may need to reach 916 GWe by 2050 to help stabilize global temperature increases under Net Zero Emissions (NZE) scenarios [54].

Beyond commercial nuclear plants, about 220 research reactors are active across over 50 countries. These reactors play vital roles in scientific research, medical isotope production, and industrial applications. Marine propulsion is another area where nuclear power has long been significant, particularly in naval submarines and large surface vessels. Currently, over 160 nuclear-powered ships (mostly submarines) are in service, with Russia and the USA operating the majority. Russia also operates a fleet of nuclear-powered icebreakers and has pioneered the use of a floating nuclear power plant at Pevek in the Arctic, which provides energy to remote areas [54].

The nuclear industry's transition from military origins to a crucial component of the global energy landscape showcases its adaptability and importance. Nuclear energy not only supplies low-carbon electricity but also supports medical, industrial, and space exploration applications. As the world confronts the dual challenges of energy demand growth and carbon reduction, nuclear technology remains an indispensable asset in achieving a sustainable and secure energy future [54].

Benefits of SMRs in Modern Power Generation

Small Modular Reactors (SMRs) represent a transformative approach to modern power generation, offering numerous benefits that address the challenges faced by traditional nuclear power plants.

One of the most significant advantages of SMRs is their enhanced safety features. SMRs are designed with passive safety systems that utilize natural forces such as gravity and convection to cool the reactor without the need for external power or operator intervention during emergencies. This design minimizes the risk of catastrophic failures, as evidenced by studies indicating that SMRs can safely shut down and cool themselves for extended periods without active systems [5, 29]. Furthermore, the smaller size of SMRs inherently limits the quantity of radioactive material present, thereby reducing the potential impact of any incident [55]. These safety characteristics make SMRs suitable for deployment in diverse locations, including urban areas and remote sites where rapid emergency response may be challenging [11, 18].

The modular design of SMRs allows for scalability in energy generation, making them adaptable to varying energy demands. Multiple SMRs can be deployed incrementally, which alleviates financial burdens associated with large-scale nuclear projects [13, 56]. This flexibility is particularly advantageous for regions with unpredictable energy needs or smaller markets, as utilities can expand capacity gradually as demand increases [14, 57]. Additionally, the factory-based production of SMRs facilitates quicker construction timelines and improved quality control, further enhancing their economic viability [13, 18].

SMRs require significantly lower initial capital investments compared to traditional nuclear power plants, which often deter potential investors due to high upfront costs [13]. The smaller footprint and modular nature of SMRs allow for more manageable financial commitments, making nuclear energy accessible to countries and utilities with limited capital resources [14, 56]. This reduced financial risk is particularly appealing for stakeholders looking to diversify their energy portfolios without incurring massive expenditures [13, 57].

SMRs are particularly well-suited for remote and off-grid applications, where traditional energy infrastructure is lacking. Many remote areas currently rely on fossil fuels, which can be costly and environmentally damaging [8, 58]. SMRs can provide stable, low-carbon power in these regions, reducing reliance on high-emission energy sources like diesel generators [11, 18]. Their compact design allows for transportation to rugged or isolated locations, making them a practical solution for military bases, island nations, and mining operations [13, 57].

Compared to traditional reactors, SMRs require less land and water, which expands the potential sites for their deployment [13, 18]. Many SMR designs utilize air cooling systems or require minimal water for cooling, making them suitable for arid regions [5, 29]. This reduced resource requirement allows for closer proximity to urban centres, aligning with sustainable development goals and minimizing environmental impacts [13, 59].

SMRs are versatile and can serve multiple roles beyond electricity generation. They can provide process heat for industrial applications such as hydrogen production and desalination [18, 57]. The high-temperature output of certain SMR designs makes them particularly suitable for these applications, offering a sustainable source of fresh water and energy [11, 56]. This versatility enhances the overall utility of SMRs in addressing diverse energy needs [3, 58].

The environmental impact of SMRs is significantly lower than that of traditional fossil fuel plants. SMRs produce no direct carbon emissions during operation, making them a crucial component in global efforts to reduce greenhouse gas emissions [18, 57]. Additionally, some SMR designs are optimized to minimize nuclear waste generation, further enhancing their environmental credentials [60, 61]. By providing reliable baseload power, SMRs can complement renewable energy sources and contribute to a more sustainable energy mix [13, 59].

SMRs can enhance grid stability, particularly in systems with high levels of renewable energy integration. Their ability to perform load-following functions allows them to adjust output in real-time, complementing the intermittent nature of renewable sources like wind and solar [58, 62]. This operational flexibility is essential for maintaining grid reliability and ensuring a continuous energy supply [18, 59].

The modular nature of SMRs allows for faster deployment compared to traditional nuclear reactors, which often require extensive construction timelines [18, 60]. SMRs can be assembled on-site after being constructed in a factory, significantly reducing the time needed to bring them online [13, 57]. This rapid deployment is advantageous for meeting urgent power needs and scaling up nuclear capacity as part of energy transition strategies [14, 56].

Finally, SMRs are positioned to play a vital role in supporting global decarbonization goals. As countries seek to reduce reliance on fossil fuels, SMRs offer a reliable, low-carbon alternative that can be deployed incrementally [57, 61]. Their ability to provide stable energy while minimizing carbon emissions makes them a strategic asset in achieving sustainable energy objectives [3, 18].

Small Modular Reactors present a compelling solution for modern power generation, characterized by enhanced safety, flexibility, and adaptability. Their ability to meet diverse energy demands while supporting global decarbonization efforts positions them as a transformative technology in the pursuit of a resilient, low-carbon energy future.

SMRs vs. Traditional Nuclear Reactors

Small Modular Reactors (SMRs) and traditional nuclear reactors both aim to provide nuclear power, yet they exhibit significant differences in their design, operation, deployment, and economic considerations. As the nuclear energy landscape evolves, these distinctions reveal unique advantages and challenges for both technologies.

SMRs are characterized by their smaller size, typically generating less than 300 MWe, with some microreactors producing as little as 5 MWe. This compact design allows for deployment in remote areas or regions with limited grid capacity, making SMRs particularly suitable for decentralized power needs [63, 64]. In contrast, traditional nuclear reactors are larger, generally producing between 1,000 to 1,600 MWe per unit, and are designed for centralized energy production, often supplying electricity to large urban areas [14, 15]. The scalability of SMRs allows for flexibility in energy production, as multiple units can be combined to meet increasing demand, whereas traditional reactors are less adaptable to fluctuating energy needs [19, 65].

The construction of SMRs emphasizes modularity, with components manufactured off-site and assembled on-site, significantly reducing construction time and labour costs [64, 65]. This modular approach enables faster deployment, as SMRs can be factory-built and transported by various means (Peng et al., 2023). Conversely, traditional reactors require extensive on-site construction, which can take over a decade due to their size and complexity, involving significant infrastructure and labour [14, 15]. The quicker deployment of SMRs can mitigate project delays and capital costs, while traditional reactors benefit from economies of scale, potentially offsetting high initial construction costs over time [14, 19].

SMRs generally have lower initial capital costs due to their smaller size and modular construction, allowing for more flexible financing options [19, 66]. They can be incrementally scaled and built as demand increases, providing a staggered cash flow model [66]. Traditional reactors, however, require substantial upfront investment and long-term financing, which can create higher financial risks, particularly given their lengthy construction timelines [14, 15]. The lower financial barrier to entry for SMRs makes them appealing for smaller utilities or countries with limited resources, while traditional reactors may achieve lower long-term energy costs per MWe due to their larger capacities [19, 64].

Many SMRs incorporate advanced passive safety features, such as gravity-driven cooling systems, which reduce reliance on active safety systems and human intervention [16]. Their simpler designs typically operate at lower pressures, enhancing safety margins and reducing the likelihood of catastrophic failures [16, 63]. Traditional reactors, while equipped with extensive safety systems, often rely on both active and passive mechanisms, necessitating more complex safety protocols due to their operation at high pressures and temperatures [15, 67]. The inherent safety features of SMRs can lower operating costs and risks associated with human error, while traditional reactors may incur higher compliance costs, especially as plants age [19, 64].

SMRs are designed for greater operational flexibility, capable of serving various applications beyond electricity generation, such as water desalination and hydrogen production [19, 64]. They can also adjust power output based on grid demand, making them compatible with renewable energy sources [63, 65]. In contrast, traditional reactors are primarily optimized for base-load electricity generation and are less adaptable to varying demands, limiting their ability to load-follow [14, 15]. This flexibility positions SMRs favourably in hybrid energy systems, particularly in smaller grids where they can support intermittent renewable sources [19, 64].

The smaller physical footprint of SMRs allows for more versatile site selection, including deployment in remote locations with limited water resources [19, 65]. Some SMRs can even be sited underground, enhancing security and minimizing environmental impact [63, 64]. Traditional reactors, however, require significant infrastructure and proximity to large water sources for cooling, which restricts their deployment options and increases their environmental footprint [14, 15]. The ability of SMRs to operate in diverse environments with less ecological disruption presents a significant advantage over traditional reactors [19, 64].

SMRs often utilize advanced fuel types, such as low-enriched uranium or molten salt, which can lead to longer operating cycles and reduced fuel consumption [19, 65]. Some designs even allow for the recycling of spent fuel, potentially decreasing the total waste generated [63, 64]. Traditional reactors typically use enriched uranium and produce substantial amounts of nuclear waste, necessitating extensive long-term storage solutions [14, 15]. The advanced fuel cycles of SMRs may simplify waste management and reduce the need for long-term storage facilities [19, 64].

The regulatory landscape for SMRs is still evolving, with many countries developing frameworks to address the unique safety protocols and deployment strategies associated with these reactors [64, 67]. While traditional reactors operate within established regulatory frameworks, the extensive licensing requirements can lead to long and costly approval processes [15, 67]. The novel designs of SMRs may face regulatory hurdles, but some countries are beginning to implement streamlined processes to facilitate their approval [19, 64].

Key Players and Developers in the SMR Industry

Small Modular Reactors (SMRs) are emerging as a transformative force in modern nuclear power, bringing with them numerous benefits that position them as key players in the future of sustainable and secure energy production. Development of SMRs in Western countries has accelerated due to substantial private investment from both large corporations and small, entrepreneurial firms. This investment shift from government-led initiatives to private sector-led R&D signifies a broader transition in nuclear innovation, characterized by a focus on clean energy deployment and the mitigation of financial risks associated with traditional large-scale reactors [68].

One of the main economic advantages of SMRs is their reduced financial risk. Traditional nuclear plants require significant upfront investments due to their large size and extended construction times, but SMRs, being smaller, modular, and factory-produced, offer a much lower capital requirement. A 2011 report by the University of Chicago for the U.S. Department of Energy highlighted that smaller reactors could lower financial risks, making them more competitive with other energy sources. Their simpler, modular design enables mass production in factory settings, improving quality control and reducing on-site construction costs. By leveraging series production, SMRs achieve economies of scale, making them financially accessible to both developed and developing countries [68].

SMRs are designed with inherent safety features that distinguish them from traditional reactors. Their smaller size results in a higher surface area-to-volume ratio, enabling more efficient passive heat removal and reducing the need for complex safety systems. Many SMRs are designed to be installed below ground, offering additional protection against natural and man-made threats, including earthquakes, tsunamis, and even aircraft impacts. Studies by the American Nuclear Society in 2010 showed that certain safety measures required for large reactors are not necessary in smaller designs, allowing SMRs to operate with simplified safety systems. This reduction in complexity decreases operational risk and enhances the overall safety profile of nuclear power [68].

While SMRs offer numerous benefits, licensing remains a complex issue. Many regulators still apply the same licensing costs and requirements to SMRs as they do to larger plants, creating a financial challenge. In response, some countries have introduced pre-licensing design review processes, like Canada's Canadian Nuclear Safety Commission (CNSC), which allows developers to identify and resolve licensing barriers early. For instance, Phase 1 of CNSC's review involves 5,000 hours of regulatory staff time, while Phase 2 doubles this commitment to thoroughly assess the system-level design. Such processes aim to streamline regulatory approvals, making it easier for SMRs to achieve compliance and faster deployment [68].

SMRs are versatile and adaptable to various settings and purposes beyond electricity generation, such as industrial heating, desalination, and remote power supply. Their smaller size and modularity make them ideal for smaller electrical grids, particularly in countries with less nuclear infrastructure or limited access to large water sources. They can also serve regions with challenging geography, like islands or remote areas, where traditional plants are impractical. Furthermore, SMRs are designed for modular assembly, which enables easier and faster construction and allows for phased capacity expansion as demand grows, providing a flexible energy solution that adjusts to local needs [68].

The World Nuclear Association has highlighted the potential of SMRs for global standardization, which could reduce costs and streamline deployment across different countries. Their small size and advanced passive safety systems make them suitable for use in smaller or developing nations with limited nuclear experience. For instance, SMRs could be fully assembled in controlled factory environments and transported to their installation sites, significantly enhancing construction quality and efficiency. As a low-carbon energy source, SMRs can directly replace fossil fuel plants, thereby contributing to global decarbonization goals. Their reduced reliance on external cooling water also makes them suitable for arid and remote regions, reducing the environmental impact [68].

A critical factor in the success of SMRs is the availability of suitable fuel, especially high-assay low-enriched uranium (HALEU), which contains U-235 enriched to nearly 20%. The U.S. Nuclear Infrastructure Council has advocated for a domestic supply of HALEU to support the development of advanced reactors, as most civil enrichment plants do not exceed 5% enrichment. Recognizing this need, the U.S. Department of Energy (DOE) contracted Centrus Energy to establish a cascade of centrifuges dedicated to producing HALEU, while Urenco USA has committed to establishing a production line in New Mexico. This fuel supply initiative is crucial, as it allows SMR developers to avoid reliance on foreign-enriched uranium, facilitating research and deployment within the United States [68].

The DOE has played a pivotal role in promoting SMR development by offering funding and fostering partnerships. In 2012, the DOE launched the SMR Licensing Technical Support Program, allocating $452 million over five years to advance SMR licensing and construction. This support has extended to collaborations with private companies, such as the Carbon-Free Power Project, which involves constructing demonstration SMR plants in the western U.S. The DOE also introduced the Advanced Reactor Demonstration Program (ARDP) in 2020 to fund the construction of advanced reactors, including SMRs, within seven years. Such programs accelerate SMR development, enable public-private partnerships, and enhance the U.S.'s role in advancing nuclear technology globally [68].

Globally, SMR development is supported by extensive research and testing, with international organizations like the International Atomic Energy Agency (IAEA) playing a significant role. The IAEA's ongoing SMR assessment programs aim to foster the design of innovative reactors, like the Multi-Application Small Light Water Reactor (MASLWR), which integrates steam generators and natural circulation cooling. By 2020, over 70 SMR designs had been catalogued by the IAEA, illustrating the diversity and potential applications of these reactors. Canada's Nuclear Safety Commission (CNSC) is also actively involved in licensing reviews for SMRs, evaluating designs that range from liquid metal and molten salt to gas-cooled reactors, reflecting a broad commitment to developing flexible and sustainable nuclear energy [68].

The modularity and reduced size of SMRs enable economies of scale in their production, akin to the manufacturing of other modular technologies. As more SMRs are produced, standardized processes, components, and workforce training can lead to further cost reductions. Additionally, international initiatives, such as those in the United Kingdom, Canada, and China, aim to create markets where SMRs can achieve high-quality, cost-effective production. By focusing on both factory assembly and site-specific deployment, SMRs allow countries without extensive nuclear infrastructure to gain access to nuclear energy, democratizing nuclear power across a more diverse array of nations and applications [68].

With the demand for new generating capacity expected to rise globally, SMRs are well-positioned to meet this need in an environmentally sustainable way. Their role in replacing aging fossil fuel plants, stabilizing grids with high renewable penetration, and providing energy security is gaining recognition worldwide. As countries commit to achieving net-zero carbon emissions, SMRs represent a scalable, clean energy option that can support varied national energy strategies. Furthermore, SMRs are versatile enough to support specialized applications, such as floating power stations or district heating in northern climates, expanding their relevance in diverse geopolitical contexts [68].

SMRs have the potential to reshape the nuclear power industry, bringing clean, reliable energy to a broader audience while minimizing environmental impact and financial risk. With ongoing innovations in fuel supply, international support, and regulatory reform, SMRs are poised to play a crucial role in a low-carbon, secure energy future. Through collaboration and technological advancement, SMRs are likely to become an integral part of global efforts to achieve sustainable energy independence and climate goals [68].

Light Water Reactors

Light water reactors (LWRs) are a well-established nuclear technology, characterized by using ordinary water as both a coolant and moderator. These reactors are highly efficient and safe, benefitting from decades of operational experience in both civilian power plants and military applications. In an LWR, uranium fuel, typically enriched to less than 5% U-235, is assembled in rods within the reactor core, and the reactor uses water to slow down, or "moderate," neutrons, thus making nuclear reactions more efficient. The water also acts as a coolant, transferring heat generated in the core to a steam generator, where it ultimately drives turbines for electricity production. LWRs are renowned for their relatively low technological risk, as they rely on proven safety protocols and components [68].

The development of small light water reactors (sLWRs) has its roots in the military applications of the 1950s, where the U.S. developed compact pressurized water reactors (PWRs) for naval vessels. One significant civilian adaptation from this early development was the Big Rock Point BWR (boiling water reactor), a small-scale reactor with a capacity of 67 MWe, which operated successfully from the 1960s until its decommissioning in 1997. More recently, the U.S. Nuclear Regulatory Commission (NRC) has shifted focus to small modular reactors (SMRs) based on LWR technology, collaborating with companies like NuScale and Holtec. These modern designs aim to harness the benefits of traditional LWRs in a smaller, modular format, potentially simplifying regulatory hurdles, lowering costs, and improving deployment flexibility.

Several LWR-based SMR designs are currently in development or deployment phases worldwide, with Russia, China, and the U.S. leading the innovation. Russia's floating nuclear power plants, including the KLT-40S and RITM-200M reactors, are among the pioneering designs. The KLT-40S, derived from icebreaker reactors, produces 35 MWe and can also supply thermal power for district heating or desalination, making it versatile for remote or coastal locations. Similarly, Russia's RITM series, including the RITM-200M, are compact, high-safety reactors intended for floating platforms and onshore installations, with operational lifetimes of up to 60 years. China's approach with the ACP100 reactor, intended for similar applications, also highlights the global interest in LWR-based SMRs for decentralized and off-grid power solutions [68].

In the U.S., companies like NuScale and Holtec are at the forefront of LWR-based SMR innovation. NuScale's Power Module, which has passed significant regulatory milestones with the NRC, is a 77 MWe reactor that features a simplified, integral design with natural convection cooling and a 24-month refuelling cycle. Holtec's SMR-160, meanwhile, is a 160 MWe reactor with a straight tube steam generator and passive cooling features, also designed for underground installation, which enhances security. Both designs emphasize factory construction and modular deployment, which reduces on-site assembly time and improves quality control. The NRC is poised to leverage international experience to enhance regulatory frameworks for these and other LWR-based SMR designs.

These modern LWR-based SMRs are designed with enhanced safety features, including passive cooling systems, compact architecture, and modularity, allowing for their deployment in varied settings, from floating platforms to underground facilities. While conventional cooling is required for the steam condensers in most designs, innovative alternatives, like NuScale's air-cooled condenser option, offer water-saving solutions for arid regions. Given the flexibility in application and robust safety, LWR-based SMRs hold promise for replacing fossil fuel plants, reducing carbon emissions, and supporting decentralized energy needs in urban, rural, and remote locations alike [68].

Heavy Water Reactors

Heavy Water Reactors (HWRs) are a unique type of nuclear reactor that uses heavy water (deuterium oxide, D_2O) as both a moderator and coolant. This choice of heavy water allows HWRs to utilize natural uranium (with low levels of U-235) as fuel without requiring extensive enrichment, making them particularly useful in countries where access to enriched uranium is limited. India has been a pioneer in developing pressurized heavy water reactors (PHWRs) for power generation, starting with collaborations and evolving into a self-sustaining indigenous program. In the HWR design, uranium fuel is contained in horizontal pressure tubes, enabling on-power refuelling, which reduces downtime and increases reactor availability. This continuous refuelling capability is a distinct advantage, allowing the reactors to operate with minimal interruption and meet ongoing energy demands reliably.

The PHWR-220 is one of India's oldest and most standardized HWR models, with sixteen reactors currently operational. With a thermal output of 800 MWt and an electrical output of 220 MWe, the PHWR-220 reactors have a long operational history in India. The first of these reactors, Rajasthan 1, was constructed in partnership with Atomic Energy of Canada Ltd (AECL) and started operations in 1972. After this initial phase, India's Nuclear Power Corporation (NPCIL) continued to develop and enhance these reactors, leading to several design advancements across subsequent installations. The early reactors, such as those at Rajasthan, featured basic dousing systems and single containment structures. However, later models saw the introduction of safety upgrades, such as suppression pools and partial double containment, with standardized reactors at Narora and beyond also incorporating double containment, suppression pools, and a calandria filled with heavy water, housed in a water-filled calandria vault to further enhance safety and reliability. These reactors typically have a burn-up of around 15 GWd/t and can be refuelled while operational, reducing the need for extensive shutdowns and improving overall reactor performance [68].

The Advanced Heavy Water Reactor (AHWR) is an innovative next-generation HWR developed by India's Bhaba Atomic Research Centre (BARC). This design marks a shift toward using thorium, an abundant resource in India, as a primary fuel. The AHWR aims to maximize thorium's potential through an in-situ conversion process, where thorium-232 transmutes into fissile uranium-233 within the reactor core. For export, a variant of the AHWR has been developed using low-enriched uranium (LEU) combined with thorium, which minimizes the need for plutonium as a starting fuel component. In this version, about 39% of the power will be derived from thorium, while the burn-up level is projected at an impressive 64 GWd/t, indicating efficient fuel utilization and waste reduction. Fuel enrichment in this export-oriented AHWR will reach up to 19.75%, with an average fissile content of 4.21% in the uranium-thorium fuel mix, making it a versatile design compatible with international markets.

Structurally, the AHWR-300 LEU features vertical pressure tubes and a light water coolant system, where water is circulated by convection under high pressure, producing steam directly within the pressure tubes. This configuration enables high-efficiency steam generation and, by extension, power generation, with a nominal capacity of 300 MWe and a net output of 284 MWe. Currently in its basic design phase, the AHWR-300 LEU represents a significant advancement in HWR technology, combining sustainability with innovative fuel cycles tailored to leverage local resources, positioning it as a promising reactor model for both domestic and international deployment [68].

High-Temperature Gas-Cooled Reactors

High-temperature gas-cooled reactors (HTRs) represent an advanced class of nuclear reactors designed to achieve higher efficiency and operational flexibility. Using graphite as a moderator and either helium, carbon dioxide, or nitrogen as the primary coolant, HTRs can operate at temperatures of 700-950°C and potentially up to 1000°C, making them suitable for various applications beyond electricity generation, such as industrial processes and hydrogen production. HTRs are distinguished by their use of tristructural-isotropic (TRISO) fuel particles, which consist of a uranium kernel surrounded by layers of carbon and silicon carbide, offering a robust containment for fission products and remaining stable at temperatures exceeding 1600°C. These reactors are characterized by a negative temperature coefficient of reactivity, meaning the fission reaction slows as temperature rises, and they can passively dissipate decay heat, enhancing their inherent safety [68].

HTRs can employ two different fuel configurations: block-type and pebble bed. In block-type designs, TRISO particles are embedded in hexagonal graphite blocks. In pebble-bed designs, the fuel is contained in billiard ball-sized graphite pebbles, each with thousands of TRISO particles. Despite the high volume of used fuel, HTRs have advantages in terms of reduced radiotoxicity and decay heat due to their high fuel burn-up. However, the core's large graphite content means the fuel is spread over a more significant volume compared to light-water reactors (LWRs). These reactors can also utilize thorium as fuel, a feature with potential advantages for energy sustainability and nuclear non-proliferation [68].

China's HTR-PM project, a commercial-scale HTR demonstration plant, is currently the most advanced HTR project. It builds upon China's HTR-10 experimental reactor, which demonstrated the reactor's safety by allowing it to reach equilibrium during an induced shutdown, proving the safety of TRISO fuel under high temperatures without active cooling. The HTR-PM consists of twin 250 MWt reactors connected to a single steam turbine producing 210 MWe. China Huaneng Group leads this project with substantial government backing, aiming to eventually build more extensive HTR-PM600 plants for large-scale power production [68].

Several HTR designs have been developed internationally, with diverse capabilities and purposes. Japan's High-Temperature Test Reactor (HTTR) achieved coolant temperatures as high as 950°C, aiming to produce hydrogen from water via thermochemical processes. Japan's GTHTR-300C, an evolution of HTTR, utilizes high-efficiency gas turbines and can produce electricity at nearly 50% thermal efficiency. X-energy in the USA is developing the Xe-100, a modular 80 MWe pebble-bed HTR, targeting utility applications and industrial heat supply. The U-Battery concept, developed by Urenco, envisions a highly compact, 4 MWe modular HTR design suitable for remote installations, further showcasing the versatility of HTR technology [68].

In addition to the direct generation of electricity, HTRs are well-suited for high-temperature applications and could potentially drive Brayton cycle gas turbines, which operate at near 50% thermal efficiency. However, achieving such efficiency requires very pure helium coolant to avoid corrosion and managing graphite dust generated within the reactor. Given the relatively low density of power output compared to traditional reactors, HTRs present advantages in load-following, meaning they can efficiently adapt to varying electricity demands without significant efficiency loss [68].

Fast Neutron Reactors

Fast neutron reactors (FNRs) represent a significant evolution in nuclear reactor design, with the potential to unlock a vastly greater portion of uranium's energy content compared to traditional reactors. Unlike conventional reactors that use slow or "thermal" neutrons and require a moderator, FNRs operate without a moderator, leveraging the high energy of fast neutrons to sustain fission. This approach allows FNRs to utilize nearly all of the energy within uranium, rather than the mere 1% typically accessed by light water reactors (LWRs). Fast reactors generally use liquid metals such as sodium, lead, or lead-bismuth as coolants due to their excellent thermal conductivity, high boiling points, and ability to operate at atmospheric pressure, which enhances safety by reducing the risk of pressure-related failures [68].

The passive safety features inherent in FNRs add a layer of reliability and robustness. A natural reactivity feedback mechanism enables these reactors to self-regulate; if coolant flow decreases, the core temperature rises, thereby slowing the fission reaction. This automatic power adjustment reduces reliance on external control mechanisms and increases resilience against potential malfunctions. The long operational life of FNRs, with refuelling intervals that can stretch up to 20 years, makes them particularly appealing for remote installations and decentralized power systems. This operational longevity is supported by fuel types enriched to approximately 15-20%, including uranium nitride and uranium-plutonium alloys, which are robust enough to withstand the intense environment within FNRs.

FNR technology has primarily advanced in two main coolant configurations: liquid metal-cooled and gas-cooled systems. Liquid metal coolants like sodium and lead-bismuth bring unique advantages and challenges. Sodium, though flammable and reactive with water, provides high thermal efficiency and simplifies engineering by operating at near-atmospheric pressure. Lead and lead-bismuth coolants, while corrosive to some metals, offer the benefit of being non-reactive with air and water, further enhancing safety in the event of leaks. Gas-cooled fast reactors, such as the Energy Multiplier Module (EM2) and Fast Modular Reactor (FMR), use helium or supercritical carbon dioxide, with some designs reaching temperatures up to 900°C. These high-temperature outputs make FNRs adaptable for both power generation and industrial applications, such as hydrogen production and desalination [68].

One notable class of FNRs includes sodium-cooled reactors like the U.S.-developed PRISM and Russia's BREST-300, which leverage sodium's excellent heat transfer properties. PRISM, for instance, is a modular reactor designed for safe operation through passive cooling. It uses electrometallurgical reprocessing to recycle spent fuel, potentially reducing the volume of nuclear waste. GE Hitachi's PRISM is proposed as a solution for disposing of surplus plutonium stocks, such as in the United Kingdom, where it could be used to transmute plutonium into less harmful forms while producing electricity.

Lead-cooled designs, such as Russia's SVBR-100 and the Westinghouse LFR, offer additional advantages for safety and longevity. Lead-cooled reactors can operate at atmospheric pressure and resist high temperatures, making them particularly suitable for high-efficiency energy production. For example, Russia's BREST-300, an integral part of the Pilot Demonstration Energy Complex, is designed to close the nuclear fuel cycle by using on-site facilities for recycling used fuel. The SVBR-100, although initially developed for submarine use, is now adapted for civilian power generation, with a focus on modular deployment and cost-effectiveness [68].

FNRs, particularly in smaller modular forms, are designed for factory production and transportation to remote sites, emphasizing simplicity and ease of installation. Many designs allow for the entire reactor unit to be returned to a regional facility at the end of its life, potentially easing waste management. For instance, Gen4 Energy's compact Gen4 Module uses lead-bismuth cooling in a portable, sealed configuration, ideal for isolated or off-grid locations.

In conclusion, fast neutron reactors hold promise for a sustainable nuclear future. By expanding uranium utilization, enabling closed fuel cycles, and offering flexible, high-efficiency designs, FNRs address both environmental concerns and energy security needs. The diversity of FNR technology—ranging from sodium-cooled to lead and gas-cooled designs—provides adaptability across varied energy demands, promising a versatile, robust solution for both large and small-scale energy needs [68].

Molten Salt Reactors

Molten Salt Reactors (MSRs) stand out for their ability to operate at high temperatures and low pressures, using molten fluoride salts as a primary coolant. These salts, primarily lithium-beryllium fluoride and lithium fluoride, remain liquid at temperatures up to 1400°C without requiring pressurization, unlike conventional Pressurized Water Reactors (PWRs) that operate at around 315°C under 150 atmospheres of pressure. This higher operational temperature can potentially increase thermodynamic efficiency and provide process heat for industrial applications. Some fast-spectrum MSRs also use chloride salts as coolants. Generally, in MSRs, the nuclear fuel is dissolved in the molten salt, allowing for a unique approach to energy production and nuclear waste management.

The concept of MSRs originated in the 1960s in the United States, where the Molten Salt Reactor Experiment (MSRE) at Oak Ridge National Laboratory ran successfully from 1965 to 1969. This prototype used uranium-235 dissolved in molten lithium, beryllium, and zirconium fluoride salts with a graphite moderator and achieved efficient energy production. However, the project was eventually shelved due to funding shifts and technological challenges. Today, renewed interest in MSRs has emerged worldwide, with notable research efforts in the USA, Japan, China, Russia, and France, as well as their inclusion in the Generation IV nuclear reactor designs for future energy solutions.

In a typical MSR, the fuel salt mixture comprises lithium and beryllium fluoride salts with dissolved enriched uranium (usually U-235 or U-233 fluorides). The reactor core incorporates unclad graphite as a moderator, facilitating the salt's flow at temperatures around 700°C. This setup allows for a flexible neutron spectrum, often in the intermediate range, which offers the possibility of breeding, although not at the efficiency of fast reactors. In a homogeneous design, thorium can be dissolved directly into the fuel salt, enabling the reactor to generate fissile uranium-233 in situ. For a two-fluid, or heterogeneous, design, separate loops for fertile thorium salt and fissile uranium salt allow for a continuous breeding cycle. Furthermore, these reactors can remove fission products from the salt in real-time, thus facilitating continuous operation and minimizing the build-up of long-lived actinides[68].

MSRs possess inherent safety features, such as a negative temperature coefficient of reactivity, meaning that as the fuel salt heats up, reactivity decreases, naturally controlling the fission reaction. Additionally, the liquid fuel provides a strong negative void coefficient, contributing further to stability. Most MSRs operate at near atmospheric pressure, minimizing the risk of explosive releases in the event of a leak. Control of reactivity can be achieved by adjusting the rate of coolant salt flow, while primary reactivity control typically involves manipulating the temperature of the secondary coolant. In the event of overheating, many MSRs incorporate freeze plugs that allow the primary salt to drain into a subcritical, passively cooled dump tank.

The MSR design offers numerous benefits, including high fuel burn-up rates, minimal production of long-lived radioactive waste, and the potential for high thermal efficiency due to elevated operational temperatures. Unlike conventional reactors that require significant quantities of uranium or plutonium, MSRs can operate on thorium—a more abundant and potentially less problematic nuclear fuel source. Furthermore, MSRs produce fewer actinides than traditional uranium-fuelled reactors, reducing the risk of weapons proliferation and lowering long-term waste management burdens.

MSRs face several technical challenges, particularly regarding material stability and tritium management. The primary salt requires lithium-7, a specific isotope of lithium, to avoid tritium production, but it is costly and not abundantly available. Even with purified lithium-7, tritium formation is an ongoing issue, necessitating advanced containment and recovery systems. Additionally, high-temperature fluoride salts are corrosive, demanding special materials that can withstand these extreme conditions over long periods [68].

Modern MSR concepts include various adaptations for specialized applications. For instance, the Liquid Fluoride Thorium Reactor (LFTR) design incorporates a fertile blanket of thorium to breed uranium-233, providing an efficient and safe thorium fuel cycle. The Integral Molten Salt Reactor (IMSR), developed by Terrestrial Energy, uses a single replaceable core unit, facilitating ease of operation and maintenance. The Stable Salt Reactor (SSR) by Moltex, another variation, employs fuel assemblies filled with molten chloride salts and stationary coolant salts, allowing for cost-effective use of standard industrial pumps.

MSRs' unique features make them adaptable for coupling with renewable energy sources. Some MSR developers, like Moltex, have introduced concepts such as molten salt heat storage (GridReserve), which stores excess heat in nitrate salt storage tanks during low demand and releases it during peak demand. This feature enables MSRs to act as a reliable backup to intermittent renewable energy sources like solar and wind, offering flexibility to the grid [68].

While MSRs have seen limited commercial deployment due to regulatory and technological challenges, their potential advantages in efficiency, waste management, and fuel flexibility position them as a promising technology for future nuclear energy solutions.

Aqueous Homogeneous Reactors

Aqueous Homogeneous Reactors (AHRs) are a unique type of nuclear reactor in which the fuel is dissolved directly into the moderator as a liquid, usually in the form of an aqueous solution. Typically, these reactors use uranium nitrate or uranium sulphate in water, with the uranium being either low-enriched or highly enriched. The solution of fuel and moderator circulates through the reactor core, where nuclear fission takes place, producing energy and creating a self-sustaining chain reaction. One of the notable advantages of AHRs is their self-regulating capability: as the reactor heats up, the fuel solution expands, decreasing the concentration of uranium and thereby reducing reactivity. This negative feedback mechanism enhances the safety profile of AHRs by naturally adjusting the reaction rate in response to changes in temperature [68].

Historically, around 30 AHRs have been built, primarily as research reactors. Their ability to operate continuously while removing fission products from the circulating solution is a key benefit. This feature not only helps maintain a steady rate of fission by minimizing the build-up of neutron-absorbing byproducts but also allows the reactor to achieve higher fuel utilization. In the Netherlands, for example, a 1 MWt AHR operated from 1974 to 1977, utilizing a mixture of thorium and highly enriched uranium (HEU) as Mixed Oxide (MOX) fuel. This setup demonstrated the reactor's suitability for research purposes, particularly in isotope production and neutron generation, which benefit from its stable neutron flux and ability to run uninterrupted.

AHRs have also been a subject of theoretical research aimed at determining the smallest feasible configuration for a nuclear reactor. In a 2006 study, scientists theorized that a miniature AHR could be powered by a solution of americium-242m (Am-242m) nitrate. This reactor design, measuring only 19 cm in diameter with a total mass of just under 5 kg, would contain 0.7 kg of Am-242m as its nuclear fuel. Such a compact reactor would produce a power output of a few kilowatts, making it suitable for specialized applications where small size and high neutron flux are essential. Potential uses include high-intensity neutron sources for scientific research and power sources for space missions, where compact, reliable energy sources are critical [68].

The simplicity of AHRs, combined with their ability to manage fission products continuously, makes them attractive for specific uses, yet their reliance on a liquid fuel poses some challenges. For instance, maintaining the solution's chemical stability, managing the effects of radiolytic decomposition (where water molecules break down under radiation), and handling the corrosive effects on reactor materials require careful design and monitoring. Despite these challenges, AHRs remain an important tool in nuclear research, providing valuable insights into reactor design, isotope production, and neutron source technology. Their unique structure and operational features continue to inspire advancements in compact reactor technology and portable nuclear energy sources, particularly for environments requiring small, high-efficiency reactors [68].

Modular Construction Using Small Reactor Units

Modular construction with small reactor units represents a shift from traditional large-scale nuclear plants to a more flexible, scalable approach that involves building multiple small reactors instead of one large reactor. This concept has been explored by Westinghouse and IRIS partners with their IRIS (International Reactor Innovative and Secure) design, a 330 MWe reactor. The main advantage of this modular approach is that it leverages the economy of serial production rather than the conventional economy of scale associated with large reactors. By standardizing small reactors, many of the components can be prefabricated in controlled factory environments, allowing faster and more consistent construction, improved quality control, and lower costs. Westinghouse has projected that the first IRIS unit could be built in three years, with subsequent units taking only two years each [68].

One of the key benefits of modular construction is the flexibility it offers in site layout and operation. Each small reactor unit can be constructed separately with sufficient physical space to allow work on subsequent units while the previous ones are operational. This phased construction approach reduces the overall site footprint: even with multiple small reactors, the land required can be comparable to or even smaller than a single, larger reactor unit with the same total power output. For example, a site with three 330 MWe IRIS units (totalling around 1000 MWe) would take up a similar or smaller footprint than a traditional 1000 MWe reactor. This flexibility enables energy companies to scale up production in a way that fits the demand and financial capabilities of the project, adapting to both site constraints and grid limitations [68].

The term "Small Modular Reactor" (SMR) describes this modular, flexible approach, but it is often applied broadly to any small reactor design, even if they are not designed with modular operation in mind. SMRs designed for modular construction allow developers to build plants incrementally, adding capacity with each new reactor while maintaining continuous revenue generation. Once a module is complete and begins producing electricity, it generates revenue that can help fund the construction of additional modules. Westinghouse has shown that this approach can result in significantly lower capital outlay and reduced financial risk. For instance, they estimated that building 1000 MWe capacity using three IRIS units in a phased approach would require a peak negative cash flow of less than $700 million, a third of the estimated cost of constructing a single large 1000 MWe unit in one phase.

This modular approach has particular advantages for developing countries and regions with smaller electric grids, where a 1000+ MWe reactor might overwhelm the grid's capacity. Instead of investing in a massive reactor that would demand substantial upfront costs and carry a longer construction timeline, countries could add capacity in small increments, aligning with grid limitations and financial resources. Moreover, SMRs can be sited in locations closer to load centres or remote areas where large reactors might be impractical. This incremental buildout model could transform the economic and operational landscape of nuclear energy, making it accessible and scalable for a wider range of applications and regions [68].

Industry Key Players

The Small Modular Reactor (SMR) industry is experiencing significant momentum, attracting major investments and commitments from established nuclear companies and innovative startups worldwide. These industry players are developing a wide range of SMR technologies, tailored for diverse applications from electricity generation to industrial heat and desalination. SMRs vary significantly in design, size, and operational focus, offering flexible solutions to meet specific energy demands. Here are some prominent developers and their SMR technologies:

1. Westinghouse Electric Company

Westinghouse has designed the eVinci Micro Reactor, a solid-state microreactor primarily intended for remote and off-grid applications. With a capacity of 5 MWe and a design emphasizing minimal refuelling, the eVinci is tailored for flexibility and ease of transport. Westinghouse, with its extensive history in nuclear energy, is focusing on factory-assembled reactors that can provide reliable power across a variety of settings.

2. NuScale Power

NuScale Power is advancing the NuScale SMR (VOYGR plant configuration), a Light Water Reactor (LWR) design offering 77 MWe per module, scalable to 12 modules per plant for a total output of 924 MWe. NuScale's design has been certified by the U.S. Nuclear Regulatory Commission (NRC), positioning it as a leader in the U.S. for SMR deployment. This reactor's modular structure allows for a flexible configuration that can grow with power needs.

3. Terrestrial Energy

Terrestrial Energy has developed the Integral Molten Salt Reactor (IMSR), a high-temperature molten salt reactor focused on industrial process heat and electricity generation. Its design offers high thermal efficiency and is suitable for industrial applications. Planned for deployment in North America, Terrestrial Energy's IMSR combines high operational temperatures with adaptable configurations for multiple industry uses.

4. Rolls-Royce SMR Ltd

In the UK, Rolls-Royce SMR Ltd is designing a 470 MWe Pressurized Water Reactor (PWR) specifically for the UK market. This reactor aims to support the country's low-carbon energy goals and is backed by government funding. Rolls-Royce's SMR leverages modular construction, reducing costs and allowing for export to other regions, with a focus on building scalable, factory-assembled reactors.

5. GE Hitachi Nuclear Energy

GE Hitachi's BWRX-300 is a Boiling Water Reactor (BWR) rated at 300 MWe, based on established BWR technology for utility-scale deployment. The BWRX-300 is being evaluated by regulatory bodies in Canada and the United States, with early interest from Ontario Power Generation. This simplified design aims for efficient utility-scale generation, making it suitable for widespread deployment by the late 2020s.

6. X-Energy

The Xe-100 from X-Energy is a High-Temperature Gas-cooled Reactor (HTGR) with each module delivering 80 MWe, scalable up to 320 MWe in a modular plant configuration. Its pebble-bed design uses TRISO fuel, making it ideal for applications requiring load-following capability, such as industrial heating or electricity generation. X-Energy's design is focused on high-temperature process heat, appealing to both grid and industrial clients.

7. Holtec International

Holtec's SMR-160 is a 160 MWe Light Water Reactor designed for small-scale electricity production and water desalination. This SMR emphasizes passive safety features, enabling deployment across diverse geographic and economic environments. Holtec's compact and versatile design targets emerging markets, offering a solution adaptable to various locations and demands.

8. China National Nuclear Corporation (CNNC)

China's ACP100 (Linglong One) is a 125 MWe Pressurized Water Reactor (PWR) suited for smaller grids, island locations, and industrial facilities. With construction underway on Hainan Island, ACP100 has become one of the first SMRs globally to reach the construction stage. Backed by the Chinese government, CNNC's design aims to provide a stable and secure energy source for various applications.

9. Rosatom (Russia)

Rosatom's RITM-200 is a PWR initially developed for icebreaker vessels, now adapted for land-based and floating power plants. Each unit produces 55 MWe and is designed for applications in remote Arctic locations, including electricity generation and desalination. Rosatom has deployed RITM-200 reactors on Russian icebreakers and is working to expand its technology to other isolated areas.

10. Moltex Energy

The Stable Salt Reactor – Wasteburner (SSR-W) from Moltex Energy is a molten salt reactor optimized for waste recycling and clean energy production. Capable of using spent nuclear fuel, this design reduces nuclear waste while generating electricity. Moltex is promoting the SSR-W in Canada, with plans to establish a demonstration plant in New Brunswick, offering a unique solution for nuclear waste management.

11. Korea Atomic Energy Research Institute (KAERI)

KAERI's SMART (System-integrated Modular Advanced Reactor) is a 100 MWe PWR targeting smaller grid systems, desalination, and combined heat and power applications. This design is attracting interest from countries with smaller power grids and regions facing water scarcity. SMART's flexible deployment options make it a viable choice for both developing countries and specific industrial applications.

12. Ultra Safe Nuclear Corporation (USNC)

The Micro Modular Reactor (MMR) by USNC is a High-Temperature Gas-cooled Reactor delivering 5 MWe, suitable for remote power generation, industrial heat applications, and hydrogen production. This microreactor is focused on providing a compact solution for off-grid and remote settings, with plans to deploy at Canada's Chalk River Laboratories by 2026, making it one of the few reactors tailored for isolated locations.

13. Toshiba

Toshiba's 4S (Super-Safe, Small & Simple) is a sodium-cooled fast reactor designed for remote areas with minimal refuelling needs, producing between 10 MWe and 50 MWe. The 4S reactor offers a long-term power solution with minimal maintenance, suited for isolated communities and locations where consistent, low-maintenance power is essential.

14. Newscale (Japan)

Newscale is developing modular Light Water Reactors (LWRs) under the NuScale Power Module (VOYGR plant) configuration. These units are designed for flexibility, aiming to provide comprehensive and adaptable power solutions suitable for diverse applications worldwide.

Industry Trends and Future Outlook

The SMR industry is evolving quickly, with developers focusing on innovative designs that offer benefits such as high-temperature operation, passive safety features, fuel flexibility, and suitability for industrial and off-grid applications. The adaptability of SMRs, coupled with lower initial investment costs and shorter construction times, is driving global interest and support from both governments and the private sector. These reactors are particularly attractive in regions with grid constraints, remote energy needs, or environmental goals. As more SMR designs approach regulatory approval and commercial deployment, the industry is poised to make significant contributions to the global energy landscape, offering cleaner, scalable, and more flexible nuclear solutions.

Chapter 2
Feasibility and Site Selection for SMR Power Stations

Key Factors in SMR Site Selection

A Small Modular Reactor (SMR) power plant requires a range of physical buildings, structures, and infrastructure, designed to optimize the modular nature of these reactors, ensure safety, and provide efficient operation and maintenance. Due to their smaller size and modular approach, SMR plants generally have a more compact footprint than traditional nuclear plants, with infrastructure tailored to support modular components and streamlined deployment.

The infrastructure surrounding SMR plants is also designed to support their unique operational requirements. This includes specialized buildings for reactor containment, maintenance facilities, and systems for waste management and cooling [63]. The compact nature of SMRs allows for more efficient use of land and resources, making them suitable for deployment in regions with limited grid capacity or where traditional nuclear plants would be impractical [3, 11].

Physical buildings, structures, and infrastructure required for SMR power stations include:

1. Reactor Building

The reactor building houses the core SMR units and is typically designed to support modular assembly. This building includes:

- **Reactor Containment Structures**: Each SMR module is often enclosed within a separate containment vessel to isolate it from the environment and prevent radiation leaks.

- **Shielding and Radiation Protection**: This includes thick concrete walls and radiation-resistant materials surrounding each reactor module, providing essential protection for workers and the surrounding area.

- **Seismic and Environmental Protection**: Due to the smaller and sometimes more compact nature of SMRs, the reactor building incorporates seismic isolation, flood barriers, and storm resilience features. In some designs, SMRs are partially or fully below ground for added protection.

2. Control and Monitoring Building

This facility houses the plant's control room, where operators monitor and manage the SMR's operations, safety, and performance. Key elements include:

- **Control Room**: Equipped with digital and automated systems that provide real-time data on the reactor's status, temperature, coolant levels, and power output.

- **Safety and Emergency Response Systems**: Redundant safety and backup systems are integrated here, allowing operators to quickly respond to any unexpected conditions. Some SMRs have passive safety features, so these areas may require less physical infrastructure compared to larger plants.

- **Data Storage and Cybersecurity Facilities**: To ensure secure operation, control buildings may have robust cybersecurity and data management infrastructure, especially if the SMR is linked to a broader energy grid or monitored remotely.

3. Cooling Systems and Heat Exchangers

SMRs are often designed with innovative cooling systems that operate at atmospheric or near-atmospheric pressure. These include:

- **Primary and Secondary Coolant Loops**: Depending on the SMR design, cooling may involve water, gas, molten salt, or liquid metal, each requiring specific infrastructure. Secondary loops may connect to external heat exchangers or cooling towers.

- **Cooling Towers and Water Supply Systems**: SMRs designed for water-based cooling require cooling towers, though on a smaller scale than traditional plants. For water-cooled SMRs, the plant may need a secure, consistent water source or incorporate air cooling as an alternative.

- **Heat Rejection Systems**: Some SMR designs use air-cooled condensers or dry cooling to reduce the water footprint, which requires dedicated infrastructure for effective thermal dissipation.

4. Spent Fuel Storage and Waste Management Facility

SMRs produce spent fuel and other radioactive waste, necessitating safe and secure storage facilities.

- **Onsite Spent Fuel Pool**: Immediately following reactor operation, spent fuel is stored in a cooled pool for several years until its radioactivity decreases.

- **Dry Cask Storage Facilities**: For long-term storage, SMRs use secure concrete or steel casks to contain spent fuel. This area is heavily shielded and monitored for radiation.

- **Waste Processing and Packaging Areas**: These areas handle lower-level radioactive waste, packaging and preparing it for transportation or onsite storage.

5. Turbine and Generator Buildings

SMRs designed for power generation feature a turbine building, though smaller than those in conventional plants, and may include:

- **Turbine Generators**: Turbines convert steam from the reactor's heat into electricity, typically using smaller and more compact equipment due to the SMR's output.

- **Steam Condensers**: Essential for converting steam back into water for reuse, condensers are connected to the turbine and are either water- or air-cooled depending on the plant's design and available resources.

- **Electrical Switchgear Rooms**: These rooms house electrical infrastructure to manage the generated electricity and ensure stable distribution to the grid.

6. Auxiliary and Maintenance Buildings

These buildings support ongoing plant operation, maintenance, and administration:

- **Maintenance and Repair Workshops**: Facilities for handling and maintaining SMR components, which may require specialized tools or remote handling capabilities.

- **Warehouse and Storage Areas**: Used to store parts, equipment, and maintenance tools required for modular replacement and regular upkeep.

- **Employee Facilities and Administrative Offices**: Includes workspaces for plant personnel, meeting rooms, and emergency response stations, often located in a safe zone away from reactor operations.

7. Security and Access Control Infrastructure

Nuclear plants require robust security measures to prevent unauthorized access and ensure safety.

- **Perimeter Security and Access Control**: The site is typically surrounded by high-security fencing, cameras, and access points staffed by security personnel.

- **Emergency Response and Medical Facilities**: Onsite first-aid and emergency response teams are prepared to respond to any incident.

8. Transmission and Grid Connection Infrastructure

To distribute the generated power, SMRs require connection to the grid:

- **Switchyard and Transmission Lines**: These carry electricity from the plant to the local grid. Some SMRs also have black start capabilities, allowing them to start independently and provide power during a grid outage.

- **Grid Interconnection Systems**: Grid management equipment is crucial for balancing output from modular units with grid demand, especially if the plant is designed for flexible operation.

The infrastructure for SMRs is streamlined compared to traditional nuclear plants, thanks to smaller reactor modules, factory-built components, and modular construction techniques. This design approach facilitates rapid assembly, reduces the overall site footprint, and allows for flexible expansion based on demand. As a result, SMR plants are adaptable, scalable, and efficient, aligning with a variety of energy and industrial applications while adhering to stringent safety and regulatory standards.

Power Plant Layout

The layout of a Small Modular Reactor (SMR) power plant is crafted to optimize modularity, safety, and efficiency, reflecting the core attributes of SMR technology. With a more compact and streamlined setup than traditional nuclear plants, SMR facilities benefit from modular components and an organized structure tailored to smaller reactors. Below is a detailed description of the essential zones within a typical SMR-based power plant.

Reactor Zone

In the reactor zone, each SMR module is housed in a dedicated containment building, designed to protect against external threats like earthquakes, floods, or aircraft impact. These containment structures are often below ground, adding extra protection. In multi-module plants, multiple reactor bays are found within the containment area, each bay housing a separate SMR unit. This arrangement allows easy access for maintenance, refuelling, and modular replacement as necessary. The primary cooling loop, responsible for circulating coolant through the reactor, is also housed here. This loop contains pumps, valves, and pipes, all customized to the specific coolant type of the SMR, such as water, gas, or molten salt.

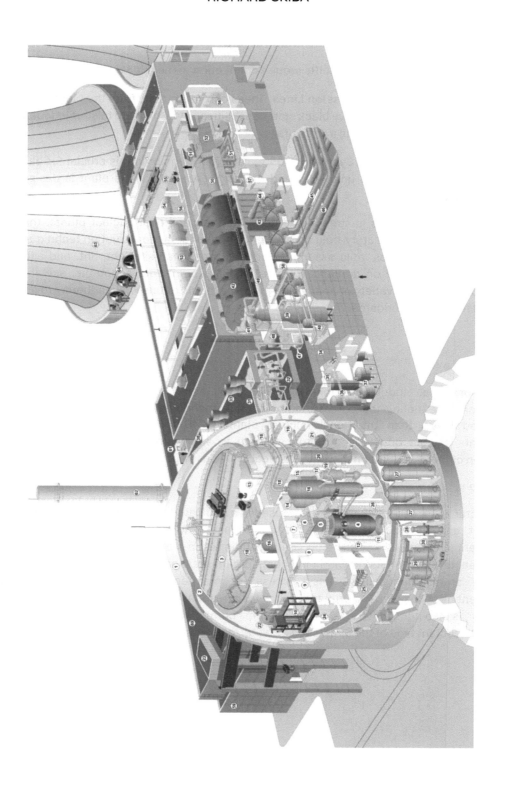

LEGEND

Reactor building:

1 – Concrete containment shell
2 – Steel containment shell
3 – Polar gantry crane
4 – Reactor pressure vessel
5 – Control rod drive mechanisms
6 – Cable bridge
7 – Reactor well
8 – Parking area for reactor internals
9 – Fuel storage pool
10 – Refueling machine
11 – Inner (biological) shield
12 – Support (biological) shield
13 – Parking area for reactor pressure vessel head
14 – Pressurizer
15 – Pressurizer relief tank
16 – Steam generators (4)
17 – Reactor coolant pumps (4)
18 – Main steam lines
19 – Feedwater lines
20 – Pressure accumulators (4)
21 – Personnel lock
22 – Equipment lock
23 – Gantry with hoist
24 – New fuel storage

25 – Transmitter room
26 – Guide tubes for neutron flux measurement (6)
27 – Flooding tanks (8)
28 – Residual heat coolers (4)
29 – Nuclear intercoolers (4)
30 – Safety injection pumps (4)
31 – Main steam and feedwater valve compartments
32 – Main steam valves
33 – Exhaust steam silencers

Auxiliary building:

34 – Radioactive wastewater treatment plant
35 – Air intake for reactor and auxiliary building
36 – Radioactive wastewater monitoring tanks
37 – Radioactive wastewater concentrate tanks
38 – Coolant storage and treatment plant

Turbine building:

39 – Steam superheaters (2)
40 – Condensate collection tanks (2)
41 – High-pressure turbine
42 – Low-pressure turbines (3)
43 – Condensers (3)
44 – Main steam bypass line

45 – Main cooling water outlet line
46 – Main cooling water inlet line
47 – Turbine oil drain tank
48 – Turbine oil supply tank
49 – Cross-over piping
50 – Generator
51 – Generator leads
52 – Exciter
53 – Feedwater tank
54 – Sprinkler tank
55 – Turbine building crane
56 – Pillar cranes
57 – Generator supports with shock absorbers
58 – Turbine supports with shock absorbers

Operation and switchgear building:

59 – Emergency diesel power supply
60 – Ventilation systems
61 – Control room
62 – Vent stack
63 – Cooling towers (two per block)
64 – Ventilators

Figure 25: Diagram of a nuclear power plant with a pressurized water reactor, based on the Biblis Nuclear Power Plant. Kaidor, CC BY-SA 4.0, via Wikimedia Commons.

Control and Operations Zone

The control room serves as the operational centre of the plant, where control systems for all reactor units are managed. Operators can monitor essential parameters, including reactor conditions, coolant flow, temperature, radiation levels, and electricity output. Often equipped with advanced digital interfaces, the control room is specifically designed to manage the unique modular features of SMRs. Adjacent to this is the Emergency Operations Center, equipped with backup systems, communication tools, and emergency protocols to ensure safety in case of incidents. Additionally, there are cybersecurity and data storage rooms to protect operational data and implement cybersecurity measures.

Cooling and Heat Exchanger Zone

This zone houses the heat exchangers, where heat from the primary coolant is transferred to a secondary coolant loop. For water-cooled SMRs, this secondary loop circulates water through cooling towers or air-cooled condensers. For designs with alternative cooling systems, like molten salt or gas, specialized equipment is used. In water-cooled SMRs, cooling towers or air-cooled condensers are compact, designed to fit the streamlined footprint of an SMR plant. Dedicated stations with pumps, valves, and control systems manage the coolant flow in both the primary and secondary loops.

Turbine and Power Generation Zone

The turbine building, though smaller than in traditional plants, houses turbine generators powered by steam from the secondary coolant loop. In modular plants, multiple turbine units may be included to accommodate scalable power generation. Steam condensers in this zone convert the steam back to water, allowing it to be reused in water-cooled SMRs. For designs utilizing air or gas cycles, the steam cycle layout may differ. Located near the turbine area, the electrical switchgear room houses infrastructure for routing power to the grid, managing load balancing, voltage regulation, and power distribution.

Spent Fuel and Waste Management Zone

Upon removal from the reactor, spent fuel rods are transferred to the spent fuel pool, located within the containment area or a separate structure in multi-module plants. Following cooling, used fuel is stored in dry casks within a secure storage area. The waste treatment and packaging facility processes low-level radioactive waste, packaging and storing it in accordance with regulatory standards for later transportation.

Auxiliary and Support Infrastructure Zone

This zone includes maintenance and assembly workshops necessary for the modular nature of SMRs, providing tools, equipment, and space to assemble or replace components as needed. Warehouse and storage facilities store reactor parts, fuel assemblies, and other essential equipment. Additionally, the zone includes administrative offices, meeting rooms, training areas, and emergency response facilities to support plant staff.

Security and Access Control Zone

The plant is surrounded by a secure perimeter with high fencing, surveillance cameras, and access-controlled entry points. Staffed 24/7, the Security Control Center oversees all security activities within the plant. Visitor and employee checkpoints are also present, where personnel undergo ID verification and radiation screening if required.

Transmission and Grid Connection Zone

The switchyard connects the SMR-generated power to the regional grid, with transformers, circuit breakers, and other components to manage power flow. Grid interconnection systems stabilize output, allowing the plant to synchronize with grid demands, and include backup generators and black start capabilities for stability during outages.

Summary Layout Considerations

The SMR layout, as per the example shown in Figure 26, emphasizes compactness, modular assembly, and operational efficiency. Each reactor unit operates and is maintained independently, supporting phased expansion where additional units can be added without disrupting operations. The SMR plant's design allows for efficient power generation, enhanced security, and a reduced environmental footprint, making it a flexible, modern solution for nuclear power.

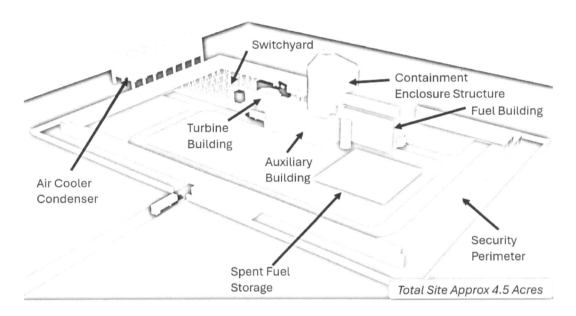

Figure 26: Example SMR Power Station layout.

The physical layout of a Small Modular Reactor (SMR) power plant is meticulously designed to maximize the advantages of modularity, safety, and efficiency. Compared to traditional nuclear power plants, SMR facilities are more compact, utilizing streamlined arrangements that enhance construction speed and lower overall costs. The site layout is broken down into several distinct zones, each dedicated to specific functions, ensuring smooth operation and efficient management.

Reactor Zone: At the heart of the SMR power plant is the Reactor Containment Building, where each SMR module is housed. These containment structures are often built underground or partially below ground level, offering protection against external threats like natural disasters and enhancing radiation shielding. Thick reinforced concrete walls provide robust structural integrity, minimizing risks from earthquakes or other potential hazards. For multi-module plants, the Modular Reactor Bays are arranged in a grid or clustered configuration, allowing each reactor to operate independently. This modular design enables staggered construction and refuelling, ensuring continuous operation without interruptions. Additionally, the Primary Cooling Loop Infrastructure is located within the reactor zone, where coolant circulates through specialized pumps and heat exchangers. Depending on the SMR's design, alternative coolants like molten salt, gas, or lead may be used, with the infrastructure tailored accordingly.

Control and Operations Zone: The Control Room serves as the command centre of the plant, managing the operation of all reactor modules. It is housed in a separate, shielded building to protect operators from radiation exposure. Advanced digital control systems enable centralized monitoring, allowing a small team to efficiently manage multiple reactors simultaneously. Adjacent to the control room is the Emergency Operations Center (EOC), which is equipped with backup systems and communication tools to handle incidents swiftly. Given the reliance on digital technology, Cybersecurity and IT Rooms are crucial for protecting the plant's digital infrastructure against cyber threats, ensuring the integrity of the control systems and data.

Cooling and Heat Exchanger Zone: The Heat Exchangers and Secondary Loop Systems play a critical role in transferring heat from the reactor's primary coolant to the secondary loop. This zone is equipped with specialized heat exchangers optimized for water, gas, or molten salt systems, depending on the specific SMR design. For SMRs using water as a coolant, Cooling Towers or Air-Cooled Condensers are strategically placed nearby to minimize heat loss and maximize cooling efficiency. Pump and Valve Stations within this zone ensure smooth and controlled circulation of the coolant, with easy access for maintenance to prevent system failures.

Turbine and Power Generation Zone: The Turbine Building houses the turbines and generators that convert thermal energy into electricity. SMR designs allow for more compact turbines, reducing the size of this building while maintaining high efficiency. The turbine building is located close to the reactor zone to minimize the length of steam pipes, reducing energy losses. Additionally, the Electrical Switchgear Room nearby manages the distribution of electricity to the grid, ensuring consistent power output and load management.

Spent Fuel and Waste Management Zone: To manage nuclear waste, each SMR unit is equipped with Spent Fuel Pools for the temporary storage of used fuel rods immediately after removal from the reactor. These pools are shielded and securely constructed to prevent radiation leakage. Once cooled, the fuel is transferred to the Dry Cask Storage Area for long-term storage in robust, monitored casks. A dedicated Waste Treatment Facility handles the processing and packaging of low-level radioactive waste for safe disposal, in compliance with environmental regulations.

Auxiliary and Support Infrastructure Zone: Supporting the plant's operations, Maintenance and Workshops are on-site for the assembly, repair, and replacement of modular components. These facilities allow for quick turnarounds, reducing downtime. Warehouse and Storage Facilities securely house reactor components, fuel assemblies, and other essential equipment. Additionally, Administrative and Employee Facilities provide office space, meeting rooms, and training centres for plant personnel, ensuring smooth administrative operations and staff well-being.

Security and Access Control Zone: The entire SMR site is protected by a Perimeter Security system, featuring controlled access points, surveillance cameras, and high-security fencing to prevent unauthorized entry. A dedicated Security Control Centre monitors the site 24/7, ensuring prompt responses to any security breaches. Visitor and Employee Checkpoints enforce stringent security protocols, including radiation screening where necessary.

Transmission and Grid Connection Zone: The Switchyard connects the SMR plant to the regional power grid, with transformers and circuit breakers managing electricity transmission. Located near the turbine building, this zone ensures efficient power delivery to the grid. Grid Stabilization Systems are also integrated, including backup generators and uninterruptible power supplies (UPS), ensuring continuous operation even during grid outages.

The SMR power plant layout is designed with a focus on compactness, modularity, and operational safety. The modular approach allows for flexible expansion, enabling the addition of new reactor units as needed without disrupting existing operations. The use of below-ground containment buildings, along with dedicated zones for waste management and security, enhances both safety and resilience. By carefully organizing each zone and leveraging modern digital technologies, SMR power plants achieve a balance between operational efficiency, safety, and scalability. This thoughtful layout supports the deployment of SMRs as a versatile and cost-effective solution for meeting global energy demands in diverse environments.

SMR Site Selection

Choosing the appropriate site for a Small Modular Reactor (SMR) power plant is a multifaceted process that necessitates careful consideration of various factors to ensure safety, efficiency, and compliance with regulatory standards. The modular design and smaller size of SMRs provide distinct advantages over traditional large-scale nuclear reactors, particularly in terms of site selection flexibility. This flexibility is crucial as it allows for a more diverse range of potential locations, which can be tailored to meet specific energy demands and environmental considerations.

The site selection process for SMRs involves a systematic evaluation of numerous criteria, including geographical, economic, and logistical factors. Site selection for SMRs is a two-stage process: the first stage involves identifying candidate locations based on objective data, while the second stage refines these locations by considering subjective factors such as public perception and political will [69]. This dual approach ensures that both technical and social dimensions are addressed, enhancing the likelihood of successful project implementation.

In addition to the technical and analytical aspects, regulatory compliance plays a pivotal role in the site selection for SMRs. The design of SMRs incorporates inherent safety features that can simplify regulatory processes compared to traditional reactors. The regulatory landscape is evolving to accommodate the unique characteristics of SMRs, which may lead to more streamlined licensing processes [70]. This regulatory flexibility is particularly beneficial in regions where energy demands are increasing, and there is a pressing need for clean energy solutions.

Furthermore, the environmental benefits of SMRs cannot be overlooked. SMRs represent a promising option for mitigating the environmental pressures associated with fossil fuel consumption, making them an attractive choice for site selection in areas facing significant ecological challenges [8]. The smaller footprint and lower emissions associated with SMRs enhance their appeal as a sustainable energy source.

Nuclear Regulatory Compliance

One of the most critical aspects of developing a Small Modular Reactor (SMR) power plant is ensuring full compliance with nuclear regulatory requirements. Before construction can begin, the site must receive approval from the relevant nuclear regulatory authority, such as the U.S. Nuclear Regulatory Commission (NRC), the International Atomic Energy Agency (IAEA), or equivalent national bodies in other countries. These regulatory agencies set stringent standards to ensure that the operation of nuclear facilities is safe, secure, and environmentally responsible. Compliance with these standards involves demonstrating that the design, construction, and operation of the SMR meet rigorous safety criteria, including radiation protection, reactor integrity, and emergency preparedness.

The regulatory process for SMRs, while similar to that of traditional reactors, also considers the unique aspects of their design, such as modularity, smaller footprint, and passive safety features. Developers must submit detailed safety analysis reports, risk assessments, and design documentation to prove that the SMR can withstand various external threats, such as natural disasters, cyber-attacks, or accidents. The approval process can take several years, as regulatory bodies perform exhaustive evaluations to confirm that all safety measures are adequately addressed. Failure to comply with regulatory standards can lead to costly delays, fines, or even the denial of a license to operate, making regulatory compliance a critical component of SMR project planning.

Environmental Impact Assessments (EIA)

In addition to regulatory compliance, a comprehensive Environmental Impact Assessment (EIA) is required before any SMR project can proceed. An EIA aims to evaluate the potential effects of the proposed power plant on the surrounding environment. This involves a thorough analysis of how the plant's operations might affect air and water quality, soil integrity, local wildlife, and vegetation. The EIA process typically begins with a scoping study to identify all potential environmental risks, followed by detailed assessments using scientific data and predictive modelling.

One key area of focus in an EIA is water usage and thermal discharge, especially if the SMR uses water-based cooling systems. The assessment must demonstrate that the plant's operations will not negatively impact nearby water bodies, such as rivers or lakes, by raising water temperatures or introducing contaminants. Additionally, the EIA examines the management of radioactive waste, ensuring that any by-products from the reactor are handled, stored, and disposed of in a way that minimizes environmental harm. The impact on local air quality is also assessed, especially in terms of emissions from auxiliary systems like backup diesel generators.

Once completed, the EIA report is typically subject to public review and consultation. Regulatory agencies may require developers to address any concerns raised by the public or environmental advocacy groups before granting approval. This process not only ensures that the SMR project aligns with environmental sustainability goals but also helps build public trust and acceptance. By identifying potential environmental risks early on, the EIA helps developers implement effective mitigation strategies, ensuring that the SMR power plant operates with minimal ecological impact.

Earthquake and Fault Line Assessment

One of the fundamental considerations in selecting a site for a Small Modular Reactor (SMR) power plant is assessing the geological and seismic stability of the location. The plant must be situated in an area that is not prone to significant earthquake activity or located near active geological fault lines. While SMRs are often designed with enhanced seismic resilience due to their smaller size and modular construction, ensuring the underlying ground is stable remains a critical safety requirement. The design of SMRs includes features like flexible foundations, shock absorbers, and reinforced structures that can withstand ground shaking; however, placing the plant in a seismically stable region is a crucial preventive measure.

During site evaluation, developers conduct detailed geological surveys to map out the presence of fault lines, past seismic activity, and the potential for future earthquakes. This includes analysing historical data and using geotechnical instrumentation to assess the likelihood of seismic events in the area. If a site shows signs of active fault lines or a history of frequent tremors, it may be deemed unsuitable for SMR construction. The goal is to mitigate risks that could compromise the structural integrity of the reactor and its containment systems, which are essential to preventing radiation leaks during a seismic event. By choosing a site with minimal seismic risk, developers can focus on designing efficient and cost-effective structures while still meeting stringent safety standards.

Soil Conditions

In addition to seismic assessments, thorough soil analysis is essential to ensure the site can adequately support the SMR's infrastructure over its operational lifespan. The stability of the soil directly impacts the safety and durability of the reactor's foundation. For an SMR power plant, which includes heavy components like the reactor containment structures, cooling systems, and auxiliary buildings, it is vital that the soil can handle these loads without significant subsidence or shifting. Soil subsidence could lead to structural misalignments, increased stress on reactor components, or, in severe cases, compromise the integrity of the reactor's containment.

Geotechnical engineers conduct soil tests to determine properties such as soil type, density, bearing capacity, and moisture content. These tests help identify any potential issues like soil liquefaction, which can occur during earthquakes, particularly in sandy or loose soils. If the soil is found to be unstable or unsuitable, additional measures such as soil compaction, grouting, or even constructing deep pile foundations may be required to enhance stability. The goal is to create a robust foundation that minimizes the risk of settlement or movement, which is especially critical for nuclear facilities where even minor structural shifts can have serious safety implications.

By carefully assessing both the geological and soil conditions of a potential site, developers can ensure that the SMR power plant is built on a stable foundation, reducing the risk of structural damage over time. This comprehensive evaluation not only enhances the safety and reliability of the plant but also contributes to its long-term operational efficiency and cost-effectiveness. The combination of seismic assessments and soil analysis is a proactive approach to managing risks, ensuring that the SMR can operate safely even under challenging environmental conditions.

Cooling Water Availability

For Small Modular Reactors (SMRs) that utilize water-based cooling systems, proximity to a reliable water source is a key factor in site selection. The cooling process in these reactors relies on water to dissipate the significant heat generated during nuclear fission. This means that the plant needs to have access to a steady supply of water, often sourced from nearby rivers, lakes, or coastal areas. A dependable water source ensures the efficient functioning of the heat exchange systems, which are crucial for maintaining safe operating temperatures within the reactor core. The availability of abundant cooling water is not only essential for normal operations but also critical in emergency situations where the reactor may need additional cooling to prevent overheating.

Locating an SMR power plant near natural water bodies can provide additional benefits, such as cost savings on infrastructure, since the proximity reduces the need for extensive piping systems to transport water over long distances. However, developers must also consider the environmental impact of using natural water sources. This includes assessing potential thermal pollution, which can adversely affect aquatic ecosystems if the discharged water is significantly warmer than the ambient temperature. Additionally, regulatory approvals are required to ensure that water withdrawal and discharge do not disrupt local water levels or harm the surrounding environment. Thus, thorough environmental impact assessments are conducted to balance the need for cooling water with ecological sustainability.

Alternative Cooling Options

In regions where water sources are scarce or unreliable, SMRs with alternative cooling technologies offer a significant advantage. Some SMRs are designed to use air-cooling or dry cooling systems, which eliminate the dependence on large quantities of water. These systems utilize air or other fluids to dissipate heat, making them particularly suitable for arid, remote, or landlocked areas where access to water is limited. The use of air-cooled condensers, for instance, allows SMRs to operate efficiently without drawing from natural water sources, thereby expanding the range of potential sites where these reactors can be deployed.

The flexibility of using air-cooling systems provides several benefits, including reducing the environmental impact associated with water withdrawal and thermal discharge. This not only helps conserve precious water resources but also simplifies the regulatory process, as there are fewer concerns related to water use permits and environmental compliance. Air-cooled SMRs can also be more resilient to climate change, as they are less affected by fluctuations in water availability due to droughts or changes in precipitation patterns. However, air cooling systems may require more space and have slightly lower thermal efficiency compared to water-cooled systems, which could influence the overall design and cost considerations for the power plant.

By incorporating both water-based and alternative cooling options into their designs, SMRs can be adapted to a wide range of environmental conditions. This versatility allows SMR technology to be deployed in diverse geographical locations, from coastal regions with abundant water supplies to inland areas facing water scarcity. This adaptability is one of the reasons SMRs are seen as a promising solution for providing clean, reliable energy in various contexts, including remote communities, arid regions, and developing countries with limited infrastructure.

Proximity to Electrical Grid

For a Small Modular Reactor (SMR) power plant, proximity to existing electrical grid infrastructure is a crucial factor in site selection. Being close to high-voltage transmission lines allows for a more efficient and cost-effective connection between the power plant and the national or regional grid. This proximity minimizes the need for extensive construction of new transmission lines, which can be costly, time-consuming, and subject to regulatory approvals. By reducing the distance to existing power lines, SMR developers can significantly lower initial capital expenses, enhance project feasibility, and accelerate the time required to bring the plant online.

Furthermore, a closer connection to the grid helps optimize power transmission efficiency. The longer electricity has to travel through transmission lines, the greater the energy losses due to resistance, which can affect the overall efficiency of power delivery. By selecting a site near established electrical infrastructure, SMR operators can reduce these losses, ensuring that more of the generated electricity reaches end-users. This is particularly important for maintaining the economic viability of SMR projects, where the goal is often to provide flexible, low-cost energy solutions. Additionally, grid connectivity near urban or industrial centres allows SMRs to directly support areas with high electricity demand, making them an attractive option for grid stabilization and peak load management.

Accessibility to Remote Areas

One of the key advantages of SMRs is their potential for deployment in remote or off-grid locations, where traditional large-scale nuclear reactors or fossil fuel power plants are not feasible. SMRs are designed to be modular, compact, and capable of operating independently, making them ideal for providing reliable power in isolated regions, such as mining sites, remote communities, or island nations. However, even in these off-grid scenarios, strategic planning is necessary to ensure that the electricity generated can be efficiently distributed to the intended users.

In remote areas, the absence of existing grid infrastructure poses unique challenges. While SMRs can operate independently, they still require robust systems for local power distribution. This may involve setting up microgrids or localized distribution networks to channel electricity to nearby industries or communities. In such cases, developers must consider the logistics of building new transmission lines, substations, and distribution centres. These projects may require collaboration with local authorities and utility companies to ensure reliable power delivery, especially if the area lacks established energy infrastructure.

The modular and flexible nature of SMRs makes them particularly suited for phased deployment in these regions. An initial module can provide immediate power to critical facilities or industries, with additional modules added as demand grows or as new transmission infrastructure is developed. This approach reduces the upfront investment and allows for gradual expansion, making it easier to adapt to changing energy needs over time. Moreover, SMRs can be paired with renewable energy sources, such as wind or solar, to create hybrid power systems that enhance energy resilience in remote areas.

Logistics for Construction and Maintenance

The accessibility of a Small Modular Reactor (SMR) power plant site is a critical factor, particularly given the modular nature of SMR construction. Unlike traditional nuclear power plants, which often require extensive on-site construction over many years, SMRs are designed to have their components fabricated in factories and then transported to the site for assembly. This modular approach significantly reduces construction time and costs, but it also means that the chosen site must be accessible for transporting large, prefabricated modules, heavy equipment, and nuclear fuel.

To ensure efficient delivery, sites need to be well-connected by robust transportation infrastructure, such as highways, rail lines, or waterways. For instance, rail access can facilitate the transportation of large reactor modules and heavy machinery, reducing the need for special road transport, which can be logistically challenging and expensive. Similarly, access to navigable waterways is advantageous, particularly for coastal or river-adjacent sites, as barges can transport oversized components that would be difficult to move overland. Roads leading to the site must also be capable of handling heavy transport vehicles, necessitating proper road quality and capacity. This infrastructure is not only essential during the initial construction phase but also crucial for ongoing maintenance, periodic refuelling, and any future upgrades to the plant.

Moreover, the modular nature of SMRs allows for future scalability. Sites with good transportation access can more easily accommodate additional modules if the plant expands its capacity. Therefore, transportation logistics are not just a one-time consideration but are integral to the long-term viability and flexibility of the power plant.

Emergency Response Access

Another vital consideration in site selection is ensuring that the SMR power plant is accessible to emergency response teams. Safety is a top priority for nuclear facilities, and while SMRs are designed with enhanced safety features, including passive cooling systems and below-ground containment structures, it is still crucial to be prepared for any potential incidents. Quick and reliable access for emergency services, such as fire brigades, medical teams, and hazardous materials response units, can make a significant difference in managing emergencies effectively.

The site must be reachable by well-maintained roads that can support emergency vehicles, even in adverse weather conditions. Proximity to nearby towns or cities with established emergency services is beneficial, as it ensures faster response times. Additionally, having alternative routes for access can be crucial in case the primary access route is blocked during a natural disaster or other emergencies. For more remote SMR installations, it may be necessary to establish dedicated emergency infrastructure on-site, such as helipads or specialized emergency response units.

Furthermore, an efficient layout within the plant itself can aid in emergency management. Clear, unobstructed pathways, designated emergency exits, and accessible control rooms are essential for the safe evacuation of personnel and for responders to quickly reach critical areas. Regular drills and coordination with local emergency services can also ensure that all parties are prepared to handle any situation that may arise, thereby enhancing the overall safety and security of the facility.

In summary, accessibility and transportation infrastructure play a critical role in both the construction and operational phases of an SMR power plant. The ability to efficiently transport modular components, equipment, and fuel, coupled with the capacity for rapid emergency response, are essential factors in ensuring the successful deployment and safe operation of SMRs. By carefully evaluating these logistical and safety considerations, developers can optimize site selection to maximize efficiency and minimize risks.

Safety Zones and Population Proximity

When selecting a site for a Small Modular Reactor (SMR) power plant, one of the key considerations is its proximity to population centres. Although SMRs are designed with enhanced safety features, such as passive cooling systems and robust containment structures, there are still regulatory requirements that mandate safety buffer zones around nuclear installations. These safety zones are established to protect the surrounding population in the rare event of a radiation leak or other incidents. Unlike traditional large nuclear reactors that require extensive exclusion zones due to their size and potential risk factors, SMRs can often operate with smaller buffer zones thanks to their inherent safety design. However, these zones must still comply with local and international nuclear safety regulations, such as those set by the International Atomic Energy Agency (IAEA) or regional bodies like the U.S. Nuclear Regulatory Commission (NRC).

Ideally, an SMR site should be located in areas with low population density to reduce the potential impact on nearby communities. By choosing a site away from densely populated urban centres, the risk to human health and safety is minimized in the unlikely event of an emergency. Additionally, a more isolated location ensures that there are fewer logistical and regulatory hurdles related to public safety, making it easier to secure approvals for construction and operation. However, this need for distance must be balanced with other factors, such as accessibility for transportation of materials and emergency response capabilities.

Economic Considerations and Industrial Applications

While safety is a primary concern, there are also significant economic factors to consider when selecting the location of an SMR power plant. In some cases, sites that are relatively close to industrial zones or growing urban areas are preferred, especially when the SMR is intended to supply not just electricity but also process heat for industrial applications. Proximity to industrial facilities can make an SMR an attractive option for industries that require consistent, high-temperature heat for their processes, such as chemical production, steel manufacturing, or water desalination. By co-locating an SMR near industrial hubs, the power plant can efficiently deliver heat and electricity, reducing energy transmission losses and lowering costs for industrial partners.

In addition, locating an SMR near urban areas experiencing growth can support local economic development. The presence of a reliable, low-carbon energy source can attract businesses, support new developments, and create jobs, thus boosting the local economy. However, while proximity to population centres can provide economic benefits, it must be carefully managed to ensure that safety standards are not compromised. This often involves comprehensive risk assessments and environmental impact studies to determine the feasibility of placing an SMR near densely populated areas.

Balancing Safety and Economic Benefits

The decision on where to locate an SMR power plant involves a careful balance between ensuring safety and maximizing economic benefits. While isolated sites are generally preferred for safety reasons, particularly in reducing the risk to human populations, there are strong incentives to place SMRs closer to where their output is most needed. By selecting sites that are near industrial zones or expanding cities, operators can capitalize on the demand for stable, low-carbon energy sources to power industries and support urban growth.

Moreover, modern SMR designs are developed with advanced safety features that can mitigate the risks associated with proximity to population centres. Features like underground containment structures, passive cooling systems, and automated shutdown mechanisms significantly reduce the likelihood of a catastrophic event. As a result, regulatory bodies may permit smaller exclusion zones for SMRs compared to traditional reactors, allowing for greater flexibility in site selection.

Ultimately, the decision on proximity to population centres hinges on a thorough analysis of both safety and economic factors. By ensuring that sites are both safe and economically viable, developers can optimize the placement of SMRs to serve the dual goals of sustainable energy production and regional economic development.

Temperature and Weather Conditions

When selecting a site for a Small Modular Reactor (SMR) power plant, understanding the local climate and environmental conditions is crucial for ensuring the plant's efficient operation and long-term safety. Extreme weather conditions, such as high ambient temperatures, heavy snowfall, or frequent storms, can significantly impact the plant's performance and reliability. For instance, high temperatures can reduce the efficiency of cooling systems, particularly for water-cooled reactors that rely on maintaining a consistent coolant temperature to avoid overheating. SMRs are often designed with robust passive cooling mechanisms, but prolonged exposure to extreme heat can still strain these systems, potentially reducing the plant's efficiency and increasing operational costs.

Conversely, cold climates pose their own challenges. Heavy snowfall can obstruct access roads, complicate maintenance activities, and even affect cooling systems if snow accumulates on air-cooled condensers. In regions that experience freezing temperatures, there's also the risk of coolant pipes freezing, which can disrupt the reactor's cooling process. Additionally, severe storms, including hurricanes and typhoons, can pose risks to infrastructure, potentially damaging external structures like cooling towers and power transmission lines. Therefore, the design of an SMR site must take into account the specific weather patterns of the area, ensuring that the plant is resilient against these conditions. This might involve reinforcing buildings, installing storm barriers, or designing cooling systems that can adapt to varying temperatures.

Flood Risk Assessment

In addition to general weather conditions, a thorough flood risk assessment is essential, especially for sites located near rivers, lakes, or coastal areas. Flooding poses one of the most significant risks to nuclear power plants, as it can lead to water ingress into critical areas, potentially compromising reactor safety systems and leading to unplanned shutdowns. Floodwaters can damage electrical systems, disable cooling pumps, and even erode the soil supporting key structures, which could result in subsidence or structural instability. Given the smaller footprint of SMR plants, they are somewhat more vulnerable to localized flooding compared to large nuclear plants, making proper site evaluation even more critical.

Flood risk assessments involve studying historical flood data, evaluating the site's elevation, and analysing the potential impact of rising sea levels and extreme weather events exacerbated by climate change. If a site is found to be at risk, additional flood protection measures may be required, such as constructing levees, flood walls, or elevated platforms to house critical infrastructure. In coastal areas, considerations must also be given to storm surges, which can lead to rapid inundation and increased flood levels. By incorporating these protective measures, developers can reduce the risk of flood damage, ensuring the safety and longevity of the SMR facility.

Mitigating Environmental Risks

To optimize the selection of an SMR site, it is essential to account for the local climate and environmental factors early in the planning process. This proactive approach helps in designing reactors that are resilient to adverse weather conditions, reducing the likelihood of disruptions or safety incidents. For example, in regions prone to high temperatures, advanced cooling systems like air-cooled condensers or dry cooling technologies can be employed to minimize dependence on water sources. For colder regions, heating systems and insulation can be incorporated into the design to protect cooling circuits from freezing.

Additionally, continuous monitoring of environmental conditions around the site, such as air and water temperatures, precipitation levels, and potential flood warnings, allows operators to respond swiftly to changing conditions. By integrating weather-resilient designs and maintaining a robust emergency response plan, SMR plants can ensure stable and safe operations even in challenging environmental conditions. This careful planning not only enhances the plant's operational efficiency but also helps in gaining regulatory approval, as adherence to stringent environmental and safety standards is often a prerequisite for project licensing.

Land Footprint Requirements

One of the significant advantages of Small Modular Reactors (SMRs) over traditional nuclear reactors is their reduced land footprint. SMRs are designed to be compact, requiring much less space than large-scale nuclear power plants, which typically occupy extensive areas to accommodate their reactors, cooling towers, waste storage, and other infrastructure. The smaller size and modular nature of SMRs mean that they can fit into tighter spaces, making them suitable for locations that might otherwise be deemed too small for conventional reactors. This reduced footprint opens up opportunities for siting SMRs in areas that were previously used for industrial purposes, known as brownfield sites. These locations often come with existing infrastructure, such as access roads, power lines, and water sources, which can reduce construction costs and time. Additionally, using brownfield sites can revitalize underutilized or abandoned industrial areas, contributing to local economic development while avoiding the environmental impact of developing greenfield sites.

Land Ownership and Zoning

Another crucial factor in selecting a site for an SMR power plant is ensuring that the land is appropriately zoned for industrial or energy generation purposes. Land ownership and zoning regulations can vary significantly by region, which can impact the feasibility of a proposed project. Ensuring that the selected site has the proper zoning designation is essential for avoiding legal complications and project delays. Land acquisition should be straightforward, ideally involving minimal bureaucratic hurdles. For example, if the land is owned by the government or a utility company, securing the necessary permissions might be easier than if the land is privately owned or has mixed ownership.

Moreover, securing a site with the right zoning reduces the risk of opposition from local communities and regulatory bodies. However, if rezoning is required, the process can be lengthy and may involve public consultations, environmental assessments, and negotiations with local authorities. Early identification of potential zoning issues and proactive engagement with stakeholders can help streamline the process. In some cases, working with local governments to align the project with regional energy strategies can also facilitate smoother land acquisition and permitting processes.

Heritage Status and Indigenous Claims

A critical aspect of land use planning for SMR sites involves considering cultural and heritage factors, particularly in areas with historical or indigenous significance. Many potential sites may be subject to heritage protections or indigenous land claims, which can complicate land acquisition and development. For instance, if the land is recognized as having cultural, historical, or archaeological significance, or if it is located near sacred indigenous sites, special permissions and consultations may be required before any development can proceed.

Engaging with indigenous communities and respecting their rights and claims to land is not only a legal obligation but also an ethical one. Failure to acknowledge these rights can lead to protests, legal challenges, and project delays, potentially derailing the entire initiative. Establishing open lines of communication with indigenous groups early in the planning process can help address concerns, build trust, and, where possible, incorporate their input into project development. This approach ensures compliance with local regulations, fosters positive community relations, and reduces the risk of project delays due to legal disputes.

Strategic Land Use Considerations

Incorporating land use considerations into the planning phase of an SMR project is vital for optimizing site selection and ensuring long-term project success. By leveraging the compact footprint of SMRs, developers can explore a wider range of site options, including those with existing infrastructure or those that might not be suitable for larger power plants. However, securing the necessary land involves more than just finding available space; it requires careful consideration of legal, regulatory, and social factors. Balancing the need for efficient land use with respect for cultural and heritage considerations is essential for ensuring that SMR projects are not only technically feasible but also socially responsible and sustainable.

Ultimately, addressing land availability, ownership, zoning, and cultural considerations early in the site selection process can help avoid costly delays and foster smoother project execution. This comprehensive approach ensures that the chosen site aligns with both regulatory requirements and community expectations, contributing to the overall success and acceptance of the SMR project.

Community Support and Public Perception

Public acceptance is a cornerstone of any nuclear power project, and Small Modular Reactors (SMRs) are no exception. The perception of nuclear energy within the community can significantly influence the success or failure of a project. While SMRs are designed with enhanced safety features and are considered more secure than traditional nuclear reactors, there can still be lingering concerns about radiation, waste management, and the potential for accidents. Therefore, gaining the trust and support of the local population is crucial for a smooth approval and construction process.

To build community support, it is essential to engage stakeholders early in the planning stages. This involves transparent communication about the benefits of the project, including its safety features, environmental advantages, and potential economic gains. Hosting town hall meetings, informational sessions, and open dialogues with residents allows developers to address concerns directly, dispel myths, and build a sense of trust. Additionally, involving local stakeholders in decision-making processes can make the community feel valued and heard, which can help reduce opposition. This proactive approach can mitigate delays caused by public protests or legal challenges, ultimately making the project timeline more predictable.

Job Creation and Economic Benefits

One of the strongest arguments in favour of siting an SMR in a particular location is the potential for job creation and economic development. The construction, operation, and maintenance of SMR plants can provide a significant boost to local economies, especially in areas that may be struggling with unemployment or lack of industrial development. During the construction phase, there is a high demand for skilled labour, engineers, project managers, and contractors, leading to immediate job opportunities. These jobs are often well-paying, contributing to the economic prosperity of the region.

Beyond the construction phase, SMRs provide long-term employment for plant operators, technicians, safety inspectors, and administrative staff. Moreover, the presence of a stable and reliable energy source can attract other industries to the area, further enhancing economic growth. For instance, businesses that require substantial, uninterrupted power supplies—such as manufacturing plants or data centres—might be more inclined to set up operations in areas with SMR facilities. This can lead to the creation of additional jobs and stimulate local economic activities, such as housing, retail, and services.

Balancing Economic Benefits with Community Concerns

While the economic benefits of SMRs can be compelling, they must be carefully balanced against community concerns. For example, while job creation is a significant advantage, it may not be enough to sway public opinion if there are fears about safety or environmental impacts. Developers need to clearly articulate how the SMR will not only bring economic benefits but also prioritize the health and safety of the community. Investments in local infrastructure, such as roads, schools, and healthcare facilities, can further demonstrate a commitment to the region's well-being.

In some cases, developers may offer community benefit agreements, which include commitments to fund local projects or provide additional services that benefit residents. This could range from supporting local education initiatives to investing in green spaces and recreational facilities. By demonstrating that the project has both economic and social value, developers can foster goodwill and strengthen the case for the project's approval.

Strategic Site Selection for Socioeconomic Impact

When choosing a site for an SMR, developers often prioritize regions that can benefit most from the economic uplift that a new power plant can provide. For instance, areas with declining industries or regions that are economically depressed may be ideal candidates. Not only do these areas often have existing infrastructure that can be repurposed, but they also stand to benefit significantly from job creation and increased tax revenue.

Strategically, this approach aligns with both community interests and the project's long-term viability. By selecting a location where the socioeconomic impact is clearly positive, developers can foster a supportive environment that facilitates smoother regulatory approvals and community acceptance. This can reduce project risks and ensure the plant's successful integration into the local community.

Physical Security

Ensuring robust physical security is a top priority for any Small Modular Reactor (SMR) power plant. Given that SMRs generate nuclear energy, they are subject to stringent security measures to prevent unauthorized access and protect against potential threats. The plant's design and layout must accommodate multiple layers of physical security to safeguard the facility, personnel, and surrounding community. This includes controlled access points with security checkpoints, reinforced fencing, surveillance cameras, motion detectors, and regular patrols by security personnel. The integration of these measures ensures that the plant is well-protected from intrusions or sabotage attempts.

A key aspect of SMR security is the use of advanced surveillance technology, such as high-definition cameras, drones, and thermal imaging sensors, to monitor the perimeter and critical zones within the site. Additionally, access to sensitive areas, like the reactor containment buildings and control rooms, is heavily restricted to authorized personnel only, using biometric scanners, access cards, and encrypted digital systems. The design of SMRs, with many units being partially or fully underground, naturally enhances security by making it more difficult for unauthorized individuals to access the reactors.

Furthermore, SMRs are often equipped with passive safety features and automated shutdown mechanisms that can act as a secondary line of defence in the event of a security breach. These systems can quickly isolate reactor modules and prevent any potential damage from escalating into a larger safety incident. By combining robust physical security infrastructure with advanced technological safeguards, SMRs aim to minimize the risk of unauthorized access and ensure the protection of the facility.

Proximity to Sensitive Facilities

The site selection process for an SMR power plant must also consider its proximity to other critical infrastructure, such as military bases, airports, or national defence installations. Placing an SMR near such facilities could pose additional security risks or lead to conflicts in terms of airspace and territorial control. For example, being close to a military base could make the plant a potential target in the event of military conflicts or terrorist attacks, thereby increasing the overall security risk profile.

Similarly, the presence of airports or air traffic routes near an SMR site can raise concerns regarding potential accidents involving aircraft. While SMRs are designed to be highly resilient, including protection against aircraft impacts, it is still advisable to avoid locations where the likelihood of such incidents is elevated. By carefully selecting sites away from sensitive facilities, developers can reduce the risk of external threats and ensure that the SMR operates in a secure and stable environment.

Additionally, regulatory bodies often impose strict guidelines on the permissible distance between nuclear power plants and sensitive infrastructure. These regulations are in place to protect both the SMR facility and the critical infrastructure it might affect. Compliance with these guidelines not only enhances safety but also simplifies the regulatory approval process, making it easier for the project to move forward without delays.

Integrated Security Strategy

Beyond physical barriers and strategic site selection, an integrated security strategy is essential for SMR facilities. This includes continuous risk assessment, cybersecurity measures, and emergency response planning. SMRs rely on digital control systems for monitoring and operations, making them potentially vulnerable to cyberattacks. Therefore, a comprehensive cybersecurity framework is crucial to protect the plant's digital infrastructure and prevent disruptions.

Emergency response planning is another critical component of an SMR's security strategy. The site must be designed with designated evacuation routes, secure shelters, and coordination protocols with local emergency services. Regular drills and training exercises for plant personnel help ensure preparedness for various scenarios, from natural disasters to security breaches. This proactive approach to safety not only protects the plant but also builds trust with the surrounding community, showing that all necessary measures are in place to handle potential threats effectively.

By incorporating these comprehensive security and safety measures, SMRs can demonstrate their commitment to maintaining the highest levels of protection while delivering clean, reliable energy.

Example Suitable Locations

Note: The locations provided in this analysis are hypothetical examples based on various site selection criteria relevant to Small Modular Reactor (SMR) deployment. These illustrative examples are for informational purposes only and do not imply any endorsement, recommendation, or suggestion that these regions should pursue or permit the construction of SMR power stations. The feasibility of building an SMR power plant in any specific location would require comprehensive studies, regulatory approvals, environmental impact assessments, and community engagement. The decision to develop nuclear infrastructure remains the sole responsibility of the respective governments, regulatory bodies, and stakeholders involved.

Below are 12 suitable locations around the world for constructing Small Modular Reactor (SMR) power stations, taking into account the factors outlined above:

1. Wyoming, USA

- **Rationale:** Wyoming has a history of supporting energy projects, particularly fossil fuels, making it an ideal candidate for transitioning to nuclear energy. The state has vast, seismically stable land, making it suitable for SMR deployment. Additionally, existing infrastructure for coal power plants provides an opportunity for site repurposing, especially in communities open to new energy sources due to job losses in coal mining. The regulatory environment is supportive of nuclear initiatives, making it an attractive candidate for SMR development [71].

- **Benefits:** Strong regulatory support, abundant land, existing grid infrastructure, and a pro-nuclear local government.

2. Ontario, Canada

- **Rationale:** Ontario has a strong nuclear regulatory framework and existing nuclear power infrastructure, particularly around the Bruce Peninsula and Darlington. The region benefits from stable geological conditions, proximity to large freshwater sources (Great Lakes), and strong public acceptance of nuclear energy. These factors contribute to a favourable environment for SMR deployment [8, 13].

- **Benefits:** Established grid connectivity, access to cooling water, supportive regulatory environment, and potential for district heating applications.

3. Scotland, United Kingdom (Caithness)

- **Rationale:** Northern Scotland, particularly near Dounreay in Caithness, is already home to former nuclear facilities, providing a skilled workforce and infrastructure. The area has low population density, access to water sources, and proximity to renewable energy sources for hybrid energy systems. Additionally, the Scottish government's support for green energy initiatives aligns with the deployment of nuclear technology [8, 13].

- **Benefits:** Existing nuclear expertise, low seismic risk, remote location suitable for safety zones, and government support for green energy initiatives.

4. Western Australia, Australia

- **Rationale:** With vast expanses of sparsely populated land and stable geological conditions, Western Australia is ideal for SMR deployment. The region is rich in mining and industrial activities, which can benefit from SMR-supplied process heat and electricity. The region's mining and industrial sectors could benefit from the process heat and electricity generated by SMRs, further enhancing their economic viability [8, 13].

- **Benefits:** Abundant land, potential for industrial applications, minimal seismic activity, and supportive state government.

5. Saskatchewan, Canada

- **Rationale:** Saskatchewan is interested in nuclear energy to reduce its carbon footprint, particularly in its resource-heavy industries. The region is geologically stable and has experience with uranium mining, which aligns with nuclear development. The region's geological stability and existing uranium mining experience align well with nuclear development, supported by pro-nuclear public policies and infrastructure [13, 71].

- **Benefits:** Pro-nuclear public policy, stable geology, potential for industrial heat applications, and existing transmission infrastructure.

6. Kazakhstan (Steppe Region)

- **Rationale:** Kazakhstan has significant experience in nuclear fuel production and uranium mining. The vast steppe region offers isolated locations suitable for SMR deployment with minimal seismic activity and plenty of land. Government support for nuclear energy further enhances its attractiveness [8, 13].

- **Benefits:** Access to uranium resources, strong government support for nuclear energy, and remote sites for safety.

7. Norrbotten, Sweden

- **Rationale:** Northern Sweden, particularly in Norrbotten, is an ideal location for SMRs due to its low population density, stable geology, and cold climate, which can improve reactor cooling efficiency [8, 13]. The region also benefits from existing renewable energy projects, providing a potential synergy.

- **Benefits:** Low population density, access to cooling water, support for clean energy transition, and strong regulatory framework.

8. United Arab Emirates (Al Dhafra Region)

- **Rationale:** The UAE has already invested heavily in nuclear energy with its Barakah Nuclear Power Plant. The Al Dhafra region offers seismically stable land, access to seawater for cooling, and a growing need for desalination, which SMRs can support [8, 13].

- **Benefits:** Established regulatory framework, government support, potential for co-generation of desalinated water, and robust security measures.

9. Chile (Atacama Desert)

- **Rationale:** Chile's Atacama Desert is one of the driest places on Earth but has significant potential for SMRs, especially air-cooled designs. The country needs reliable energy sources for its mining industry and to replace its reliance on fossil fuels. The region's high demand for energy in the mining sector and the need to diversify energy sources away from fossil fuels make it a viable candidate for nuclear development [8, 13].

- **Benefits**: Stable geology, isolated location, high demand for industrial heat, and potential for energy diversification.

10. Finland (Lapland Region)

- **Rationale**: Finland has a well-established nuclear energy program and is exploring SMRs for district heating. The Lapland region, with its cold climate, sparse population, and stable geological conditions, is suitable for SMR deployment [8, 13].

- **Benefits**: Low seismic risk, supportive regulatory environment, potential for district heating, and strong public acceptance.

11. South Korea (Gyeongbuk Province)

- **Rationale**: South Korea is actively developing SMRs like the SMART reactor and has extensive experience in nuclear energy. The Gyeongbuk Province offers existing infrastructure, proximity to industrial zones, and supportive government policies. Gyeongbuk Province offers existing infrastructure and proximity to industrial zones, coupled with supportive government policies, making it a prime location for new nuclear facilities [8, 13].

- **Benefits**: Advanced nuclear technology expertise, established regulatory framework, and proximity to industrial consumers.

12. South Africa (Northern Cape)

- **Rationale**: South Africa is exploring SMRs to stabilize its energy grid. The Northern Cape, with its low population density, ample land availability, and stable geology, is an ideal site for SMRs, especially air-cooled designs. The region's low population density, ample land, and stable geological conditions make it an ideal site for SMRs, particularly air-cooled designs, supported by strong government backing for nuclear energy [8, 13].

- **Benefits**: Strong government support for nuclear energy, stable geological conditions, and potential for energy security.

Summary of Site Selection Criteria Considered

- **Regulatory Compliance**: All sites are in regions with supportive or established nuclear regulatory frameworks.

- **Geological Stability**: Selected sites are in areas with low seismic activity and stable soil conditions.

- **Cooling Water or Alternatives**: Sites near water sources are chosen where necessary, while others are suitable for air-cooled reactors.

- **Grid and Transportation**: Locations are chosen for proximity to electrical grids or remote areas where SMRs can add value.

- **Accessibility and Security**: Each site is accessible for construction, maintenance, and emergency response, with considerations for physical security.

- **Economic and Socioeconomic Factors**: Sites are in regions where SMRs can boost local economies through job creation and industrial support.

These locations represent a strategic mix of developed and developing regions, balancing technological feasibility, regulatory readiness, and local energy needs.

Selecting an appropriate site for an SMR power plant involves balancing technical, regulatory, environmental, and community considerations. The modular and flexible nature of SMRs allows them to be deployed in locations that may not be suitable for traditional reactors, such as remote or off-grid areas. However, ensuring safety, regulatory compliance, and community acceptance remains essential for successful site selection and plant operation.

Environmental Impact and Regulatory Requirements

Environmental Impact of Small Modular Reactors

Small Modular Reactors (SMRs) are increasingly recognized for their potential to provide a flexible and environmentally friendly alternative to traditional nuclear power generation. However, their deployment raises several environmental concerns that must be addressed comprehensively. This synthesis evaluates the key environmental impacts associated with SMR deployment, drawing on recent literature.

Greenhouse Gas Emissions and Air Quality

SMRs are touted for their low carbon emissions during operation, producing virtually no greenhouse gases, which positions them as a viable option for reducing carbon footprints compared to fossil fuel-based power plants [14]. The lifecycle emissions, however, must also be considered. Indirect emissions can arise from various stages, including construction, transportation of components, and the mining and processing of nuclear fuel [72]. While the operational phase of SMRs is characterized by minimal emissions, the overall environmental impact must account for these lifecycle stages, which can contribute to greenhouse gas emissions [64].

As an example of Green House Gas reduction and sustainability, the Shin-Kori Nuclear Power Plant (See Figure 27), located in Ulju County, Ulsan, along South Korea's southeastern coast, is one of the nation's most significant nuclear energy facilities. This plant plays a critical role in South Korea's strategy to diversify its energy sources while reducing greenhouse gas emissions.

It is managed by Korea Hydro & Nuclear Power Co., Ltd. (KHNP), a subsidiary of the state-owned Korea Electric Power Corporation (KEPCO). The facility is a cornerstone in South Korea's push towards achieving energy security and sustainability.

Key Features and Reactor Units:

The Shin-Kori plant consists of multiple reactor units, featuring both older and newer technologies to maximize efficiency and safety.

Shin-Kori Units 1 & 2 (OPR-1000) -

- These units utilize the OPR-1000 (Optimized Power Reactor) design, which is a domestically developed version based on earlier Westinghouse technologies.

- Each of these reactors has a generating capacity of 1,000 MWe.

- Unit 1 began commercial operations in August 2011, followed by Unit 2 in July 2012.

Shin-Kori Units 3 & 4 (APR-1400) -

- The next phase of development includes the APR-1400 (Advanced Power Reactor) units, which showcase significant advancements in reactor technology, emphasizing safety and efficiency.

- Each of these units has a capacity of 1,400 MWe, representing a major upgrade in output compared to the older reactors.

- Unit 3 was commissioned in December 2016, with Unit 4 following in August 2019.

- The APR-1400 reactors feature improved emergency core cooling systems, advanced digital control systems, and enhanced seismic resilience to withstand earthquakes.

Shin-Kori Units 5 & 6 (APR-1400) -

- The latest additions to the Shin-Kori complex, Units 5 and 6, are also based on the APR-1400 design.

- Construction for Unit 5 began in April 2017, while Unit 6 broke ground in September 2018.

- These reactors are expected to become operational by 2024 and 2025, respectively, further expanding the plant's capacity.

Figure 27: Unit 1-2 of Korea Shin-Kori Nuclear Power Plant. IAEA Imagebank, CC BY-SA 2.0, via Wikimedia Commons.

The newer reactors at Shin-Kori, especially the **APR-1400 units,** incorporate state-of-the-art technologies to enhance both performance and safety:

- **Passive Safety Systems**: These reactors are equipped with systems that can cool the reactor core without the need for active mechanical components or operator intervention, significantly increasing safety in emergency scenarios.

- **Digital Instrumentation and Control**: Advanced digital systems are used to monitor and control plant operations, enhancing reliability and improving response times to potential issues.

- **Seismic Resilience**: Designed to withstand significant seismic events, these reactors are suitable for regions with earthquake risks, ensuring robust safety margins.

The Shin-Kori Nuclear Power Plant plays a pivotal role in ensuring South Korea's energy stability and economic growth:

- **Electricity Generation**: Upon full completion, Shin-Kori will boast a total capacity of nearly **7,800 MWe,** making it one of the largest nuclear power facilities in the country.

- **Energy Independence**: By relying on nuclear power, South Korea can reduce its dependence on imported fossil fuels, bolstering energy independence.

- **Greenhouse Gas Reduction**: The low-carbon footprint of nuclear power supports South Korea's commitment to international climate agreements, such as the **Paris Agreement,** by significantly lowering greenhouse gas emissions.

Despite its technological advancements, the Shin-Kori plant has not been without its challenges:

- **Safety Concerns**: Following the Fukushima disaster in 2011, local residents and environmental groups raised concerns about potential nuclear accidents. In response, the government has conducted extensive safety reviews and implemented additional safeguards.

- **Policy Shifts:** The South Korean government's stance on nuclear energy has fluctuated over the years. During President Moon Jae-in's tenure, there was a push to phase out nuclear energy. However, subsequent administrations have recognized the need for stable, low-carbon power sources, leading to renewed support for expanding nuclear capacity.

With the construction of Units 5 and 6 nearing completion, South Korea is positioning itself as a leader in nuclear technology, particularly through the APR-1400 design. The country has successfully exported its nuclear technology, with the Barakah Nuclear Power Plant in the United Arab Emirates being a flagship international project utilizing the APR-1400 reactors.

The Shin-Kori Nuclear Power Plant exemplifies South Korea's expertise in nuclear technology and its strategic focus on energy security and sustainability. As South Korea seeks to balance its energy mix, integrating more renewables while leveraging the reliability of nuclear power, Shin-Kori remains at the forefront of these efforts. This facility not only contributes to the nation's economic stability but also underscores South Korea's commitment to achieving a low-carbon future.

Water Use and Thermal Pollution

SMRs typically require significant volumes of cooling water, which can impact local water bodies, particularly in regions experiencing water scarcity [72]. The thermal pollution resulting from the discharge of warm water back into aquatic environments can elevate local temperatures, potentially disrupting aquatic ecosystems [15]. Some designs, such as those utilizing air-cooling or dry cooling systems, aim to mitigate these impacts by reducing freshwater consumption [73]. This adaptability in cooling methods is crucial for minimizing the ecological footprint of SMRs.

Land Use and Habitat Disruption

The compact footprint of SMRs generally results in reduced land disturbance compared to traditional reactors, allowing for the potential siting of these facilities on brownfield sites, which minimizes impacts on untouched natural areas [14]. However, the construction and operation of SMRs can still disrupt local ecosystems, particularly if located near sensitive habitats [64]. Therefore, careful site selection and environmental assessments are essential to balance development with ecological preservation.

Waste Management

Like traditional nuclear reactors, SMRs generate radioactive waste, including spent nuclear fuel that poses long-term storage challenges due to its hazardous nature [72]. However, some SMR designs are engineered to produce less waste per unit of energy generated, which could alleviate some waste management concerns [64]. Additionally, certain designs, such as molten salt reactors, have the potential to recycle spent nuclear fuel, thereby reducing the overall volume of nuclear waste [14].

Nuclear Accidents and Radiological Impact

SMRs are designed with enhanced safety features, including passive cooling systems and below-ground containment, which significantly reduce the risk of accidents and radiation leaks [5]. Despite these advancements, the inherent risks associated with nuclear facilities necessitate robust emergency preparedness plans to address potential accidental releases of radioactive materials [72]. The emphasis on safety in SMR design is a critical factor in mitigating radiological impacts.

Decommissioning and Site Restoration

The modular design of SMRs may facilitate easier decommissioning processes, potentially reducing the time and environmental impact associated with dismantling the plant [64]. However, residual contamination from decommissioned sites may require ongoing monitoring and remediation efforts to ensure safety for future land use [5]. This aspect underscores the importance of planning for the entire lifecycle of SMRs, including decommissioning.

Transportation and Logistics

The factory-built nature of SMR components allows for reduced construction impacts; however, transportation introduces environmental risks, including potential fuel spills and emissions associated with logistics [5]. Moreover, the environmental impact of uranium mining and processing must be considered, as these activities can lead to habitat destruction and water pollution [64]. Thus, a comprehensive assessment of the supply chain is essential.

Social and Community Impact

The siting of SMRs near communities raises public concerns regarding radiation exposure, particularly in areas without prior nuclear infrastructure [64]. Engaging local communities early in the planning process is vital to address these concerns and minimize social impacts [14]. Effective communication and transparency can foster public trust and acceptance of SMR projects.

While SMRs present significant opportunities for reducing carbon emissions and providing reliable low-carbon energy, their deployment involves a complex array of environmental considerations. These include impacts on water resources, waste management challenges, land use, and the potential for nuclear accidents. Addressing these issues through proper planning, regulatory oversight, and community engagement is essential for ensuring that SMRs can contribute sustainably to the future energy mix.

Regulatory Requirements for Small Modular Reactors

Small Modular Reactors (SMRs) are gaining traction as a promising technology to support low-carbon energy generation. However, regulatory frameworks differ significantly across countries, reflecting varied approaches to safety, licensing, environmental protection, and public engagement. Below is a detailed overview of the key regulatory requirements for SMRs in various regions, including specific examples to illustrate how different nations are addressing the deployment of this technology.

1. United States

In the U.S., the **Nuclear Regulatory Commission (NRC)** is the primary regulatory authority overseeing nuclear reactor licensing. The NRC employs a rigorous and established multi-stage process, which includes **Design Certification (DC), Early Site Permit (ESP)**, and **Combined License (COL)** for reactor projects. The process ensures that all safety, environmental, and operational standards are met before construction begins.

Recent developments in the U.S. have highlighted the potential of SMRs:

- **NuScale Power**: In 2020, NuScale became the first company to receive design certification for its SMR from the NRC, demonstrating that SMRs can meet stringent safety requirements.

- The **Advanced Reactor Demonstration Program (ARDP)**, led by the Department of Energy (DOE), provides funding to accelerate the deployment of SMRs through public-private partnerships.

The Nuclear Regulatory Commission (NRC) serves as the principal regulatory body in the United States responsible for overseeing the licensing and regulation of nuclear reactors. Established in 1975, the NRC's mission is to ensure the safe use of radioactive materials for beneficial civilian purposes while protecting people and the environment. The NRC employs a comprehensive and multi-stage licensing process that includes Design Certification (DC), Early Site Permit (ESP), and Combined License (COL) applications. This rigorous process is designed to ensure that all safety, environmental, and operational standards are met before any construction or operation of nuclear facilities can commence.

Multi-Stage Licensing Process

1. Design Certification (DC):

The Design Certification process is a critical first step in the licensing of nuclear reactors. Under this process, reactor designs are evaluated against safety standards established by the NRC. The design certification ensures that the reactor design can be constructed and operated safely. The NRC reviews the design documentation, which includes detailed safety analyses, operational procedures, and environmental impact assessments. Once a design is certified, it can be referenced in subsequent licensing applications, thereby streamlining the process for future reactors utilizing the same design.

2. Early Site Permit (ESP):

The Early Site Permit process allows applicants to obtain approval for the proposed site of a nuclear reactor before a specific reactor design is selected. This stage assesses the suitability of the site based on environmental, safety, and emergency preparedness considerations. The ESP process can address potential site-specific issues, such as geological and hydrological factors, and can be beneficial in expediting the overall licensing process for future reactor construction.

3. Combined License (COL):

The Combined License process integrates the construction and operating license into a single application. This process allows applicants to submit a comprehensive application that includes both the reactor design and the proposed site, along with the necessary safety and environmental analyses. The COL process is designed to ensure that all regulatory requirements are satisfied before construction begins, thus minimizing delays and uncertainties associated with the licensing process.

2. Canada

Canada's regulatory framework for nuclear projects is managed by the **Canadian Nuclear Safety Commission (CNSC)**. The CNSC has introduced a **Vendor Design Review (VDR)** process that helps evaluate the safety and feasibility of new reactor designs before formal licensing. Although the VDR is voluntary, it significantly streamlines the subsequent licensing stages, making it an attractive option for developers.

Key projects in Canada include:

- **Terrestrial Energy** and **Moltex Energy**, both of which have completed Phase 1 of the VDR. Terrestrial Energy is advancing towards the construction of its Integral Molten Salt Reactor (IMSR).

- The **Canadian Nuclear Laboratories (CNL)** support SMR development through their **SMR Siting Initiative**, allowing developers to propose projects at the Chalk River site.

Canada's regulatory framework for nuclear projects is primarily overseen by the Canadian Nuclear Safety Commission (CNSC), which plays a critical role in ensuring the safety and feasibility of nuclear energy developments. A significant aspect of this framework is the Vendor Design Review (VDR) process, which, while voluntary, provides a structured pathway for evaluating new reactor designs before they enter formal licensing stages. This process not only enhances safety assessments but also streamlines subsequent licensing phases, making it an appealing option for developers looking to introduce innovative reactor technologies [8, 74].

Among the notable projects in Canada, Terrestrial Energy and Moltex Energy have successfully completed Phase 1 of the VDR. Terrestrial Energy is progressing towards the construction of its Integral Molten Salt Reactor (IMSR), which is designed to leverage the unique safety and operational advantages of molten salt technology. Molten salt reactors (MSRs) are recognized for their ability to operate at atmospheric pressure, which significantly reduces the risk of catastrophic failures associated with high-pressure systems [75, 76]. Furthermore, the inherent safety features of MSRs, such as their low vapor pressure and high thermal conductivity, contribute to their appeal as a next-generation nuclear option [77, 78].

The Canadian Nuclear Laboratories (CNL) are also actively supporting the development of small modular reactors (SMRs) through initiatives like the SMR Siting Initiative. This program facilitates the proposal of new projects at the Chalk River site, which has a long-standing history of nuclear research and operations. The Chalk River site is particularly advantageous due to its established infrastructure and regulatory familiarity, which can expedite the development process for new SMR technologies [69, 74]. The site selection for SMRs is a multifaceted process that requires careful consideration of environmental, technical, and regulatory factors, ensuring that proposed projects align with national energy goals and safety standards [8, 69].

3. United Kingdom

In the UK, nuclear projects are regulated by the **Office for Nuclear Regulation (ONR)**. The **Generic Design Assessment (GDA)** is a critical part of the UK's licensing process, assessing new reactor designs' safety, security, and environmental implications. Once a design passes the GDA, developers can move forward with site-specific licensing.

Recent advancements include:

- **Rolls-Royce SMR**: In 2021, Rolls-Royce submitted its SMR design for GDA, with significant government funding to accelerate development.

- The UK has established an **Advanced Nuclear Fund** to support SMRs as part of its strategy to achieve net-zero emissions by 2050.

In the United Kingdom, nuclear projects are overseen by the Office for Nuclear Regulation (ONR), which plays a pivotal role in ensuring that nuclear energy is developed safely and securely. A crucial component of the UK's nuclear licensing process is the Generic Design Assessment (GDA). This process evaluates the safety, security, and environmental impacts of new reactor designs before they can be constructed and operated. The GDA is essential for maintaining high safety standards in the nuclear industry, as it allows for a thorough examination of design specifications and operational protocols [79]. Once a design successfully completes the GDA, developers are then permitted to pursue site-specific licensing, which is necessary for the actual construction and operation of nuclear facilities.

Recent advancements in the UK's nuclear sector include the development of Small Modular Reactors (SMRs), particularly the Rolls-Royce SMR project. In 2021, Rolls-Royce submitted its SMR design for GDA, marking a significant step forward in the UK's nuclear capabilities. The UK government has provided substantial funding to expedite the development of this technology, recognizing its potential to contribute to the country's energy mix and climate goals [79]. The establishment of an Advanced Nuclear Fund further underscores the UK's commitment to supporting SMRs as part of its broader strategy to achieve net-zero emissions by 2050. This fund aims to facilitate innovation and investment in advanced nuclear technologies, which are seen as critical to transitioning to a low-carbon economy.

The UK government has also framed nuclear power as a vital component of its energy strategy, particularly in the context of climate change mitigation. The Climate Change Act of 2008 set ambitious targets for reducing carbon emissions, and the role of nuclear energy in achieving these targets has been increasingly recognized [80]. The political landscape surrounding nuclear energy has shifted, with a growing consensus that new nuclear power generation is necessary to meet future energy demands while minimizing environmental impacts [81]. This shift has been reflected in public discourse and policy, highlighting the importance of nuclear energy in the transition to sustainable energy systems.

4. European Union

While the **European Atomic Energy Community (EURATOM)** provides a broad regulatory framework, individual EU member states retain control over their national nuclear regulations. For instance:

- **France**: The **French Nuclear Safety Authority (ASN)** oversees nuclear projects. France is investing in SMRs, with EDF exploring the NUWARD design.

- **Poland**: The **Polish National Atomic Energy Agency (PAA)** is developing a regulatory framework to support SMR projects. In 2021, Poland signed agreements with companies like NuScale and Westinghouse to explore SMR deployment.

The European Union (EU) plays a pivotal role in the regulation and oversight of nuclear energy through the European Atomic Energy Community (EURATOM). While EURATOM establishes a broad regulatory framework for nuclear safety and security, individual EU member states maintain authority over their national nuclear regulations. This dual structure allows for a harmonized approach to nuclear safety while accommodating the specific needs and contexts of each member state.

In France, for example, the French Nuclear Safety Authority (ASN) is responsible for overseeing nuclear projects, ensuring compliance with both national and EU regulations. France has been proactive in advancing its nuclear capabilities, particularly in the development of Small Modular Reactors (SMRs). The state-owned utility company Électricité de France (EDF) is currently exploring the NUWARD design, which represents a significant investment in next-generation nuclear technology aimed at enhancing safety and efficiency in energy production. This initiative aligns with the broader objectives of EURATOM to promote research and innovation in nuclear safety across Europe.

Poland, on the other hand, is in the process of establishing a regulatory framework to support its burgeoning SMR projects. The Polish National Atomic Energy Agency (PAA) is spearheading this effort, which includes signing agreements with prominent companies such as NuScale and Westinghouse in 2021 to facilitate the exploration and potential deployment of SMRs. These partnerships are crucial for Poland as it seeks to diversify its energy sources and reduce its carbon footprint, reflecting the EU's commitment to sustainable energy practices.

The regulatory landscape established by EURATOM is further reinforced by directives such as Directive 2013/59/Euratom, which sets basic safety standards for protection against ionizing radiation. This directive underscores the EU's commitment to ensuring that all member states adhere to high safety standards in their nuclear operations. Moreover, the collaborative nature of EURATOM's initiatives, such as the European Fast Sodium Reactor Community Project, illustrates the EU's strategy of fostering cooperation among member states to enhance nuclear safety and efficiency.

5. Russia

In Russia, nuclear projects are regulated by the **Federal Service for Environmental, Technological, and Nuclear Supervision (Rostekhnadzor)**. Russia has streamlined processes for approving nuclear technologies, leveraging decades of nuclear expertise, particularly in small and mobile reactors.

- **RITM-200 SMRs**: Initially developed for icebreakers, Russia plans to expand this technology to remote regions and floating nuclear power plants, such as the **Akademik Lomonosov**, the world's first floating SMR plant.

In Russia, the regulation of nuclear projects falls under the jurisdiction of the Federal Service for Environmental, Technological, and Nuclear Supervision, known as Rostekhnadzor. This agency is responsible for overseeing the safety and compliance of nuclear facilities and technologies, ensuring that they meet stringent environmental and safety standards. The Russian approach to nuclear energy is characterized by a streamlined process for the approval of nuclear technologies, which is supported by decades of accumulated expertise in the field. This expertise is particularly evident in the development and deployment of small modular reactors (SMRs), which are increasingly seen as a viable solution for both domestic energy needs and international energy cooperation [11, 82, 83].

One of the notable advancements in Russia's nuclear technology is the RITM-200 small modular reactor. Originally designed for use in nuclear icebreakers, the RITM-200 is being adapted for broader applications, including deployment in remote regions and as part of floating nuclear power plants. The Akademik Lomonosov, which is the world's first floating SMR plant, exemplifies this innovative approach. The RITM-200's design incorporates features that enhance safety and efficiency, making it suitable for operation in isolated power systems where traditional large-scale reactors may not be feasible [82-84]. The modular nature of SMRs like the RITM-200 allows for easier construction and scalability, which is particularly advantageous for regions with limited infrastructure [3, 11].

The global interest in SMRs has surged in recent years, driven by the need for sustainable and flexible energy solutions. These reactors are designed to be built in factories and transported to sites, which significantly reduces construction times and costs compared to traditional nuclear plants. Additionally, SMRs offer enhanced safety features, such as passive safety systems that can operate without external power, thereby reducing the risk of accidents [11, 16, 30]. The economic viability of SMRs is also a critical factor, as they require lower initial capital investments and can be deployed incrementally, allowing for a gradual increase in energy capacity that aligns with demand [30, 56].

6. China

China's nuclear projects are overseen by the **National Nuclear Safety Administration (NNSA)**. The country follows a structured regulatory framework that includes site selection, construction approval, and operational licensing.

- **ACP100 (Linglong One)**: In 2021, China initiated the construction of the world's first commercial SMR at Hainan Island. This project is closely monitored by the NNSA to ensure compliance with safety and environmental standards.

China's nuclear projects are overseen by the National Nuclear Safety Administration (NNSA), which ensures that all nuclear activities adhere to stringent safety and environmental standards. The regulatory framework established by the NNSA encompasses several critical phases, including site selection, construction approval, and operational licensing. This structured approach is essential for maintaining safety and public confidence in nuclear energy, particularly in the wake of global nuclear incidents such as the Fukushima disaster (see Figure 28), which prompted a re-evaluation of nuclear safety protocols worldwide [85, 86].

Figure 28: Debris from the upper levels of Unit 4 lies beside the building. The rubble as been cut away to prepare for construction of a new cover, so that fuel can be moved from the unit's spent fuel pool to a common pool. Fukushima Daiichi Nuclear Power Plant. 18 December 2012. IAEA Imagebank, CC BY-SA 2.0, via Flickr.

One of the significant developments in China's nuclear landscape is the construction of the ACP100, also known as Linglong One, which commenced in 2021 on Hainan Island. This project is notable as it represents the world's first commercial small modular reactor (SMR). The NNSA closely monitors the ACP100 to ensure compliance with both safety regulations and environmental standards, reflecting China's commitment to advancing nuclear technology while prioritizing safety [16, 87]. The ACP100 is designed to utilize advanced safety features inherent in SMR technology, which are believed to enhance operational safety compared to traditional large reactors [30].

The regulatory framework governing nuclear projects in China is characterized by its comprehensive nature, which includes rigorous assessments at each stage of the project lifecycle. This framework is crucial for domestic projects and positions China as a potential leader in global nuclear technology, particularly in regions involved in the Belt and Road Initiative. By sharing its regulatory standards and construction experiences, China aims to foster international cooperation in nuclear energy development, thereby contributing to global efforts to combat climate change [87, 88].

Moreover, the emphasis on small modular reactors like the ACP100 aligns with a broader trend in the nuclear industry towards modularization and scalability. These reactors are designed to be manufactured in a factory setting, which can lead to reduced construction times and costs, as well as enhanced safety through standardized designs [30]. The NNSA's oversight ensures that these innovations are implemented without compromising safety, thereby addressing public concerns about nuclear energy and its associated risks [89].

7. Japan

Japan's **Nuclear Regulation Authority (NRA)** has stringent safety regulations, particularly following the Fukushima disaster. SMRs must meet enhanced seismic and safety standards to receive approval.

- Japan is actively researching SMRs for industrial and remote applications, focusing on **High-Temperature Gas-Cooled Reactors (HTGRs)** and molten salt reactors to diversify its energy portfolio.

The Fukushima disaster marked a pivotal moment in Japan's nuclear energy policy, leading to a comprehensive re-evaluation of safety protocols and regulatory frameworks. The Nuclear Regulation Authority (NRA), established in 2012, has since been responsible for enforcing stringent safety regulations to ensure the integrity of nuclear facilities across the nation. This paper will explore the implications of these regulations on the development of Small Modular Reactors (SMRs) and the broader context of Japan's energy strategy.

The NRA was formed in response to the inadequacies exposed by the Fukushima disaster, which underscored the need for a more robust regulatory body. The authority's primary objectives include:

- Safety Oversight: The NRA has implemented rigorous safety standards that nuclear facilities must meet before receiving operational approval. This includes enhanced seismic standards, which are critical given Japan's susceptibility to earthquakes.
- Public Transparency: The NRA emphasizes transparency in its operations, aiming to rebuild public trust in nuclear energy. This involves public consultations and the dissemination of safety information.
- Continuous Improvement: The NRA mandates that nuclear operators engage in continuous safety improvements, reflecting lessons learned from both domestic and international incidents.

SMRs represent a new frontier in nuclear technology, characterized by their smaller size, modular construction, and enhanced safety features. The development of SMRs in Japan is particularly relevant in the post-Fukushima context, as they offer several advantages:

- Safety Features: SMRs are designed with inherent safety features that reduce the risk of catastrophic failures. For instance, many SMRs utilize passive safety systems that rely on natural forces, such as gravity and convection, to maintain safe operations without the need for active intervention.
- Seismic Resilience: Given Japan's geological conditions, SMRs are engineered to withstand seismic events. The NRA's stringent seismic standards necessitate that these reactors undergo rigorous testing and validation to ensure their resilience.
- Flexibility in Deployment: SMRs can be deployed in remote locations or areas with limited infrastructure, making them suitable for industrial applications and energy supply in isolated regions.

Japan is actively pursuing research and development in two specific types of SMRs: High-Temperature Gas-Cooled Reactors (HTGRs) and molten salt reactors.

HTGRs utilize helium as a coolant and graphite as a moderator, allowing for high operational temperatures. This characteristic enables HTGRs to produce hydrogen efficiently, which can be utilized in various industrial processes. The Japanese government views HTGRs as a potential solution for reducing greenhouse gas emissions while providing a stable energy supply.

Molten salt reactors operate using a liquid salt mixture as both coolant and fuel. This technology offers several advantages, including improved safety due to low pressure operation and the ability to utilize a variety of fuel sources, including thorium. Japan's interest in molten salt reactors aligns with its goals of energy diversification and sustainability.

8. South Korea

The **Korea Institute of Nuclear Safety (KINS)** manages the regulatory processes for nuclear technology in South Korea. The country uses a multi-stage review process focused on safety and environmental impacts.

- South Korea's **SMART Reactor** has been licensed domestically, with plans to export the technology to countries in the Middle East, targeting regions with smaller grids and limited water resources.

The management of nuclear technology in South Korea is primarily overseen by the Korea Institute of Nuclear Safety (KINS), which implements a multi-stage review process that emphasizes safety and environmental impacts. This regulatory framework is crucial, especially as South Korea seeks to expand its nuclear energy capacity to meet the growing electricity demand. The country has recognized the need for nuclear power as a reliable and cost-effective energy source, which is essential for reducing greenhouse gas emissions compared to fossil fuels [90, 91]. However, public attitudes towards nuclear energy remain complex, influenced by historical events such as the Fukushima disaster, which initially raised safety concerns but did not deter the government's commitment to nuclear expansion [92, 93].

South Korea's innovative approach to nuclear technology is exemplified by the development of the System-integrated Modular Advanced Reactor (SMART). This reactor has been licensed for domestic use and is designed to operate effectively in regions with smaller electrical grids and limited water resources, making it particularly suitable for export to countries in the Middle East [17, 94]. The SMART reactor's design incorporates advanced safety features, such as canned motors for reactor coolant pumps, which enhance operational safety by preventing coolant leakage [17]. This focus on safety and adaptability positions South Korea as a potential leader in the global nuclear technology market, particularly in areas where traditional large-scale reactors may not be feasible.

The economic implications of South Korea's nuclear energy strategy are significant. The country aims to balance its energy mix by transitioning towards a more sustainable system while still relying on nuclear power as a critical component [91, 93]. This transition is not without challenges, as it requires careful management of existing nuclear facilities and the integration of renewable energy sources. The South Korean government has initiated policies to improve transparency and public understanding of nuclear energy, which could enhance social acceptance and support for nuclear projects [90, 95]. As the nation navigates these complexities, the interplay between energy policy, public perception, and international market opportunities will be pivotal in shaping the future of its nuclear energy landscape.

9. Australia

Australia currently has a ban on nuclear power plants. However, the **Australian Radiation Protection and Nuclear Safety Agency (ARPANSA)** is exploring the potential of SMRs for remote and mining sites, given the technology's flexibility and smaller footprint.

- Australian policymakers are considering SMRs as a way to achieve their net-zero carbon targets, particularly for remote and off-grid areas.

Australia's current energy landscape is characterized by a ban on nuclear power plants, a policy rooted in historical, environmental, and public safety concerns. However, the Australian government is now exploring the potential of Small Modular Reactors (SMRs) as a viable alternative, particularly for remote and mining sites. This exploration is driven by the flexibility and smaller footprint of SMRs, which could provide a reliable energy source in areas that are often off the main grid. The Australian Radiation Protection and Nuclear Safety Agency (ARPANSA) is actively involved in assessing the safety and regulatory frameworks necessary for the deployment of such technologies [96].

The consideration of SMRs aligns with Australia's broader commitment to achieving net-zero carbon emissions by 2050. Policymakers are increasingly recognizing that traditional energy sources, particularly fossil fuels, are incompatible with these targets. As Australia transitions from a fossil fuel-dominated energy mix to one that emphasizes renewable sources like solar and wind, the role of nuclear energy—specifically through SMRs—becomes more pertinent [97]. The potential for SMRs to support energy needs in remote areas is particularly significant, as these regions often lack access to reliable power sources, which can hinder economic development and sustainability efforts [98].

Moreover, the integration of SMRs into Australia's energy strategy is not merely about energy production; it is also about enhancing energy security and reducing greenhouse gas emissions. The Australian government is exploring various pathways to decarbonize its energy sector, and SMRs could play a crucial role in this transition by providing a stable and low-emission energy source that complements intermittent renewable energy generation [97]. This approach is consistent with global trends where countries are increasingly looking to diversify their energy portfolios to include nuclear options as part of their climate strategies [98].

10. Argentina

The **Nuclear Regulatory Authority (ARN)** of Argentina plays a pivotal role in the country's nuclear initiatives. Argentina is a leader in SMR development within Latin America:

- The **CAREM reactor**, a domestically designed SMR, is under construction, aimed at providing energy to remote areas and for potential export.

Argentina's Nuclear Regulatory Authority (ARN) plays a crucial role in overseeing the country's nuclear initiatives, particularly in the development of Small Modular Reactors (SMRs). Argentina is recognized as a leader in SMR technology within Latin America, with the CAREM reactor being a notable example. The CAREM reactor, designed and constructed domestically, aims to provide energy solutions for remote areas and has the potential for export, reflecting Argentina's commitment to advancing its nuclear capabilities while addressing energy needs in less accessible regions [37, 99].

The CAREM reactor is a Pressurized Water Reactor (PWR) type, which is one of the most widely used designs for nuclear power generation globally. This reactor design is particularly advantageous for Argentina, as it aligns with the country's strategic goals of technological autonomy and energy independence [37, 100]. The development of the CAREM reactor is part of a broader trend where countries are increasingly investing in SMR technologies due to their enhanced safety features, modular construction, and lower financial risks compared to traditional large-scale reactors [3, 101]. The modular nature of SMRs allows for incremental capacity expansion, making them suitable for regions with limited grid capacity, which is a significant consideration for Argentina's diverse geographical landscape [3, 102].

Figure 29: Construction Status of the CAREM Nuclear Power Plant (July 2019). Mariadelmar28, CC BY-SA 4.0, via Wikimedia Commons.

Furthermore, Argentina's historical context regarding nuclear technology is essential to understand its current initiatives. The country has navigated complex geopolitical dynamics, particularly in relation to Brazil, where both nations have pursued independent nuclear capabilities. This pursuit has been motivated by a desire for technological sovereignty and regional prestige, leading to cooperative agreements in the peaceful use of nuclear energy [103, 104]. Argentina's commitment to nuclear energy is also reflected in its regulatory framework, which aims to ensure safety and compliance with international standards, thereby facilitating the development and potential export of nuclear technology [105].

The ARN's regulatory role is vital in maintaining safety standards and fostering innovation within the nuclear sector. The authority's framework is designed to adapt to the evolving landscape of nuclear technology, ensuring that new developments, such as the CAREM reactor, meet stringent safety and operational criteria [106]. This adaptability is crucial as Argentina seeks to position itself as a key player in the global nuclear market, particularly in the context of SMR technology, which is gaining traction worldwide due to its potential to address energy security and environmental challenges [107].

Global Trends in SMR Regulation

- **International Cooperation**: Many countries are collaborating on regulatory frameworks to streamline SMR approvals. The **International Atomic Energy Agency (IAEA)** is facilitating these efforts, promoting best practices and safety standards.

- **Standardization Efforts**: There is an ongoing push for harmonized regulatory guidelines to support cross-border technology transfers, which could simplify the deployment of SMRs globally.

- **Public Engagement**: To address safety concerns and improve public acceptance, regulatory bodies are increasingly focusing on transparency and community involvement in the decision-making process.

These examples illustrate the diverse regulatory landscapes worldwide and how various countries are adapting their frameworks to accommodate the unique characteristics of SMRs. As the technology evolves, harmonizing and simplifying regulatory processes will be crucial for supporting the global deployment of SMRs, helping nations transition to cleaner energy sources.

International Regulations for Small Modular Reactors (SMRs)

The deployment of Small Modular Reactors (SMRs) is subject to a complex set of regulations designed to ensure safety, security, environmental protection, and public trust. These regulations vary from country to country but are influenced by international standards set by organizations like the International Atomic Energy Agency (IAEA). Below is a detailed explanation of the specific regulations that apply to SMRs, focusing on their construction, operation, and safety requirements.

1. International Atomic Energy Agency (IAEA) Guidelines

The IAEA provides a global framework for nuclear safety and regulation, offering guidance on best practices for the design, construction, and operation of SMRs. Although the IAEA does not have direct regulatory authority over national nuclear programs, its recommendations are widely adopted to harmonize safety standards across borders. Key IAEA standards relevant to SMRs include:

- **Safety of Nuclear Installations (SSR-2/1, SSR-2/2)**: These standards cover the safety requirements for the design and operation of nuclear power plants, including SMRs. They emphasize safety assessments, accident prevention, and emergency preparedness.

- **SMR-specific Guidelines (SSG-53)**: The IAEA has developed guidance specifically for SMRs, focusing on the unique characteristics of these reactors, such as modular construction, passive safety features, and scalability. This includes recommendations on site selection, licensing, and environmental impact assessments.

- **Nuclear Security Recommendations (NSS-13)**: These guidelines ensure that SMRs are protected from threats such as sabotage or terrorism. They cover physical security, cybersecurity, and the protection of nuclear materials.

2. United States: Nuclear Regulatory Commission (NRC)

In the United States, the NRC is responsible for licensing and regulating SMRs. The NRC has adapted its regulatory framework to accommodate the unique characteristics of SMRs:

- **Design Certification (DC)**: Developers must obtain design certification for their SMR models, proving that they meet all safety and technical requirements. The NuScale SMR design was the first to receive NRC certification in 2020.

- **Early Site Permit (ESP)**: The ESP allows for site approval before a specific reactor design is chosen, streamlining the licensing process.

- **Combined License (COL)**: This combines construction and operating licenses into a single process, reducing the time and cost needed to deploy SMRs.

- **Regulatory Guidance for Advanced Reactors (RG 1.233)**: The NRC has issued specific guidance to address the design and safety features unique to SMRs, including passive safety systems and modular construction.

3. Canada: Canadian Nuclear Safety Commission (CNSC)

Canada has developed a flexible regulatory process to support the deployment of SMRs:

- **Vendor Design Review (VDR)**: The VDR is a three-phase, pre-licensing assessment process that evaluates the safety of new reactor designs before formal licensing. This process is voluntary but helps expedite the licensing phase.

- **Licensing for Site Preparation, Construction, and Operation**: The CNSC requires separate licenses for site preparation, construction, and operation of SMRs. Each license application must include environmental assessments, safety analysis, and public consultation.

- **Environmental Impact Assessments (EIA)**: SMR projects must undergo comprehensive environmental reviews to assess their impact on ecosystems, water resources, and local communities.

4. United Kingdom: Office for Nuclear Regulation (ONR)

The UK's regulatory framework supports the development of SMRs through:

- **Generic Design Assessment (GDA)**: The GDA process evaluates the safety, security, and environmental impact of a new reactor design before construction. Rolls-Royce's SMR design is currently undergoing the GDA process.

- **Site Licensing**: Once the GDA is complete, developers must obtain a site-specific license for construction. This involves demonstrating that the design meets all safety and security standards for the chosen site.

- **Nuclear Industry Strategy**: The UK government has committed significant funding to support the development and deployment of SMRs, aiming to streamline the regulatory process and reduce deployment times.

5. European Union: EURATOM Framework

The European Atomic Energy Community (EURATOM) provides a regulatory framework that all EU member states must follow for nuclear safety and waste management:

- **Directive 2009/71/Euratom**: This directive focuses on nuclear safety, requiring member states to establish national regulatory frameworks aligned with international standards.

- **Radioactive Waste Management Directive (2011/70/Euratom)**: This directive ensures that SMR projects have robust waste management plans in place, covering storage, transportation, and disposal.

- **Country-Specific Regulations**: While EURATOM sets the overarching framework, individual countries like France and Poland have their own nuclear regulatory authorities to approve and oversee SMR projects.

6. Russia: Rostekhnadzor (Federal Service for Environmental, Technological, and Nuclear Supervision)

Russia has a well-established regulatory framework for nuclear power, including SMRs:

- **Streamlined Licensing for Mobile Reactors**: Russia has developed specific regulations for floating and land-based SMRs, such as the RITM-200 reactors used on icebreakers and floating power stations.

- **Construction and Operation Licensing**: Russian regulations allow for a more expedited process for projects that use proven technologies, such as the Akademik Lomonosov floating SMR plant.

7. China: National Nuclear Safety Administration (NNSA)

China's regulatory framework supports its ambitious nuclear expansion, including SMRs:

- **Stepwise Licensing Process**: China uses a phased approach, starting with site selection and feasibility studies, followed by construction permits and operational licenses.

- **ACP100 (Linglong One)**: This SMR design was the first in the world to begin construction in 2021 after receiving regulatory approval from the NNSA.

- **Focus on Innovation**: China's regulatory approach encourages the rapid deployment of innovative nuclear technologies, supported by government investment.

8. Japan: Nuclear Regulation Authority (NRA)

Japan's stringent nuclear regulations, revised after the Fukushima disaster, also apply to SMRs:

- **Enhanced Safety and Seismic Standards**: SMRs in Japan must meet rigorous safety requirements, particularly related to seismic resilience and emergency response.

- **Public Consultation and Transparency**: Japan emphasizes public engagement in the regulatory process, ensuring that local communities are informed and consulted on new SMR projects.

9. South Korea: Korea Institute of Nuclear Safety (KINS)

South Korea is actively pursuing SMR development with its SMART reactor:

- **Multi-Stage Licensing Process**: KINS uses a detailed, multi-stage review for new reactor designs, focusing on safety assessments and environmental protection.

- **Export Licensing**: South Korea's regulatory framework also includes provisions for exporting nuclear technology, which is a key aspect of its strategy to deploy SMART reactors abroad.

Assessing Geographical and Geological Considerations

Selecting a site for constructing a Small Modular Reactor (SMR) power station involves a thorough assessment of geographical and geological factors to ensure safety, stability, and operational efficiency. Unlike traditional large nuclear power plants, SMRs offer greater flexibility in site selection due to their smaller footprint and enhanced safety features. However, specific geographical and geological considerations remain critical to ensure long-term stability and mitigate risks associated with natural disasters.

Seismic Stability and Earthquake Risk Assessment

One of the foremost considerations in site selection is seismic stability. An SMR power plant must be constructed in an area with minimal seismic activity to prevent damage during earthquakes. Although SMRs are designed with improved seismic resilience, building on a stable site reduces the risk of structural damage and potential radioactive leaks.

Seismic stability is a critical consideration in the site selection process for Small Modular Reactors (SMRs). The inherent design of SMRs, which includes advanced safety features, provides a level of resilience against seismic events. However, constructing these facilities in areas with minimal seismic activity significantly reduces the risk of structural damage and potential radioactive leaks during earthquakes. This is particularly important given the catastrophic consequences that can arise from seismic events affecting nuclear facilities, as highlighted by past incidents such as Fukushima, which underscored the necessity for robust seismic safety measures in nuclear power plant design [108].

Geological surveys are conducted to identify fault lines, historical earthquake data, and ground stability. Sites near active fault lines or in areas prone to high seismic activity are generally avoided. However, if a site is in a moderate-risk zone, additional safety measures such as base isolation, shock absorbers, and reinforced foundations may be employed to enhance the reactor's resilience.

Procedure for Conducting Geological Surveys for SMR Power Station Site Selection

Conducting geological surveys is an essential step in selecting a safe and stable site for a Small Modular Reactor (SMR) power station. This detailed procedure outlines how to systematically identify fault lines, historical earthquake data, and assess ground stability. It also includes guidelines on evaluating potential risks and implementing additional safety measures in moderate-risk zones.

Step 1: Preliminary Desk Study and Data Collection
1. **Gather Existing Data:**
   o Obtain geological maps, seismological data, and historical records from national geological and seismic institutions.
   o Utilize satellite imagery, aerial photography, and remote sensing data to assess the general geology of the area.
   o Review reports from previous geological surveys, if available, for the selected region.
2. **Identify Key Geological Features:**
   o Identify known fault lines, seismic zones, and areas with historical earthquake activity.
   o Review the geological history to understand past seismic events, ground movements, and land stability.
3. **Regulatory Review:**
   o Consult local and international regulatory guidelines to understand the seismic and geological requirements for nuclear power plants.
   o Review specific seismic design standards set by agencies like the International Atomic Energy Agency (IAEA) and local regulatory bodies.

Step 2: Site Reconnaissance and Initial Assessment
1. **Conduct Site Visits:**

- Perform a physical inspection of the potential site to confirm the findings from the desk study.
- Take note of visible geological features such as rock outcrops, soil layers, slope stability, and signs of previous ground movement or landslides.

2. **Interview Local Experts**:
   - Engage with local geologists, engineers, and residents to gather anecdotal evidence of historical ground movements or unusual geological activity.

3. **Develop a Preliminary Report**:
   - Based on initial findings, compile a preliminary assessment report highlighting potential geological concerns and recommendations for further investigation.

Step 3: Detailed Geophysical Surveys

1. **Seismic Refraction and Reflection Surveys**:
   - Use seismic refraction methods to determine subsurface layers' velocities, which helps identify fault lines and ground stability.
   - Conduct seismic reflection surveys to create detailed images of the subsurface structures, such as faults, fractures, and bedrock depth.

2. **Ground Penetrating Radar (GPR)**:
   - Utilize GPR to detect subsurface anomalies, such as voids, fractures, or buried fault lines. This is particularly useful in areas with shallow soil cover.

3. **Magnetic and Gravimetric Surveys**:
   - Use magnetic surveys to detect subsurface variations caused by geological formations.
   - Perform gravimetric surveys to measure gravitational field variations, which can help identify faults, fractures, and subsurface voids.

Step 4: Borehole Drilling and Soil Sampling

1. **Borehole Drilling**:
   - Drill boreholes at strategic locations to collect soil and rock samples from various depths.
   - Install sensors in the boreholes to monitor ground stability, pore water pressure, and seismic activity over time.

2. **Geotechnical Testing**:
   - Perform laboratory tests on collected soil and rock samples to assess soil composition, shear strength, compressibility, and liquefaction potential.
   - Analyse the soil's load-bearing capacity to ensure it can support the weight of the SMR infrastructure.

Step 5: Seismic Hazard Assessment

1. **Historical Earthquake Analysis**:
   - Analyse historical earthquake data to determine the frequency, magnitude, and epicentre locations of past earthquakes in the region.
   - Use probabilistic seismic hazard analysis (PSHA) to estimate the likelihood of future seismic events.

2. **Fault Line Mapping**:
   - Map existing fault lines and determine whether they are active or dormant.

o   Assess the potential for fault reactivation due to tectonic stress or human activities.

3. **Microzonation Studies**:
   o   Conduct microzonation to divide the site into zones based on varying levels of seismic risk. This helps optimize the placement of critical infrastructure.

Step 6: Ground Stability Assessment
1. **Slope Stability Analysis**:
   o   Assess the stability of slopes near the site to prevent landslides or soil erosion that could endanger the plant's safety.
   o   Use software models to simulate slope behaviour under different conditions, such as heavy rainfall or seismic activity.
2. **Subsidence and Settlement Analysis**:
   o   Evaluate the potential for ground subsidence, especially in areas with loose soil or underlying cavities.
   o   Conduct soil compaction tests to understand settlement risks and the need for soil stabilization.

Step 7: Evaluation of Results and Risk Mitigation Measures
1. **Data Analysis**:
   o   Compile and analyse all collected data to assess the overall geological suitability of the site.
   o   Rank the site's risk factors, such as proximity to fault lines, soil instability, and historical earthquake data.
2. **Develop Mitigation Strategies**:
   o   For sites in moderate-risk zones, recommend additional safety measures such as:
       ▪   **Base Isolation**: Installing flexible bearings to isolate the reactor building from ground movements.
       ▪   **Shock Absorbers**: Using dampers to absorb seismic energy and reduce structural stress.
       ▪   **Reinforced Foundations**: Employing deep foundations or pile systems to enhance stability.
3. **Cost-Benefit Analysis**:
   o   Conduct a cost-benefit analysis to determine whether the additional construction costs for safety measures are justified by the site's strategic benefits.

Step 8: Final Reporting and Recommendations
1. **Prepare a Comprehensive Geological Survey Report**:
   o   Document all findings, analyses, and recommendations in a detailed report.
   o   Include maps, charts, and diagrams illustrating fault lines, seismic zones, soil conditions, and recommended safety measures.
2. **Present Findings to Stakeholders**:
   o   Share the report with regulatory authorities, project managers, and other stakeholders.
   o   Address any concerns or requests for additional data or analysis.

3. **Obtain Regulatory Approvals**:
   - ○ Submit the report as part of the site licensing application to relevant regulatory bodies.
   - ○ Make adjustments to the site plan if required by regulatory feedback.

Step 9: Ongoing Monitoring and Assessment
1. **Establish a Monitoring Program**:
   - ○ Install seismometers, ground movement sensors, and other monitoring equipment to continuously track the site's geological conditions.
   - ○ Set up a regular schedule for reviewing data and updating risk assessments.
2. **Implement Contingency Plans**:
   - ○ Develop contingency plans for responding to unexpected seismic events or ground instability.
   - ○ Train site personnel in emergency procedures related to geological hazards.

Geological surveys play a pivotal role in identifying suitable sites for SMR construction. These surveys assess fault lines, historical earthquake data, and ground stability to ensure that selected sites are not located near active fault lines or in regions prone to high seismic activity. In cases where potential sites fall within moderate-risk zones, additional safety measures can be implemented. Techniques such as base isolation, which involves decoupling the structure from ground motion, and the use of shock absorbers can enhance the seismic resilience of the reactor [109, 110]. Reinforced foundations are also critical in ensuring that the structural integrity of the plant is maintained during seismic events [111].

The selection process for SMR sites is multifaceted and requires a comprehensive analysis of various factors, including geological and environmental considerations. Research indicates that the site selection process should be guided by a clearly established set of criteria, which includes not only seismic stability but also community engagement and logistical feasibility [8, 69]. The integration of advanced technologies, such as Geographic Information Systems (GIS), can facilitate a more thorough analysis of potential sites, allowing for better-informed decisions that align with both technical requirements and community needs [65].

Furthermore, the design features of SMRs contribute to their resilience against seismic events. For instance, the modular construction of SMRs allows for shorter construction periods and enhanced safety protocols, which are crucial in mitigating risks associated with seismic activity [11]. The passive safety features of many SMRs, such as natural circulation of coolant and independent core cooling systems, further enhance their ability to withstand seismic shocks without compromising safety [18].

Soil Conditions and Ground Stability

The soil composition and ground stability are crucial factors that influence the structural integrity of an SMR power station. The site must have soil capable of supporting the heavy infrastructure of the reactor, containment buildings, cooling systems, and other components. Poor soil conditions, such as soft clay or sandy soils, can lead to subsidence, which may compromise the safety and stability of the plant over time.

Geotechnical investigations are necessary to evaluate soil type, bearing capacity, and groundwater levels. Sites with stable bedrock or firm soil foundations are preferred, as they can better support the weight of the reactor modules and reduce the risk of settlement or uneven ground shifting. If a site has less-than-ideal soil conditions, techniques such as soil compaction, pile foundations, or ground stabilization may be required to ensure stability.

Procedure for Conducting Geotechnical Investigations for SMR Power Station Site Selection

Geotechnical investigations are critical in evaluating the suitability of a site for the construction of a Small Modular Reactor (SMR) power station. This process involves a series of steps to assess soil type, bearing capacity, groundwater levels, and overall ground stability. By conducting thorough geotechnical assessments, developers can determine if a site is capable of supporting heavy reactor modules and infrastructure. Additionally, if a site has suboptimal conditions, appropriate ground improvement techniques can be identified to ensure long-term stability.

Step 1: Preliminary Site Assessment and Planning
1. **Review Existing Data and Reports:**
    o   Gather information on soil conditions, geological features, and previous land use from local government records, geological surveys, and historical reports.
    o   Review topographic maps, aerial imagery, and satellite data to get an overview of the site's physical characteristics.
2. **Initial Site Visit:**
    o   Conduct a preliminary site visit to assess surface conditions, identify potential access points, and locate any visible geological features such as outcrops, slopes, or signs of erosion.
    o   Interview local experts or landowners to gather anecdotal information about the site's soil stability and groundwater behaviour.
3. **Develop a Geotechnical Investigation Plan:**
    o   Create a detailed plan outlining the scope of the investigation, including the number of boreholes, soil tests, and groundwater monitoring wells needed.
    o   Identify specific areas for testing based on the site layout, focusing on locations where heavy structures such as reactor modules, cooling towers, and auxiliary buildings will be constructed.

Step 2: Soil Boring and Sampling
1. **Drilling Boreholes:**

- o Use drilling rigs to create boreholes at multiple locations across the site. Boreholes should typically reach a depth of at least 20-30 meters or until stable bedrock is encountered.
- o Space the boreholes strategically to cover critical areas, such as proposed foundation zones, access roads, and storage areas.
2. **Soil Sampling:**
   - o Extract soil samples at various depths using techniques like split-spoon sampling, Shelby tubes, or continuous coring.
   - o Store samples in labelled containers for transport to a geotechnical laboratory, ensuring they are protected from contamination or moisture changes.
3. **Rock Core Sampling (if necessary):**
   - o If bedrock is encountered, collect rock core samples to assess the quality and bearing capacity of the underlying rock layers.
   - o Analyse the core samples for fractures, weathering, and other structural characteristics.

Step 3: In-Situ Testing for Soil Properties
1. **Standard Penetration Test (SPT):**
   - o Conduct SPTs within boreholes to measure soil resistance and provide an indication of soil density and bearing capacity.
   - o Record the number of blows required to drive a split-barrel sampler into the soil, providing data on soil compaction and consistency.
2. **Cone Penetration Test (CPT):**
   - o Perform CPTs to obtain continuous profiles of soil resistance, pore water pressure, and soil stratification. This test is especially useful for identifying weak or compressible soil layers.
   - o Use CPT results to assess soil stiffness, shear strength, and potential liquefaction risks.
3. **Groundwater Level Monitoring:**
   - o Install monitoring wells in boreholes to measure groundwater levels and fluctuations over time.
   - o Collect groundwater samples to analyse chemical composition, which could affect construction materials or foundations.

Step 4: Laboratory Testing of Soil and Rock Samples
1. **Soil Classification Tests:**
   - o Conduct grain size analysis, Atterberg limits, and soil classification tests to determine the type of soil (e.g., clay, silt, sand, gravel).
   - o Identify the presence of expansive clays or loose sandy soils that may require stabilization.
2. **Bearing Capacity and Compaction Tests:**
   - o Perform unconfined compressive strength tests, triaxial tests, and direct shear tests to evaluate soil bearing capacity and shear strength.
   - o Conduct Proctor compaction tests to determine the optimal moisture content and compaction level needed to improve soil stability.
3. **Rock Quality Analysis:**

- o If rock samples were collected, conduct tests to determine rock quality, uniaxial compressive strength, and fracture density.
- o Analyse rock cores for potential weaknesses that may affect structural integrity.

Step 5: Data Analysis and Interpretation
1. **Create Soil Profiles and Cross-Sections**:
   - o Use the data collected from boreholes, in-situ tests, and lab results to create detailed soil profiles and cross-sectional diagrams of the site.
   - o Identify layers of weak soil, groundwater tables, and potential zones of instability.
2. **Assess Site Suitability**:
   - o Evaluate the site's overall geotechnical characteristics to determine if it is suitable for supporting heavy reactor modules.
   - o Calculate bearing capacity, settlement potential, and the risk of liquefaction, particularly in areas prone to seismic activity.
3. **Prepare a Geotechnical Investigation Report**:
   - o Document all findings, including soil and rock properties, groundwater conditions, and recommendations for foundation design.
   - o Include maps, charts, and diagrams to illustrate the site's geotechnical conditions.

Step 6: Recommendations for Ground Improvement (if necessary)
1. **Soil Compaction Techniques**:
   - o Recommend dynamic compaction, vibro-compaction, or soil densification methods for loose, granular soils that need strengthening.
   - o For cohesive soils, suggest preloading or surcharge techniques to consolidate the ground before construction.
2. **Pile Foundations**:
   - o If soil conditions are too weak for shallow foundations, propose using driven piles, drilled shafts, or auger-cast piles to reach stable soil layers or bedrock.
   - o Specify pile lengths, diameters, and materials based on the load-bearing requirements of the SMR modules.
3. **Ground Stabilization**:
   - o For sites with unstable soils, suggest soil stabilization techniques such as cement grouting, lime stabilization, or the use of geosynthetic reinforcements.
   - o Implement base isolation pads or shock absorbers in areas with moderate seismic risk to enhance structural resilience.

Step 7: Final Review and Approval
1. **Consult with Structural and Civil Engineers**:
   - o Collaborate with structural engineers to integrate geotechnical recommendations into the overall site design.
   - o Review foundation plans, building layouts, and construction methods to ensure compatibility with site conditions.
2. **Submit Report to Regulatory Authorities**:

> 3.   o   Prepare a comprehensive geotechnical report for submission to regulatory bodies as part of the site approval process.
>   o   Address any feedback or additional requirements from authorities before final site selection.
> 3. **Plan for Ongoing Monitoring**:
>   o   Establish a monitoring program to track soil settlement, groundwater levels, and structural stability during and after construction.
>   o   Schedule periodic geotechnical inspections to identify and address potential issues early.

Flood Risk Assessment

Proximity to water sources is often a benefit for SMR plants that rely on water cooling, but it also introduces the risk of flooding. A comprehensive flood risk assessment is essential, especially if the site is near rivers, lakes, or coastal areas. Rising sea levels, storm surges, and extreme weather events must be factored into the assessment.

For coastal sites, considerations include the likelihood of tsunamis and the impact of climate change on sea levels. For inland sites, the focus is on river flooding, flash floods, and dam break scenarios. Protective measures, such as elevated foundations, levees, and flood barriers, may be incorporated into the design to safeguard against flooding.

Topographical and Geographical Features

The topography of the selected site influences construction, operational efficiency, and safety. Flat or gently sloping land is generally preferred as it simplifies construction and reduces the need for extensive land leveling or excavation. However, areas with natural elevation can offer additional protection against flooding.

Geographical features such as hills, mountains, or valleys can impact wind patterns, cooling efficiency, and the dispersion of any emissions in case of an incident. Sites with natural windbreaks or those located in areas with stable weather patterns are preferred to minimize operational disruptions.

Proximity to Water Sources for Cooling

For SMRs that rely on water cooling, access to a reliable water source is essential. However, geological assessments are needed to ensure that water extraction will not negatively impact the local ecosystem or cause land subsidence. Sites near rivers, lakes, or coasts are preferred, but groundwater resources may also be used if surface water is unavailable.

For air-cooled or dry-cooled SMR designs, the flexibility of site selection increases, as they are less dependent on proximity to large water bodies. However, the local climate and air temperature must be considered to ensure that the cooling systems remain effective even in extreme weather conditions.

Landslide and Soil Erosion Risks

Geological assessments should also consider the risk of landslides, especially in areas with steep slopes or unstable soil layers. Landslides can be triggered by heavy rainfall, earthquakes, or human activities, leading to significant damage to infrastructure. Soil erosion, particularly in coastal or riverine areas, can undermine the stability of foundations over time.

Preventive measures, such as retaining walls, terracing, and vegetation cover, can be used to reduce the risk of landslides and soil erosion. Sites with a history of landslides or significant erosion are generally avoided unless mitigation measures can be effectively implemented.

Assessment of Underground Water Tables and Hydrogeology

The underground water table plays a crucial role in the construction and operation of SMRs. High water tables can lead to groundwater infiltration, which may compromise the integrity of underground structures and containment systems. Conversely, very low water tables may limit the availability of groundwater for cooling purposes.

Hydrogeological studies help determine the depth, flow, and quality of groundwater. These studies are particularly important for regions where water resources are scarce or heavily regulated. Ensuring that the construction of the SMR does not adversely affect local water supplies is essential for both regulatory compliance and community support.

Long-Term Geotechnical Monitoring

Given the long lifespan of nuclear reactors, it is essential to monitor the geological and geotechnical conditions of the site over time. Changes in soil stability, groundwater levels, or seismic activity can affect the safety of the plant. Continuous monitoring systems, combined with periodic site assessments, help identify potential risks early and allow for proactive maintenance.

This long-term monitoring is particularly crucial for sites that may be prone to gradual geological changes, such as coastal erosion, shifting river courses, or ground subsidence due to water extraction or natural compaction.

The construction of an SMR power station requires a detailed understanding of the site's geographical and geological characteristics to ensure safety, stability, and operational efficiency.

By thoroughly assessing factors such as seismic stability, soil conditions, flood risks, and water availability, developers can select sites that minimize risks and support the long-term success of the project. While SMRs offer greater flexibility in site selection compared to traditional nuclear reactors, careful consideration of these factors remains essential to achieving a safe and reliable energy source.

Graded Scale for Assessing Site Suitability for an SMR Power Plant

To evaluate the suitability of a site for a Small Modular Reactor (SMR) power plant, the following graded scale uses a point-based system that rates key factors. The scale ranges from **1 (Poor)** to **5 (Excellent)** for each criterion, with higher scores indicating better conditions for SMR construction and operation. The total score will help in making an informed decision on site selection.

Site Assessment Criteria and Grading Scale

Criteria	1 (Poor)	2 (Fair)	3 (Moderate)	4 (Good)	5 (Excellent)

1. Regulatory and Licensing Compliance

- **1**: Significant regulatory hurdles, complex approval processes, and low likelihood of regulatory support.
- **2**: Regulatory environment is challenging but feasible with significant adjustments.
- **3**: Moderately supportive regulatory environment; standard approval process with clear requirements.
- **4**: Positive regulatory environment; streamlined process with some incentives.
- **5**: Strong regulatory support, fast-track approval, and government incentives for SMR projects.

2. Geological and Seismic Stability

- **1**: High seismic activity, proximity to active fault lines, unstable soil conditions.
- **2**: Moderate seismic risk with some history of earthquakes; soil stability concerns.
- **3**: Low to moderate seismic activity; generally stable ground with minimal risk.
- **4**: Very low seismic risk; stable bedrock and no fault lines.
- **5**: Seismically stable, solid bedrock, and ideal soil conditions for heavy structures.

3. Proximity to Water Sources (for Water-Cooled SMRs)

- **1**: No nearby water source, difficult to secure a stable water supply.
- **2**: Limited water availability, requiring extensive infrastructure to access water.

- **3**: Water source available within a reasonable distance; some infrastructure required.
- **4**: Reliable water source nearby, easily accessible with minimal infrastructure.
- **5**: Abundant water source on-site; no additional infrastructure needed.

4. Grid Connectivity and Power Distribution

- **1**: No existing electrical infrastructure nearby; significant investment needed.
- **2**: Limited grid access; requires extensive upgrades to connect.
- **3**: Moderate distance to grid; requires some upgrades.
- **4**: Close proximity to electrical grid; minimal upgrades required.
- **5**: Direct access to high-capacity grid infrastructure with minimal connection costs.

5. Accessibility and Transportation Infrastructure

- **1**: Remote location with no access roads, rail, or ports; costly logistics.
- **2**: Difficult access; requires significant road or rail development.
- **3**: Moderately accessible; some infrastructure improvements needed.
- **4**: Good road, rail, or port access with minor enhancements required.
- **5**: Excellent access via roads, railways, and waterways; fully developed infrastructure.

6. Proximity to Population Centres

- **1**: Located in densely populated area; high risk to nearby populations.
- **2**: Moderate population density; requires extensive safety measures.
- **3**: Low population density; buffer zones can be established.
- **4**: Sparsely populated area; minimal risk with established safety zones.
- **5**: Remote location with no nearby populations, providing maximum safety.

7. Local Climate and Environmental Conditions

- **1**: Extreme weather conditions, high flood or storm risk.
- **2**: Challenging climate with potential impact on operations; some flood risk.
- **3**: Moderate weather conditions with manageable risks.
- **4**: Favourable climate with minimal operational disruptions.
- **5**: Ideal climate conditions; low risk of extreme weather or flooding.

8. Land Availability and Land Use

- **1**: Land acquisition issues, zoning restrictions, or indigenous claims.
- **2**: Complicated land-use permissions; some zoning adjustments needed.
- **3**: Land available but may require rezoning; moderate acquisition process.
- **4**: Readily available land with appropriate zoning; few obstacles.

- **5**: Ample land available with clear zoning for industrial use; easy acquisition.

9. Socioeconomic Factors

- **1**: Strong public opposition; low community support.
- **2**: Mixed public sentiment; requires significant community engagement.
- **3**: Moderate community support; manageable public concerns.
- **4**: Generally supportive community; positive economic impact expected.
- **5**: Strong community support; high demand for jobs and economic benefits.

10. Security and Safety Concerns

- **1**: High risk of security threats; close proximity to sensitive facilities.
- **2**: Moderate security risks; requires extensive protective measures.
- **3**: Manageable security risks with standard protective measures.
- **4**: Low security risk with established safety protocols.
- **5**: Highly secure area with minimal risk; easy to enforce security measures.

Scoring and Evaluation

1. **Scoring**:
   - Assign a score between **1 to 5** for each criterion based on the site assessment.
   - Total the scores for all criteria.
2. **Grading Scale**:
   - **45-50**: Excellent Site — Highly suitable with minimal additional investments needed.
   - **35-44**: Good Site — Suitable, with some improvements or mitigations required.
   - **25-34**: Moderate Site — Feasible, but will require significant adjustments and investments.
   - **15-24**: Marginal Site — Substantial challenges; may not be cost-effective without major interventions.
   - **10-14**: Poor Site — Not recommended due to severe limitations.
3. **Decision-Making**:
   - Sites scoring **35 and above** are generally recommended for further feasibility studies.
   - Sites scoring below **25** should be reconsidered unless critical factors can be mitigated cost-effectively.

This graded scale provides a systematic approach to assessing potential sites for SMR power station construction. By evaluating each site against key criteria, project developers can make informed decisions and prioritize sites that offer the best balance of safety, cost-effectiveness, and regulatory compliance.

Quantification for Each Criterion in the Graded Scale for Assessing Site Suitability

To improve objectivity and consistency, the following quantifiable metrics can be added for each criterion. These metrics will help to assign scores based on specific, measurable factors, ensuring a clear and systematic evaluation process.

1. Regulatory and Licensing Compliance

- **1 (Poor)**: Regulatory approval process estimated to take over 8 years; significant opposition from regulators.

- **2 (Fair)**: Approval process expected to take 5-8 years with substantial adjustments; moderate regulatory challenges.

- **3 (Moderate)**: Standard approval process with an estimated timeline of 3-5 years; clear regulatory requirements.

- **4 (Good)**: Streamlined process with some regulatory incentives; approval within 2-3 years.

- **5 (Excellent)**: Fast-track approval (less than 2 years); strong government support and incentives for SMR projects.

2. Geological and Seismic Stability

- **1 (Poor)**: Located within 10 km of active fault lines; historical earthquake magnitude > 7.0.

- **2 (Fair)**: Seismic risk zone with a history of earthquakes between 5.0-6.9; soil stability concerns.

- **3 (Moderate)**: Low to moderate seismic risk; minimal history of earthquakes (< 5.0).

- **4 (Good)**: Very low seismic activity; stable ground with rare minor tremors.

- **5 (Excellent)**: No significant seismic history; stable bedrock with optimal soil conditions.

3. Proximity to Water Sources (for Water-Cooled SMRs)

- **1 (Poor)**: No water source within 50 km; requires expensive infrastructure.

- **2 (Fair)**: Water source between 20-50 km; limited availability.

- **3 (Moderate):** Water source within 10-20 km; requires moderate infrastructure.

- **4 (Good):** Reliable water source within 5-10 km; easy access with minimal infrastructure.

- **5 (Excellent):** Abundant water source on-site or within 5 km; no additional infrastructure needed.

4. Grid Connectivity and Power Distribution

- **1 (Poor):** Nearest electrical grid > 50 km; requires new substations and extensive transmission lines.

- **2 (Fair):** Grid access between 20-50 km; requires significant upgrades.

- **3 (Moderate):** Moderate grid access within 10-20 km; requires some upgrades.

- **4 (Good):** Close grid access within 5-10 km; minimal upgrades needed.

- **5 (Excellent):** Direct access to high-capacity grid within 5 km; no significant upgrades needed.

5. Accessibility and Transportation Infrastructure

- **1 (Poor):** Remote location; no existing roads or railways; construction logistics highly challenging.

- **2 (Fair):** Limited access via poorly maintained roads or railways; significant infrastructure development needed.

- **3 (Moderate):** Accessible via regional roads; moderate upgrades needed.

- **4 (Good):** Good access via national roads and rail; minor enhancements required.

- **5 (Excellent):** Fully developed infrastructure with highways, rail, and ports nearby.

6. Proximity to Population Centers

- **1 (Poor):** Located within 10 km of a densely populated urban area (> 1 million residents).

- **2 (Fair):** Population density > 500,000 within 20 km; requires extensive safety zones.

- **3 (Moderate):** Located in areas with < 100,000 residents within 20 km; manageable safety zones.

- **4 (Good):** Sparsely populated area with < 50,000 residents within 20 km; minimal safety concerns.

- **5 (Excellent):** Remote area with no significant population within 30 km; maximum safety.

7. Local Climate and Environmental Conditions

- **1 (Poor)**: Frequent extreme weather events (e.g., hurricanes, heavy flooding, temperatures > 40°C).

- **2 (Fair)**: Periodic harsh weather (e.g., seasonal storms, snowfall > 1 meter/year).

- **3 (Moderate)**: Moderate climate with occasional weather disruptions.

- **4 (Good)**: Generally stable climate with minimal weather disruptions.

- **5 (Excellent)**: Mild climate with ideal conditions for year-round operation.

8. Land Availability and Land Use

- **1 (Poor)**: Complicated land ownership issues; multiple indigenous claims or zoning restrictions.

- **2 (Fair)**: Requires rezoning and lengthy acquisition process; moderate opposition.

- **3 (Moderate)**: Land available with some zoning adjustments needed; moderate acquisition timeline.

- **4 (Good)**: Readily available land with few legal obstacles; quick acquisition.

- **5 (Excellent)**: Ample land available with clear zoning for energy use; immediate acquisition.

9. Socioeconomic Factors

- **1 (Poor)**: Strong public opposition (> 80% against the project); significant community protests.

- **2 (Fair)**: Mixed public sentiment (50-80% support); requires intensive engagement.

- **3 (Moderate)**: Moderate support (50-60%); concerns manageable with community outreach.

- **4 (Good)**: Generally supportive community (> 60%); visible economic benefits.

- **5 (Excellent)**: Strong community support (> 80%); high demand for jobs and economic impact.

10. Security and Safety Concerns

- **1 (Poor)**: High-risk area with security threats (e.g., conflict zones, terrorism-prone).

- **2 (Fair)**: Proximity to sensitive facilities like military bases or airports; moderate risk.

- **3 (Moderate)**: Manageable security risks; requires standard protective measures.

- **4 (Good)**: Low security risks; well-established safety protocols.

- **5 (Excellent)**: Highly secure, remote area with minimal threats; easy to enforce safety.

Scoring and Evaluation

- **Total Score**: Sum the scores for all criteria to assess site suitability.

  o **45-50**: **Excellent Site** — Ideal for SMR construction with minimal additional investments.

  o **35-44**: **Good Site** — Suitable with some improvements required.

  o **25-34**: **Moderate Site** — Feasible but will require significant adjustments.

  o **15-24**: **Marginal Site** — Substantial challenges; not cost-effective without major interventions.

  o **10-14**: **Poor Site** — Not recommended due to severe limitations.

Decision Guidelines

- **Priority Sites**: Sites scoring **35 and above** are strong candidates for further feasibility studies and investments.

- **Caution**: Sites scoring between **25-34** should only proceed if improvements are cost-justifiable.

- **Avoid**: Sites scoring below **25** are generally unsuitable unless critical issues can be mitigated effectively.

Engaging Local Communities and Stakeholders

Engaging local communities and stakeholders is a critical step in the successful planning and construction of a Small Modular Reactor (SMR) power plant. Building trust, addressing concerns, and fostering public support are essential to ensuring smooth project execution. A well-structured engagement process can lead to greater acceptance and cooperation from the community, which is vital for the long-term success of the project. Below is a comprehensive guide on how to effectively engage stakeholders throughout the various phases of the SMR project lifecycle.

The initial step is to identify all relevant stakeholders, including local residents, community leaders, government officials, regulatory bodies, environmental organizations, and local businesses. Conducting an analysis of these stakeholders helps to understand their concerns, priorities, and interests. This understanding is crucial for crafting communication strategies that effectively address their needs. Additionally, mapping stakeholders according to their influence and the potential impact of the project on them allows for the prioritization of engagement efforts, ensuring that key stakeholders are consulted early in the process.

Engaging with marginalized groups, particularly indigenous populations, is vital as they often lack formal representation in public forums [112]. Cultural sensitivity is paramount; employing anthropologists or social scientists can aid in understanding the cultural significance and traditional land use practices of these communities [113].

Developing a clear communication plan is vital to maintaining transparency throughout the project. This plan should include a mix of communication methods such as traditional media, social media, community meetings, and direct outreach initiatives. Consistent messaging across these channels helps prevent the spread of misinformation. Developing a culturally appropriate engagement strategy is critical. This includes respecting cultural protocols and traditions when interacting with indigenous communities, such as conducting meetings led by community elders [114]. Additionally, providing materials in indigenous languages and employing interpreters can facilitate better understanding and communication [115]. Training project staff in cultural awareness ensures that they are equipped to engage meaningfully with local communities [112].

Appointing a dedicated community liaison officer can further streamline communication by serving as a reliable point of contact for the community. Utilizing multiple platforms, such as newsletters, websites, social media, and public notices, ensures that information reaches a diverse audience, fostering inclusivity and trust. Transparent communication channels are essential for effective stakeholder engagement. Appointing community liaison officers who are experienced in working with local and indigenous communities can bridge gaps between project teams and stakeholders [116]. Utilizing diverse communication platforms—such as public meetings, social media, and local radio—ensures that information reaches various segments of the community [117]. Regular updates and feedback mechanisms foster trust and demonstrate accountability [118].

Organizing public information sessions, workshops, and town hall meetings allows the project team to educate the community about SMR technology, including its safety features, environmental benefits, and economic impacts. Utilizing visual aids like models, videos, and virtual reality simulations can help simplify complex concepts and demystify the technology. Addressing common concerns and debunking myths related to radiation risks, waste management, and potential accidents through evidence-based explanations can significantly reduce public apprehension. Pre-consultation workshops can simplify complex information and address community concerns [119]. Organizing town hall meetings in accessible locations ensures maximum participation, particularly from indigenous groups, by involving their decision-makers in discussions [112].

Gathering feedback is essential to understand community sentiment and address specific concerns. Distributing surveys and questionnaires can help capture opinions on various aspects of the project, while focus groups and roundtable discussions with different stakeholder segments can provide deeper insights into their views. Maintaining an open-door policy, where residents can visit project offices or speak with representatives, promotes transparency and fosters trust, allowing for more direct engagement.

Establishing a Community Advisory Board (CAB) early in the project provides a formal platform for continuous dialogue between the project team and the community [114]. Comprising local leaders, technical experts, and other key stakeholders, the CAB serves as a bridge for communication, ensuring that community concerns are addressed promptly. Regular meetings and updates with the CAB enhance transparency and accountability, while empowering its members with access to independent experts ensures informed decision-making. Empowering indigenous groups through dedicated working groups allows them to contribute meaningfully to discussions on land use and environmental management [120].

Respecting land rights and treaties is fundamental to gaining a social license to operate. Compliance with national laws and international conventions, such as the United Nations Declaration on the Rights of Indigenous Peoples (UNDRIP), is crucial [121]. Securing Free, Prior, and Informed Consent (FPIC) from indigenous communities before project activities commence is a key aspect of this engagement [122].

Engaging independent experts to review project plans and conduct environmental assessments adds credibility to the project and builds trust among sceptical stakeholders. Making third-party findings publicly available demonstrates transparency and reinforces the project's commitment to safety and environmental protection. Independent reviews can also provide assurance that the project meets all regulatory standards, which is crucial for gaining public acceptance.

Highlighting the economic and social benefits of the SMR project can significantly enhance community support. Emphasize potential job creation, local procurement opportunities, and economic revitalization. Providing specific data on the number and types of jobs to be created can illustrate tangible benefits. Additionally, investing in community programs, such as educational scholarships, infrastructure improvements, or partnerships with local schools, demonstrates a long-term commitment to the community's well-being.

Ensuring that local and indigenous communities derive economic benefits from the project is vital. This can be achieved through job creation, local procurement, and skills training programs [116]. Investing in community projects that address local needs—such as infrastructure, healthcare, and education—can foster long-term partnerships and support [112].

Figure 30: Nuclear Energy Summit, "Powering tomorrow, Today", transition night held at the Atomium, Place de l'Atomium 1, Brussels, Belgium. 20 March 2024. IAEA Imagebank, CC BY 2.0, via Flickr.

Conducting comprehensive Environmental Impact Assessments (EIA) and sharing the results with the public helps demonstrate how the project will protect local ecosystems and water resources. Hosting emergency preparedness drills in collaboration with local authorities further reassures the community of the plant's safety protocols. Transparency in waste management plans, including how radioactive waste will be stored, managed, or recycled, helps address public concerns and clarify misconceptions.

Continuously monitoring public sentiment through sentiment analysis tools, media monitoring, and community feedback forms is essential throughout the project lifecycle [119]. Adjusting engagement strategies based on feedback ensures that the project remains aligned with community expectations. Regular updates on progress, particularly on commitments made during consultations, build long-term trust and reinforce the project team's accountability.

To ensure compliance with regulatory requirements, it is crucial to document all community engagement activities, including feedback and responses. Preparing for public hearings, which are often a requirement in many jurisdictions, involves ensuring that stakeholders are well-informed and supportive. Engaging stakeholders early and consistently helps mitigate potential legal challenges that could delay the project, thereby ensuring smoother regulatory approval.

Maintaining comprehensive records of all engagement activities is essential for transparency and regulatory compliance [113]. Publicly sharing engagement reports can illustrate how stakeholder input has influenced project decisions, further enhancing trust and community support [112].

The following table provides a guide that can be easily used for planning, tracking, and implementing community and stakeholder engagement activities during the development of an SMR power plant project:

Local Community and Stakeholder Engagement Checklist			
No.	Task	Description	Action Items
1	Stakeholder Identification & Mapping	Identify all relevant stakeholders (e.g., local residents, indigenous groups, government officials, environmental organizations, local businesses). Conduct a stakeholder analysis to understand their concerns, influence, and interests. Map stakeholders based on their level of influence and potential impact from the project.	- Identify and prioritize key groups that may need special attention (e.g., indigenous communities, marginalized populations).
2	Initial Engagement Planning	Develop a comprehensive communication and engagement strategy. Allocate budget and resources for engagement activities. Appoint a dedicated Community Liaison Officer to serve as a consistent point of contact.	- Provide training for project staff on cultural sensitivity, especially when working with indigenous communities.
3	Establishing Communication Channels	Set up multiple communication platforms (e.g., newsletters, social media, community meetings, hotlines). Develop project websites or online portals for sharing information. Translate key materials into local languages and provide interpreters for meetings if necessary.	- Establish a feedback mechanism (e.g., suggestion boxes, online surveys, community hotlines).
4	Organizing Informational & Educational Workshops	Plan public information sessions, town halls, and community meetings. Prepare visual aids (e.g., models, videos, simulations) to explain SMR technology, safety, and benefits. Conduct workshops to address specific concerns like radiation safety, environmental impacts, and waste management.	- Schedule meetings at convenient times and accessible locations for maximum participation.

Local Community and Stakeholder Engagement Checklist			
No.	**Task**	**Description**	**Action Items**
5	**Gathering Community Feedback**	Distribute surveys or questionnaires to gather input on community concerns and expectations. Conduct focus group discussions with diverse community segments (e.g., youth, seniors, environmental advocates). Set up an open-door policy for residents to visit project offices and discuss concerns.	- Document all feedback received and respond promptly to inquiries.
6	**Forming Community Advisory Boards (CAB)**	Establish a CAB with representatives from local communities, indigenous groups, and other stakeholders. Schedule regular CAB meetings to discuss project updates, address concerns, and seek advice.	- Provide CAB members with access to independent experts and resources to ensure informed recommendations.
7	**Leveraging Independent Third-Party Experts**	Engage independent experts to review environmental assessments and project plans. Share findings from third-party reviews with the community for transparency.	- Use independent assessments to address potential scepticism and build trust.
8	**Highlighting Economic & Social Benefits**	Prepare detailed information on job creation, local procurement opportunities, and economic benefits. Develop community investment programs (e.g., scholarships, infrastructure improvements, STEM education).	- Share case studies of successful SMR projects and their benefits to local communities.
9	**Addressing Environmental & Safety Concerns**	Conduct comprehensive Environmental Impact Assessments (EIAs) and share the results with stakeholders. Host emergency preparedness drills in collaboration with local authorities. Clearly explain the project's approach to nuclear waste management and safety measures.	- Develop materials to dispel myths and misinformation about nuclear technology.
10	**Continuous Monitoring & Feedback**	Use sentiment analysis tools and community feedback forms to monitor public perception. Regularly adjust engagement strategies based on feedback and evolving concerns. Schedule	- Document lessons learned to improve future engagement efforts.

No.	Task	Description	Action Items
	Local Community and Stakeholder Engagement Checklist		
		follow-up meetings with community members to address ongoing concerns.	
11	**Legal & Regulatory Compliance**	Keep detailed records of all engagement activities, meetings, agreements, and feedback. Ensure compliance with local and international regulations related to community consultation. Prepare for public hearings by engaging stakeholders early to secure their support.	- Obtain Free, Prior, and Informed Consent (FPIC) from indigenous groups if required.
12	**Post-Engagement Review & Reporting**	Conduct a post-engagement review to assess the effectiveness of communication efforts. Prepare a report summarizing stakeholder feedback and how it was addressed. Share the final report with stakeholders and make it publicly available.	- Plan long-term community engagement initiatives to maintain trust beyond the construction phase.

Case Studies: Site Selection in Different Regions

To illustrate the practical application of site selection criteria for Small Modular Reactors (SMRs), here are a few case studies from different regions around the world. Each case study highlights how various factors such as regulatory compliance, geological stability, access to cooling water, grid connectivity, and community engagement were addressed during the site selection process.

1. NuScale Power Plant - Idaho National Laboratory (INL), United States

Background:

NuScale Power chose the Idaho National Laboratory (INL) as the site for its first commercial deployment of its SMR technology. The project is part of the Carbon Free Power Project (CFPP), which aims to provide clean, reliable energy to several Western U.S. states.

Key Site Selection Factors:

- **Regulatory Compliance:** The U.S. Nuclear Regulatory Commission (NRC) has a well-established regulatory framework, and INL's history as a national laboratory facilitated the approval process.

- **Geological Stability:** INL is located in a region with low seismic activity, providing a stable foundation for the reactor modules.

- **Access to Water Sources:** The site has access to sufficient water for cooling through the Snake River Aquifer, which is essential for water-cooled reactors.

- **Grid Connectivity:** INL is well-connected to the Western power grid, facilitating efficient power distribution to participating utilities.

- **Community Support:** NuScale conducted extensive public outreach, gaining support from local stakeholders and leveraging INL's established reputation in nuclear research.

Outcome:

The project is expected to become operational by 2029 and will be the first SMR deployment in the United States, demonstrating the feasibility of SMR technology in reducing carbon emissions.

2. CAREM Reactor - Atucha Nuclear Complex, Argentina

Background:

Argentina's CAREM project, developed by the National Atomic Energy Commission (CNEA), is located at the Atucha Nuclear Complex, approximately 100 km northwest of Buenos Aires. The CAREM reactor is a domestically developed SMR designed for both power generation and remote applications.

Key Site Selection Factors:

- **Regulatory Compliance:** Argentina's Nuclear Regulatory Authority (ARN) provided support, ensuring that the site met local and international safety standards.

- **Geological and Seismic Considerations:** The Atucha site is situated in a seismically stable area with low earthquake risk, making it ideal for nuclear installations.

- **Proximity to Water Sources:** The plant is located near the Paraná River, providing an abundant source of cooling water.

- **Land Availability:** The site is already zoned for nuclear use, reducing regulatory and land acquisition challenges.

- **Community Engagement:** CNEA conducted community meetings and public consultations to address concerns about safety and environmental impact.

Outcome:

The CAREM project aims to be fully operational by 2027, making it one of the first SMRs in Latin America. It serves as a model for deploying SMRs in developing countries.

3. Rolls-Royce SMR - Trawsfynydd Site, Wales, United Kingdom

Background:

Rolls-Royce has identified the Trawsfynydd site, a former Magnox nuclear power station in Wales, as a potential location for its first SMR deployment in the UK. The site was chosen for its existing infrastructure and nuclear heritage.

Key Site Selection Factors:

- **Regulatory Compliance:** The UK's Office for Nuclear Regulation (ONR) is supporting the project through its Generic Design Assessment (GDA) process, streamlining regulatory approvals.

- **Geological Stability:** The Trawsfynydd site is located in a geologically stable region with no significant fault lines, reducing the risk of seismic activity.

- **Existing Infrastructure:** The site already has grid connections, access roads, and other infrastructure, significantly reducing construction costs and timelines.

- **Community Support:** Given its history as a nuclear power site, the local community is familiar with the nuclear industry, and initial consultations showed strong public support.

- **Economic Benefits:** The project is expected to create jobs and stimulate economic activity in the region, which has faced economic decline since the closure of the original plant.

Outcome:

If approved, the Trawsfynydd SMR project could be operational by the early 2030s, contributing to the UK's goal of achieving net-zero emissions by 2050.

4. RITM-200 Reactor - Pevek, Chukotka, Russia

Background:

Russia's state-owned nuclear corporation, Rosatom, deployed its RITM-200 reactors on the floating nuclear power plant Akademik Lomonosov in Pevek, Chukotka. This remote Arctic location was selected to provide reliable energy to the region's isolated communities and industries.

Key Site Selection Factors:

- **Geological and Seismic Stability:** The Chukotka region is geologically stable, with minimal risk of earthquakes or soil subsidence.

- **Cooling Water Availability:** Being located on the coast, the floating plant has access to seawater for cooling.

- **Grid Connectivity:** The plant provides power to the isolated regional grid, which previously relied on diesel generators.

- **Accessibility:** Pevek is accessible by sea during the summer months, allowing the modular components to be transported via specialized vessels.

- **Community Engagement:** Rosatom engaged with local communities and indigenous groups to address concerns about environmental impacts and radiation safety.

Outcome:

The Akademik Lomonosov plant is currently operational and has demonstrated the potential of SMRs to supply power in remote and harsh environments.

5. ACP100 (Linglong One) - Changjiang Nuclear Power Plant, Hainan Island, China

Background:

China's first land-based SMR, the ACP100, also known as Linglong One, is under construction at the Changjiang Nuclear Power Plant site on Hainan Island. The project is led by the China National Nuclear Corporation (CNNC).

Key Site Selection Factors:

- **Regulatory Compliance:** The National Nuclear Safety Administration (NNSA) is overseeing the project, ensuring compliance with stringent safety standards.

- **Geological Stability:** The Changjiang site is located in a low-seismicity zone, reducing the risks associated with earthquakes.

- **Proximity to Water Sources:** The plant is situated near the South China Sea, providing ample water for cooling.

- **Climate Considerations:** The site has favourable weather conditions, with low risk of extreme weather events such as typhoons.

- **Community Support:** CNNC has engaged in public outreach efforts to address concerns and gain local support for the project.

Outcome:

The ACP100 reactor is expected to be operational by 2026, marking a significant step in China's efforts to deploy SMR technology for both domestic and international markets.

These case studies demonstrate how different regions adapt site selection strategies based on local conditions, regulatory environments, and community engagement. By assessing factors such as geological stability, access to water, regulatory compliance, and public perception, project developers can optimize site selection for successful SMR deployment.

Chapter 3
Design and Engineering of SMR Power Stations

Core Design Principles of SMRs

Small Modular Reactors (SMRs) are a new generation of nuclear reactors designed with several key principles in mind. These principles focus on enhancing safety, efficiency, flexibility, and cost-effectiveness compared to traditional large-scale nuclear reactors. The core design principles that guide the development of SMRs include:

1. **Modularity and Scalability**

    o   **Modular Construction**: SMRs are designed to be factory-built as modular units. These modules can be transported to the site and assembled quickly, reducing construction time and costs.

    o   **Scalable Deployment**: The modular design allows for phased construction, enabling utilities to add capacity incrementally based on demand. This flexibility is ideal for both small grids and growing energy needs.

    o   **Prefabrication**: Factory-based prefabrication ensures high-quality control, reduces on-site labour, and minimizes construction risks, leading to faster deployment.

2. **Enhanced Safety Features**

    o   **Passive Safety Systems**: SMRs rely on passive safety systems that do not require active controls, power, or human intervention to operate. This includes natural circulation for cooling, gravity-fed water reservoirs, and passive heat dissipation.

- o **Inherent Safety**: The reactors are designed to be inherently safe, with features such as negative temperature coefficients, which naturally reduce reactor power as temperatures increase.

- o **Underground and Seismic Resilience**: Many SMRs are designed to be installed underground or partially buried, providing additional protection against natural disasters, aircraft impacts, or sabotage.

3. **Compact and Simplified Design**

- o **Reduced Footprint**: SMRs have a much smaller physical footprint compared to traditional reactors, making them suitable for a wider range of locations, including remote areas and industrial zones.

- o **Simplified Systems**: The design focuses on simplifying reactor components and reducing the number of parts, which enhances reliability and lowers maintenance requirements.

- o **Integrated Design**: SMRs often use integrated reactor vessels, where primary components like the reactor core, steam generator, and pressurizer are contained within a single structure to reduce complexity and enhance safety.

4. **Fuel Efficiency and Waste Reduction**

- o **Higher Fuel Burnup**: SMRs are designed for higher fuel efficiency, allowing for longer fuel cycles and reducing the frequency of refuelling.

- o **Closed Fuel Cycle**: Some SMRs, particularly those using molten salt or fast reactor designs, aim to utilize a closed fuel cycle to recycle spent fuel and reduce nuclear waste.

- o **Waste Minimization**: SMRs produce less nuclear waste per unit of energy generated compared to large reactors, making waste management more manageable and less costly.

5. **Flexibility in Fuel Types**

- o **Diverse Fuel Options**: SMRs can be designed to use a variety of fuel types, including low-enriched uranium (LEU), thorium, and mixed oxide (MOX) fuels.

- o **Hybrid Designs**: Some designs incorporate advanced fuel types like TRISO (Tri-structural Isotropic) fuel, which is more resilient to high temperatures and radiation.

o   **Fuel Security**: SMRs are designed to optimize fuel use and reduce dependency on frequent fuel deliveries, making them suitable for remote or isolated locations.

6. **Economic Competitiveness**

o   **Lower Initial Capital Costs**: The smaller size and modular construction of SMRs reduce upfront capital costs, making nuclear power more accessible to a wider range of markets, including developing countries.

o   **Faster Return on Investment**: The shorter construction time and quicker deployment of SMRs enable faster revenue generation compared to traditional large reactors.

o   **Reduced Operating Costs**: The simplified design, reduced staffing needs, and lower maintenance requirements of SMRs contribute to lower overall operating costs.

7. **Flexibility in Applications**

o   **Decentralized Power Generation**: SMRs can be deployed in decentralized grids, providing power to remote areas, islands, or locations with limited grid connectivity.

o   **Process Heat and Industrial Applications**: High-temperature SMRs can be used for industrial processes like hydrogen production, desalination, or district heating.

o   **Load Following Capabilities**: SMRs are designed to be more flexible in adjusting power output, making them suitable for integration with intermittent renewable energy sources like wind and solar.

8. **Enhanced Regulatory Compliance**

o   **Standardization**: The modular nature of SMRs allows for design standardization, which can streamline regulatory approval processes and reduce licensing times.

o   **Licensing Innovations**: SMR developers work closely with regulatory bodies to develop new licensing frameworks that accommodate the unique features of SMRs, ensuring safety without unnecessary delays.

o   **Global Collaboration**: The development of international standards and harmonized regulations can help accelerate the global deployment of SMRs.

9. **Environmental Sustainability**

o   **Reduced Greenhouse Gas Emissions**: SMRs provide a low-carbon energy source, contributing to global efforts to combat climate change.

- o **Minimal Environmental Impact**: The smaller size and lower fuel requirements reduce the environmental footprint of SMR plants compared to traditional nuclear reactors.

- o **Adaptability for Brownfield Sites**: SMRs can be sited on brownfield or previously developed industrial sites, reducing the need for new land use and minimizing ecological disruption.

The design principles for the nuclear core in Small Modular Reactors (SMRs) are fundamentally rooted in enhancing safety, efficiency, and adaptability to various operational contexts. These principles are informed by the unique characteristics of SMRs, which are typically defined by their smaller size, modular construction, and advanced safety features.

One of the primary design principles is the incorporation of passive safety systems. SMRs are engineered to utilize natural physical laws, such as gravity and natural circulation, to ensure safety without the need for active mechanical systems. This is particularly evident in designs like the Westinghouse SMR, which integrates passive safety features derived from larger reactor designs, such as the AP1000 [28, 123]. These systems are crucial for maintaining core integrity during accidents, as they can function without external power or operator intervention, thereby reducing the risk of core meltdowns and radiological releases [1].

Another significant principle is the long operational life of the reactor core, which is designed to minimize the frequency of refuelling. For instance, some SMR designs feature cores that can operate for over 18 years without needing fresh fuel [124]. This long core life is achieved through advanced fuel management strategies, such as the use of burnable absorbers and optimized fuel compositions, which help control reactivity and extend the fuel cycle [125, 126]. The ability to operate for extended periods without refuelling not only enhances operational efficiency but also reduces the logistical challenges associated with fuel supply and management.

The adaptability of SMRs to various energy demands and grid conditions is another critical design principle. SMRs are designed to be flexible in their operation, capable of adjusting output to match the variable demands of the grid, especially in conjunction with renewable energy sources [18]. This flexibility is facilitated by features such as turbine bypass systems and battery energy storage, which allow SMRs to maintain stability and reliability in power supply [18]. Moreover, their modular nature enables them to be deployed in multiple units, providing scalability to meet specific energy needs in diverse geographical and economic contexts [42].

Furthermore, the choice of fuel and coolant materials in SMR designs reflects a commitment to safety and efficiency. For example, some designs utilize low-enriched uranium or thorium fuels, which can be more proliferation-resistant and environmentally friendly [127]. The use of advanced coolants, such as supercritical carbon dioxide or molten salts, enhances thermal efficiency and allows for higher operational temperatures, which can improve the overall efficiency of electricity generation [63, 128, 129].

Engineering Design Details of a Small Modular Reactor (SMR) Power Station

The engineering design of a Small Modular Reactor (SMR) power station incorporates advanced nuclear technology, modular construction techniques, and innovative safety systems to optimize efficiency, safety, and scalability. Below is a detailed breakdown of the key engineering components, systems, and design considerations involved in constructing and operating an SMR power station.

Reactor Core and Containment Structure

The design of the reactor core and its containment structure is a critical aspect of Small Modular Reactors (SMRs), reflecting a combination of safety, efficiency, and modern nuclear technology advancements. The compact nature of SMRs allows for innovative design choices that differentiate them from traditional large-scale reactors, with a focus on minimizing risks while maximizing energy output.

Reactor Core Design: The core of an SMR is engineered to optimize fuel efficiency and safety. Low-enriched uranium (LEU), typically enriched to below 20%, is commonly used in most SMR designs. This fuel composition strikes a balance between achieving a sustained nuclear chain reaction and adhering to non-proliferation standards. The reactor core is configured with fuel assemblies, where each assembly consists of multiple fuel rods containing a ceramic uranium dioxide (UO_2) core. The ceramic nature of UO_2 provides excellent thermal stability and radiation resistance, ensuring the integrity of the fuel even under high-temperature conditions.

Each fuel rod is encased in zirconium alloy cladding, which serves as a barrier to prevent the release of radioactive fission products. The cladding is designed to withstand high radiation doses and the corrosive environment within the reactor. Additionally, the fuel assemblies are optimized for longer burn-up cycles, which means that they can remain in the reactor for extended periods before needing replacement. This not only reduces operational downtime but also decreases the volume of spent fuel generated, thereby improving the overall efficiency of the reactor.

In terms of moderators and coolants, many SMRs use light water reactors (LWR) configurations, where water serves as both the moderator (to slow down neutrons) and the coolant (to carry away heat). However, some advanced SMR designs employ alternative coolants like molten salt, helium, or lead, which enable higher operating temperatures and, consequently, greater thermal efficiency. These alternative coolants are particularly effective in enhancing the heat transfer properties of the reactor, making them suitable for high-temperature applications like hydrogen production or industrial process heat.

Containment Structure: The containment structure is an essential safety feature in any nuclear power plant, and in SMRs, it is designed to be both compact and robust. The reactor core is housed within a reinforced concrete and steel containment vessel, which is engineered to withstand internal pressures as well as external threats, such as earthquakes, floods, or potential aircraft impacts. This robust construction ensures that in the unlikely event of an internal failure, radioactive materials are contained within the vessel, preventing any release into the environment.

One of the unique features of many SMRs is their below-ground containment structures. By placing the reactor vessel partially or entirely underground, the design adds an extra layer of protection against external threats, including natural disasters or potential terrorist attacks. This below-ground setup also provides additional shielding from radiation, thus reducing the risk to plant personnel and nearby communities.

A key component of the containment structure in SMRs is the integration of passive safety features. Unlike traditional reactors that rely heavily on active systems (such as pumps and external power sources) to maintain cooling, SMRs are designed with passive systems that operate without human intervention or external power. For example, in the event of an emergency shutdown, natural convection, gravity-fed cooling systems, and heat exchangers can automatically activate to dissipate heat from the reactor core, preventing overheating. This design significantly enhances the reactor's safety profile by reducing the likelihood of catastrophic failures, such as core meltdowns.

The combination of robust containment, passive safety systems, and advanced materials used in SMR design ensures that these reactors are not only safer but also more resilient to various operational challenges. The design of the reactor core and containment structure is pivotal in making SMRs a viable, scalable, and sustainable option for the future of nuclear energy, especially in regions where traditional large-scale reactors may not be feasible due to space, cost, or regulatory constraints.

Cooling Systems and Heat Transfer in SMR Design

The design of cooling systems in Small Modular Reactors (SMRs) is a crucial aspect that ensures both operational efficiency and safety. Cooling systems in SMRs are engineered to manage heat transfer efficiently, even under emergency conditions, by utilizing a combination of traditional and innovative technologies. The compact and modular nature of SMRs allows for advanced cooling methods, enhancing their reliability and resilience compared to traditional large reactors.

Primary Cooling System: The primary cooling system plays a fundamental role in transferring heat away from the reactor core to maintain safe operating temperatures. In traditional water-cooled SMRs, the primary loop uses pressurized water as the coolant. This water circulates through the reactor core, absorbing the heat generated from nuclear fission. The heated water, kept under high pressure to prevent it from boiling, is then transported to a heat exchanger where the heat is transferred to the secondary loop.

In contrast, advanced SMR designs—such as those using molten salt or gas as coolants—operate at significantly higher temperatures, often reaching 700-900°C. These high-temperature coolants circulate through specially designed channels within the reactor core, extracting heat more efficiently than water-based systems. Molten salt coolants, for example, have excellent thermal conductivity and can remain in liquid form at very high temperatures, making them ideal for reactors that aim to achieve higher thermal efficiencies. Similarly, gas-cooled reactors utilize helium, which is chemically inert and has a high heat capacity, allowing for effective heat transfer at elevated temperatures.

A unique feature of many SMRs is their ability to use natural circulation in the primary cooling loop. This means that in the event of a power loss or pump failure, the coolant can still flow through the system without the need for mechanical pumps. Natural circulation relies on the principles of buoyancy, where heated coolant rises and cooler liquid falls, creating a self-sustaining flow. This passive mechanism greatly enhances the safety of the reactor, particularly in emergency situations where active systems might fail.

Secondary Cooling System: The secondary cooling system is responsible for converting the heat extracted from the reactor core into useful energy. After the primary coolant absorbs heat from the reactor core, it transfers this heat to the secondary loop via a heat exchanger. In the secondary loop, water is converted into steam, which drives a turbine connected to an electricity generator.

To reduce dependency on water resources, many SMRs are designed with dry cooling systems or air-cooled condensers. Unlike traditional cooling towers that require substantial amounts of water, these systems use air to condense the steam back into water, making them ideal for deployment in arid regions or areas with limited water availability. By using air-cooled systems, SMRs can be sited in locations where traditional reactors would be impractical due to water scarcity, thus expanding their potential deployment zones.

Passive Cooling Systems: One of the standout safety features in SMRs is the integration of passive cooling systems. These systems are designed to operate without the need for external power sources or active mechanical components, relying instead on natural physical processes to manage residual heat after the reactor has been shut down.

Passive decay heat removal systems are a key component of this design. After the reactor is turned off, fission reactions cease, but residual decay heat continues to be produced. To prevent overheating, SMRs are equipped with systems that utilize natural convection, gravity-fed water tanks, and passive heat exchangers. For instance, in some designs, gravity-driven water tanks release coolant into the system if temperatures exceed a certain threshold, allowing for rapid cooling without operator intervention.

Another approach used in passive systems is the deployment of air or natural draft heat exchangers, which dissipate heat through natural airflow. These systems require no moving parts, making them highly reliable even in scenarios where external power sources are unavailable. The inclusion of passive cooling features not only enhances safety but also reduces the need for extensive backup systems, thus lowering the overall cost and complexity of the reactor design.

Turbine and Power Generation System in SMRs

The turbine and power generation system is a crucial component of a Small Modular Reactor (SMR) power plant, converting the thermal energy generated by the reactor into electricity. SMRs are designed to be highly efficient and adaptable, with a focus on compactness and modularity. This section provides a detailed explanation of how SMRs utilize advanced turbine technology and power conversion systems to optimize electricity generation.

Steam Turbine Operation: The power generation process in SMRs begins with the steam turbine, which is driven by steam produced in the secondary cooling loop. After the reactor core heats the primary coolant, this heat is transferred to the secondary loop through a heat exchanger. In water-cooled SMRs, the heat causes water in the secondary loop to boil, generating steam. This high-pressure steam is then directed into a steam turbine, which is connected to a generator. The turbine blades are turned by the steam's force, converting thermal energy into mechanical energy, which in turn drives the generator to produce electricity.

SMRs utilize compact turbine-generator units, designed to fit within the smaller footprint of the plant. Unlike traditional large-scale nuclear reactors, which require expansive turbine halls, SMR turbine systems are optimized for space efficiency. This compact design reduces both construction time and costs, allowing SMRs to be deployed in diverse locations, including remote or off-grid areas where space is limited.

Power Conversion Efficiency: A key advantage of SMRs is their potential for achieving high thermal efficiency, often in the range of 35-40% or higher. This efficiency is made possible by the reactor's ability to operate at elevated temperatures, especially in advanced designs like gas-cooled or molten salt reactors. The higher operating temperatures improve the Rankine cycle efficiency, which is the thermodynamic cycle used in steam turbines for converting heat into work.

For example, gas-cooled reactors, which use helium as a coolant, or molten salt reactors can achieve temperatures up to 700-900°C, significantly higher than traditional water-cooled reactors. This allows SMRs to extract more energy from each unit of fuel, enhancing overall efficiency and reducing the volume of nuclear waste produced. Additionally, the use of alternative coolants with superior heat transfer properties enables more efficient steam generation, resulting in higher electricity output with less fuel consumption.

Electrical Systems and Power Distribution: Beyond the turbine system, SMRs are equipped with sophisticated electrical infrastructure to manage power distribution and ensure seamless integration with the electrical grid. The plant includes electrical switchgear, transformers, and synchronization equipment to regulate power output, ensuring stable delivery to the grid. These components are essential for maintaining voltage levels, managing load variations, and ensuring compatibility with existing grid infrastructure.

A critical aspect of SMR design is the inclusion of redundant power supplies to maintain essential plant functions during grid disruptions. This typically involves diesel generators and uninterruptible power supplies (UPS) systems, which provide backup power to critical systems such as the reactor cooling pumps, control systems, and safety mechanisms. In the event of a grid outage, these backup systems can kick in almost instantly, ensuring the reactor remains in a safe state.

SMRs are also designed with black-start capabilities, allowing them to restart independently without relying on external grid power. This feature makes SMRs particularly valuable for remote or isolated regions where grid reliability may be a concern. The ability to support microgrid applications and off-grid power generation also makes SMRs a strategic asset for regions with limited access to stable electricity.

Control and Instrumentation Systems in SMRs

The control and instrumentation systems of Small Modular Reactors (SMRs) play a pivotal role in ensuring the safe and efficient operation of the reactor and associated power generation systems. These systems integrate cutting-edge digital technology to provide real-time monitoring, automation, and safety controls. This section explains how these systems are designed to optimize performance while ensuring robust protection against potential threats.

Centralized Control Room: The heart of an SMR's operational oversight is its centralized control room. Unlike traditional nuclear power plants that often rely on a combination of analogue and digital controls, SMRs are designed with fully integrated digital control systems. These advanced systems allow operators to monitor and control the reactor, turbine, cooling systems, and other essential plant components from a single, centralized location.

The control room is equipped with real-time data displays that provide continuous information on critical parameters such as reactor core temperature, coolant flow rates, pressure levels, and radiation levels. These data feeds are processed by sophisticated algorithms that can identify abnormal patterns or deviations, allowing operators to react promptly to any changes. In addition to manual control, the system includes automated functions to optimize reactor performance, adjust power output, and enhance efficiency.

One of the key features of SMR control systems is the incorporation of automated safety systems. These systems are designed to automatically initiate protective actions if they detect anomalies such as coolant loss, pressure deviations, or temperature spikes. For instance, if the system detects a significant drop in coolant flow or a rise in core temperature beyond safe limits, it can automatically trigger a reactor shutdown, isolating the reactor core and initiating passive cooling mechanisms. This automation enhances the plant's ability to respond swiftly to potential safety threats, minimizing the risk of operator error during critical situations.

Cybersecurity Measures: Given the digital nature of SMRs, cybersecurity is a critical aspect of their control and instrumentation systems. With increasing global concerns over cyber threats, SMR plants are designed with robust cybersecurity infrastructure to protect their digital assets from malicious attacks that could compromise safety or disrupt operations.

The SMR's digital control systems are fortified with multiple layers of cybersecurity defences, including firewalls, intrusion detection systems (IDS), and encrypted communication channels. Firewalls act as the first line of defence by filtering incoming and outgoing network traffic, while intrusion detection systems continuously monitor the plant's network for unusual activities that could indicate a cyberattack. In the event of suspicious activity, automated alerts are generated to notify the control room operators, enabling a rapid response to potential threats.

Additionally, encrypted communication protocols are used to secure data transfers between control systems and remote monitoring stations. This ensures that sensitive operational data, such as reactor status or safety logs, remains confidential and protected from interception. Periodic cybersecurity audits and penetration tests are conducted to identify vulnerabilities and implement corrective measures, ensuring that the plant's digital infrastructure remains resilient against evolving cyber threats.

To further safeguard the plant, access to critical control systems is tightly regulated. Multi-factor authentication (MFA) and secure login procedures are required for operators and maintenance personnel to access sensitive systems. This prevents unauthorized access and ensures that only trained and certified personnel can interact with the reactor's control mechanisms.

Waste Management and Fuel Handling in SMR Power Plants

Managing nuclear waste and handling spent fuel are crucial aspects of the operational safety and sustainability of Small Modular Reactors (SMRs). Given their compact size, advanced design, and enhanced safety features, SMRs are engineered to minimize waste production and simplify fuel handling processes. Below, we explore the key components of waste management and fuel handling in SMRs.

Spent Fuel Storage: The management of spent nuclear fuel is a critical step in ensuring the safety and environmental sustainability of SMR operations. When nuclear fuel reaches the end of its useful life in the reactor, it remains highly radioactive and generates significant heat. To safely handle this, spent fuel is first transferred to on-site spent fuel pools. These pools are specially designed to cool the fuel by submerging it in water, which helps dissipate heat while also providing a barrier against radiation. The spent fuel typically remains in these pools for several years until it cools down sufficiently.

After the initial cooling period, the spent fuel is moved to dry cask storage for longer-term management. Dry casks are robust, airtight containers made of steel and concrete, designed to safely store spent fuel for decades. The use of dry cask storage not only frees up space in the spent fuel pools but also reduces the risk associated with keeping large quantities of spent fuel in one location. This staged approach to fuel storage enhances safety by allowing for flexible, modular expansion of storage capacity as needed.

Advanced SMR designs, such as molten salt reactors, take waste management a step further by integrating on-site fuel reprocessing or recycling capabilities. These reactors can process spent fuel to extract usable materials, thereby reducing the overall volume of high-level waste that needs long-term storage. This capability could potentially allow SMRs to operate in a more sustainable and closed fuel cycle, where less waste is generated, and more of the fuel's energy content is utilized.

Waste Minimization: One of the significant advantages of SMRs is their ability to minimize radioactive waste production compared to traditional large-scale reactors. This is achieved through several design improvements:

1. **Higher Fuel Efficiency**: SMRs are designed to achieve longer burn-up cycles, meaning they can extract more energy from the nuclear fuel before it is spent. This efficiency results in less frequent refuelling and a reduced volume of spent fuel.

2. **Reduced Production of Long-Lived Isotopes**: By optimizing neutron economy and using advanced fuel compositions, SMRs can minimize the generation of long-lived radioactive isotopes. This reduces the long-term environmental impact of nuclear waste, making disposal more manageable.

3. **Waste-Burning Capabilities**: Some advanced SMR designs, particularly **molten salt reactors**, are capable of using spent fuel from conventional reactors as part of their fuel mix. These reactors can "burn" the long-lived actinides found in spent fuel, thereby converting existing nuclear waste into usable energy. This not only reduces the stockpile of spent fuel but also lessens the burden on geological repositories needed for waste disposal.

Fuel Handling and Safety Measures: Fuel handling in SMRs is designed to be safer and more efficient due to their smaller size and modular construction. The fuel used in SMRs, such as low-enriched uranium (LEU) or thorium, is typically loaded into sealed fuel assemblies that are easier to handle and transport. This modular approach reduces the risk of accidental exposure to radiation during the refuelling process.

In many SMR designs, refuelling can be done without shutting down the entire plant. For example, in reactors with online refuelling capabilities, such as pebble bed reactors, fresh fuel can be introduced and spent fuel removed continuously, which minimizes downtime. This feature not only improves operational efficiency but also reduces the risk of accidents associated with traditional refuelling processes.

For molten salt reactors, the fuel is already in a liquid state and is continuously circulated through the reactor. This allows for real-time removal of fission products and reduces the accumulation of radioactive waste. The fuel can be reprocessed on-site to extract valuable isotopes, which can then be reused, further minimizing waste generation.

Long-Term Waste Disposal Considerations: Despite the advantages of SMRs in waste reduction, there remains the challenge of long-term disposal for the remaining high-level waste. While spent fuel from SMRs is less in volume, it still requires secure storage for thousands of years due to its radioactive nature. Countries developing SMRs are exploring geological repositories as a permanent solution for the disposal of high-level waste. Additionally, international efforts are ongoing to develop advanced waste management technologies, such as deep borehole disposal, which could provide safer and more cost-effective alternatives.

SMRs incorporate advanced waste management strategies that significantly reduce the environmental impact and safety risks associated with nuclear power. By utilizing longer burn-up cycles, incorporating passive safety features, and exploring innovative fuel recycling methods, SMRs are paving the way for a more sustainable and efficient approach to nuclear energy production.

Structural Engineering and Site Layout in SMR Power Plants

Building a Small Modular Reactor (SMR) power plant involves advanced structural engineering and careful site planning to ensure safety, efficiency, and resilience. The design principles focus on modular construction, robust foundations, and a well-organized site layout, allowing for faster deployment, cost efficiency, and improved safety. Below, we delve into the specifics of these critical elements.

Modular Construction: One of the core design features of SMR power plants is their modular construction approach. This method involves prefabricating major components in controlled factory environments before transporting them to the construction site. By using factory-fabricated modules, the construction process is streamlined, reducing the overall time and cost associated with building the plant. Prefabrication also ensures higher quality control, as modules are produced under stringent factory conditions with less exposure to weather and environmental factors.

In addition to speeding up the construction timeline, modular components are designed to fit together seamlessly once on-site, minimizing the complexities typically associated with traditional nuclear plant construction. This method also reduces labour risks by limiting the amount of work required in potentially hazardous environments, allowing for safer assembly in a controlled setting. The prefabricated modules can include everything from reactor containment units to cooling systems and control room components, making the SMR a highly adaptable solution for various locations.

Foundations and Structural Support: The structural engineering of an SMR power plant prioritizes resilience against natural disasters, especially in regions prone to earthquakes or soil instability. The foundations are engineered to be robust, employing techniques such as base isolation and deep piling to absorb seismic shocks. This isolation system helps to decouple the reactor structures from ground movements, thereby reducing the impact of an earthquake on the critical components of the plant.

In areas where the soil may not be naturally stable, ground improvement techniques like soil compaction or the use of reinforced concrete piles are employed. These measures ensure that the foundation can support the heavy weight of the reactor modules and associated infrastructure, preventing issues like subsidence or uneven ground shifting over time. The reinforced structures used in SMRs are designed to handle both internal pressures from reactor operations and external forces, providing a high level of protection.

Site Layout: The site layout of an SMR power plant is meticulously planned to optimize safety, operational efficiency, and accessibility. The plant is organized into distinct zones, each serving a specific function, which helps to streamline operations and enhance safety protocols. The primary zones typically include:

1. **Reactor Containment Area**: This is the central zone where the reactor modules are housed. It includes reinforced containment structures to prevent the release of radioactive materials in the event of an incident. Often, these structures are partially or fully underground for added protection against external threats like aircraft impact or natural disasters.

2. **Control and Operations Zone**: The control room and associated operations buildings are situated near the reactor zone but are shielded to protect personnel from radiation. This zone includes the centralized control systems for monitoring reactor operations, safety mechanisms, and communication infrastructure.

3. **Cooling Infrastructure**: The cooling systems, including cooling towers, heat exchangers, or air-cooled condensers, are strategically placed to maximize efficiency. The proximity of cooling systems to the reactor zone reduces heat loss and improves energy transfer rates.

4. **Turbine Hall**: The **turbine hall** is designed to house steam turbines and generators, converting thermal energy into electricity. This area is compact yet efficiently laid out to optimize space while ensuring easy access for maintenance.

5. **Waste Management Area**: The waste management zone is dedicated to the **storage and handling of spent fuel** and radioactive waste. It includes secure spent fuel pools, dry cask storage, and waste processing facilities, all designed to minimize environmental impact.

6. **Security Perimeter**: The entire site is enclosed within a secure perimeter with surveillance systems, access control points, and barriers to prevent unauthorized entry. This helps protect critical infrastructure from physical threats and ensures compliance with regulatory safety standards.

Access Roads and Transportation: Effective site layout also requires well-planned access roads and transportation infrastructure. These are essential for transporting large prefabricated modules, reactor fuel, and heavy equipment to the site. The design includes wide access routes capable of accommodating oversized transport vehicles, ensuring that modules can be delivered without delays.

Additionally, the layout is designed to facilitate the rapid deployment of emergency response teams if needed. Dedicated access points for first responders, as well as designated evacuation routes, are incorporated into the overall site design. This ensures that, in the event of an emergency, personnel can respond swiftly and safely.

By focusing on these structural and site planning elements, SMR power plants achieve a balance of safety, efficiency, and flexibility. The modular design allows for scalability, meaning additional reactor modules can be added over time to meet growing energy demands. The robust engineering and careful site planning not only optimize performance but also ensure the resilience of the plant against both natural and human-induced threats, making SMRs a viable solution for sustainable energy in diverse environments.

Safety Systems and Emergency Preparedness in SMR Power Plants

The safety of Small Modular Reactors (SMRs) is a paramount consideration in their design and operation. SMRs are engineered with multiple layers of both passive and active safety systems to ensure a high level of protection against accidents, natural disasters, and other emergencies. The design focuses on redundancy, resilience, and automated responses to mitigate potential risks. Below is an in-depth explanation of how safety systems and emergency preparedness are integrated into SMR power plant operations.

Passive and Active Safety Systems: A significant advantage of SMR technology is its incorporation of both passive and active safety systems. Unlike traditional reactors, which often rely on active systems that require external power sources or human intervention, SMRs are designed to be largely self-sufficient in emergencies.

Passive safety systems leverage natural physical processes, such as gravity, natural convection, and thermal expansion, to manage reactor cooling and shutdown operations. For instance, the Emergency Core Cooling System (ECCS) is engineered to activate automatically in the event of a reactor overheating. It utilizes gravity-fed water tanks and natural convection to circulate coolant through the reactor core, even if the power supply is compromised. These passive systems are designed to function without the need for mechanical pumps, external power, or operator action, significantly reducing the risk of human error during a crisis.

In addition to passive systems, active safety systems are integrated to provide additional layers of protection. These systems, such as backup diesel generators and automated control valves, are designed to monitor reactor conditions and intervene if necessary. For example, if an anomaly is detected in coolant flow or pressure, automated shutdown mechanisms are triggered to prevent reactor damage. This dual approach ensures that SMRs are highly resilient to both internal malfunctions and external threats.

Seismic and Flood Protection: SMR power plants are often sited in areas that may be exposed to natural disasters such as earthquakes and floods. As a result, they are equipped with advanced seismic and flood protection measures to safeguard against these risks.

To enhance seismic resilience, the plant is designed with base isolation systems and seismic dampers. Base isolation involves placing the reactor building on specialized bearings that absorb seismic shocks, effectively decoupling the structure from ground movements. This prevents damage to critical components during an earthquake. Additionally, seismic dampers act as shock absorbers, further reducing the impact of vibrations on the reactor core and cooling systems.

Flood protection is another critical aspect of SMR safety. The design includes elevated platforms and waterproof barriers to protect key infrastructure from rising water levels. The reactor containment area is often built above the highest historical flood levels to prevent water ingress during extreme weather events. Drainage systems and watertight doors are also installed to minimize the risk of flooding and ensure the plant remains operational in adverse conditions.

Emergency Response Plans: A comprehensive emergency preparedness plan is essential for any nuclear facility, and SMRs are no exception. These plans are designed to protect both the plant personnel and the surrounding community in the unlikely event of an incident.

SMR facilities include emergency response centres equipped with real-time monitoring and communication systems to coordinate responses during a crisis. These centres serve as the command hub for managing incidents, enabling operators to quickly assess the situation, initiate safety protocols, and communicate with local authorities.

To facilitate a swift evacuation if needed, the site is equipped with clearly marked evacuation routes and assembly points. Drills and training exercises are regularly conducted to ensure that all plant personnel are familiar with emergency procedures. Additionally, communication systems are set up to provide timely updates to the local community, ensuring transparency and maintaining public trust.

In collaboration with local authorities, the plant also has emergency response agreements in place to mobilize external resources, such as firefighting units, medical teams, and law enforcement, if necessary. By integrating internal safety measures with external support systems, SMR power plants are prepared to handle a wide range of emergency scenarios.

Environmental Impact Mitigation in SMR Power Plants

Small Modular Reactors (SMRs) have been designed with a focus on minimizing their environmental impact. As the world grapples with climate change and increasing energy demands, SMRs offer a cleaner alternative to fossil fuel power generation. However, like any energy source, they still have environmental implications that must be managed carefully. Below is a detailed explanation of how SMRs mitigate their environmental impact through strategic design and technology.

Minimal Land Use: One of the most significant environmental advantages of SMRs is their minimal land footprint compared to traditional nuclear reactors and other large-scale power plants. Due to their smaller size and modular construction, SMRs can be built on brownfield sites—previously developed industrial areas that may not be suitable for other forms of development. This reduces the need to clear new land, thereby preserving natural habitats and reducing the impact on local ecosystems.

The compact design of SMRs also makes them ideal for areas with limited space, such as densely populated regions or remote locations where land availability is constrained. By requiring less land, SMRs help protect forests, wetlands, and other critical ecosystems from being converted for energy infrastructure. Additionally, this smaller footprint allows for more flexibility in site selection, reducing potential conflicts with existing land use and heritage sites.

Low Carbon Emissions: SMRs play a crucial role in the transition to a low-carbon energy future. Unlike fossil fuel-based power plants, SMRs generate electricity with virtually no carbon emissions during operation. This makes them an attractive option for countries looking to meet their climate commitments, such as the goals set by the Paris Agreement.

By providing a reliable and consistent source of baseload power, SMRs can support the integration of intermittent renewable energy sources like wind and solar. This combination helps to stabilize the grid, reduce the need for fossil fuel backup, and ultimately lower overall carbon emissions. Additionally, since SMRs are often designed for longer fuel cycles with high burn-up rates, they are more efficient in using nuclear fuel, which further contributes to reducing the carbon footprint associated with nuclear energy production.

Cooling Water Management: Water usage is a critical environmental concern for many power plants, particularly those that rely on water for cooling. Traditional nuclear reactors often require large amounts of water to dissipate heat, which can lead to issues such as thermal pollution, where heated water discharged into rivers or lakes disrupts aquatic ecosystems.

SMRs, on the other hand, are designed to optimize water use and minimize environmental impact. For water-cooled SMRs, cooling towers and air-cooled condensers are employed to reduce the need for continuous water intake. These systems are highly efficient and help prevent the release of heated water back into natural water bodies, thus protecting aquatic life and reducing thermal pollution.

In regions where water availability is limited, SMRs can also be designed with dry cooling systems that rely on air instead of water to dissipate heat. This innovation not only conserves water resources but also allows SMRs to be sited in arid or water-scarce regions, expanding the range of suitable locations for these reactors. By prioritizing efficient water management, SMRs contribute to sustainable water use while maintaining operational efficiency.

Figure 31: A Fast-Breeder Test Reactor, Kalpakkam Nuclear Complex, India. IAEA Imagebank, CC BY-SA 2.0, via Flickr.

Safety Features and Redundancies in SMR Design

The design of Small Modular Reactors (SMRs) incorporates a variety of safety features and redundancies that enhance their operational reliability and public acceptance. These features are critical in addressing the inherent risks associated with nuclear power generation, particularly in the aftermath of significant nuclear incidents such as Fukushima. The integration of advanced safety systems, passive safety mechanisms, and modular design principles are fundamental to the SMR concept, ensuring that these reactors can operate safely under a wide range of conditions.

One of the primary safety features of SMRs is their inherent safety characteristics, which stem from their low power density and large heat capacity. This design allows for a greater margin of safety during operational anomalies, such as Loss of Coolant Accidents (LOCAs) and core-melting scenarios. For instance, the modular high-temperature gas-cooled reactor (MHTGR) utilizes helium as a coolant, which has a high heat capacity and low density, contributing to its ability to manage heat effectively even in emergency situations [1]. Furthermore, the design of SMRs often incorporates passive safety systems that rely on natural physical laws, such as gravity and natural circulation, to ensure reactor cooling without the need for active mechanical systems [123, 130]. This reliance on passive systems significantly reduces the likelihood of human error and mechanical failure, which are common causes of nuclear accidents.

Case Study: Modular High-Temperature Gas-Cooled Reactor (MHTGR)

A practical example of these inherent safety features can be seen in the Modular High-Temperature Gas-Cooled Reactor (MHTGR). This reactor uses helium as a coolant, which is an inert gas with excellent thermal properties. Helium's high heat capacity and low density contribute to efficient heat management, even under accident conditions. Unlike water, helium does not become radioactive upon exposure to neutron radiation, and it cannot chemically react with the reactor materials, thus minimizing risks associated with coolant loss or chemical reactions. In the event of a loss of forced cooling, the MHTGR can still passively dissipate heat through natural convection and conduction, leveraging the large thermal inertia of its graphite moderator and fuel blocks.

In addition to passive safety features, SMRs are designed with multiple redundancies to enhance their safety profile. These redundancies include backup systems for critical components, such as emergency cooling systems and power supply systems. For example, the Westinghouse SMR design integrates an integral pressurized water reactor (iPWR) system that houses all components within a single pressure vessel, thereby minimizing the potential for leaks and failures [28]. Moreover, the incorporation of battery energy storage systems allows SMRs to maintain operational integrity during power outages, ensuring that essential safety functions can be performed even in the absence of external power [18]. This layered approach to safety, combining inherent design features with redundant systems, creates a robust framework for managing potential accidents.

The development of SMRs has also been informed by lessons learned from past nuclear incidents. Enhanced safety features have been integrated into SMR designs to address specific vulnerabilities identified in traditional nuclear reactors. For instance, the incorporation of advanced fuel technologies and improved instrumentation enhances the ability to monitor and control reactor conditions, thereby reducing the risk of overheating and fuel failure [131]. Additionally, the design of SMRs often includes advanced containment structures that are capable of withstanding extreme external events, such as earthquakes and tsunamis, further bolstering their safety credentials [57]. These design considerations reflect a proactive approach to safety, emphasizing the importance of learning from historical precedents to inform future reactor designs.

The modular nature of SMRs also contributes to their safety and operational flexibility. By allowing for incremental capacity expansion, SMRs can be deployed in a manner that aligns with local energy demands, reducing the risks associated with overbuilding and underutilization of nuclear facilities [3]. This adaptability is particularly beneficial in regions with limited grid capacity, where smaller, modular units can be integrated into existing energy systems without overwhelming local infrastructure [102]. Furthermore, the ability to manufacture SMRs in a factory setting allows for greater quality control and consistency in construction, which is critical for ensuring safety standards are met throughout the production process [29].

Techniques for Managing Loss of Coolant Accidents (LOCAs) and Core-Melting Scenarios in SMRs

Small Modular Reactors (SMRs) are designed with multiple layers of safety features to handle Loss of Coolant Accidents (LOCAs) and core-melting scenarios effectively. These techniques are a combination of passive safety mechanisms, advanced materials, and innovative engineering designs that significantly reduce the likelihood of catastrophic failures. Here's a detailed explanation of the techniques used to manage these situations:

Emergency Core Cooling Systems (ECCS)

Emergency Core Cooling Systems (ECCS) are a fundamental safety feature designed to protect Small Modular Reactors (SMRs) during a Loss of Coolant Accident (LOCA), which may occur due to a leak, pipe rupture, or other mechanical failures that result in the sudden loss of coolant from the primary system. In such events, the reactor's cooling capacity is compromised, and without intervention, the core could overheat, potentially leading to severe damage. To address this risk, SMRs incorporate both active and passive ECCS mechanisms that ensure continuous cooling and prevent core damage, even without external power or manual intervention.

Gravity-fed Coolant Injection: One of the key features of SMR ECCS is the gravity-fed coolant injection system. In the event of a coolant loss, large, elevated water tanks release coolant directly into the reactor core through gravity alone. This design eliminates the need for pumps or electrical power, which may be unavailable during an emergency. The advantage of gravity-fed systems is their simplicity and reliability: as long as there is water in the tanks, the coolant will flow into the core, driven purely by gravity. This feature is especially beneficial in scenarios where external power sources are compromised, such as during natural disasters or severe system failures. By ensuring a continuous flow of coolant to the reactor core, the gravity-fed system helps prevent overheating and potential core damage.

Passive Heat Exchangers: Another critical component of the ECCS in SMRs is the use of passive heat exchangers connected to natural convection loops. These heat exchangers are designed to transfer residual heat from the reactor core to an external cooling system without the need for mechanical pumps. The process relies on thermal gradients to drive the circulation of coolant, where hot coolant rises and cooler fluid sinks, creating a continuous flow. This natural convection mechanism allows the reactor to dissipate excess heat even if the primary cooling system fails. The passive heat exchangers are particularly effective in managing decay heat, which continues to be generated by the reactor fuel even after the reactor has been shut down. By removing this residual heat, the heat exchangers play a crucial role in preventing a rise in core temperature and maintaining overall reactor safety.

Pressure Release Valves: In addition to gravity-fed systems and passive heat exchangers, SMRs equipped with pressurized water reactors (PWRs) also include pressure release valves as part of their ECCS. These valves are designed to automatically release pressure from the reactor vessel if it reaches critical levels, thereby preventing further damage to the system. During a LOCA, the rapid loss of coolant can cause a significant build-up of steam pressure within the reactor vessel. If left unchecked, this pressure could lead to structural failure. The pressure release valves act as a safeguard by releasing steam in a controlled manner, thereby protecting the integrity of the containment vessel and reducing the risk of catastrophic failure.

Key Advantage - Ensuring Continuous Cooling and Preventing Core Damage: The integration of these passive ECCS components ensures that SMRs can remain safe even under the most challenging conditions. Unlike traditional nuclear reactors that often rely on complex active systems and external power sources, SMRs are designed to be self-sustaining in emergency scenarios. The ability to cool the reactor core without pumps or electrical power is a significant advantage, as it eliminates potential failure points associated with mechanical systems. In a worst-case scenario, the combination of gravity-fed coolant injection, passive heat exchangers, and pressure release valves ensures that the reactor core remains covered with coolant, thereby preventing overheating, fuel damage, and potential radioactive release.

This layered approach to emergency core cooling highlights the robust safety measures embedded within SMR designs, making them an attractive option for enhancing nuclear safety while meeting global energy demands.

Natural Circulation Cooling

Small Modular Reactors (SMRs) are designed to leverage **natural circulation cooling**, a passive cooling technique that eliminates the need for mechanical pumps. This innovative approach is a cornerstone of SMR safety, ensuring that the reactor core remains adequately cooled even in emergency situations where external power sources may be compromised. By relying on fundamental physical principles such as thermal gradients and fluid dynamics, natural circulation enhances the reactor's resilience, making SMRs inherently safer than traditional nuclear reactors that depend on active cooling systems.

Thermal Siphon Effect: At the heart of natural circulation cooling is the thermal siphon effect, a process driven by differences in the density of the coolant as it heats up and cools down. When the reactor core generates heat, it raises the temperature of the coolant, typically water, causing it to expand and become less dense. This hot, less dense coolant naturally rises through the reactor vessel toward the heat exchangers located at a higher elevation. As the hot coolant passes through these heat exchangers, it releases its heat to a secondary cooling loop or directly to the atmosphere (in air-cooled systems). The coolant, now cooled, becomes denser and sinks back down to the bottom of the reactor vessel, where it re-enters the core to absorb heat once again.

This continuous convection loop is entirely driven by the differences in fluid density and does not require any external power or mechanical components to maintain flow. The thermal siphon effect is particularly effective in SMRs because of their compact size and optimized design, which allows for efficient heat transfer and circulation over shorter distances compared to large-scale reactors.

No External Power Required: One of the most significant advantages of natural circulation cooling is that it operates independently of external power sources. In the event of a power outage, equipment failure, or a severe incident like a Loss of Coolant Accident (LOCA), the reactor can continue to cool itself passively. This eliminates the need for emergency diesel generators, backup pumps, or other mechanical systems that could fail under stress or during catastrophic events. The simplicity and reliability of natural circulation make it an ideal feature for SMRs, which are often deployed in remote or off-grid locations where maintaining external power supplies can be challenging.

The reliance on natural circulation significantly reduces the risk of human error or mechanical failure, which are common factors in nuclear accidents. By removing the dependency on pumps and electrical components, SMRs equipped with natural circulation cooling can maintain core stability for extended periods, even in the absence of operator intervention. This feature is especially critical during emergencies, providing additional time for operators to assess the situation and implement safety measures if needed.

Key Advantage: Reliable Cooling Under Extreme Conditions: The key advantage of natural circulation cooling is its ability to maintain the reactor core at safe temperatures, even during a total loss of power or catastrophic system failures. This passive cooling mechanism ensures that the reactor core remains submerged in coolant, effectively dissipating residual heat and preventing overheating. The natural convection process is self-regulating; as the temperature in the core increases, the thermal siphon effect becomes more pronounced, accelerating the flow of coolant and enhancing heat removal. Conversely, as the core cools down, the circulation rate decreases, preventing overcooling.

This built-in self-regulation makes natural circulation a highly effective and fail-safe method for managing reactor temperatures, contributing to the overall safety and reliability of SMR designs. By minimizing the need for active components and reducing reliance on human intervention, natural circulation cooling aligns with the safety-first philosophy of modern nuclear engineering, offering a robust solution to ensure reactor stability under all conditions.

Decay Heat Removal Systems

Even after a reactor has been shut down, the fuel within the core continues to generate heat due to the radioactive decay of fission products. This residual heat, known as decay heat, can reach significant levels and, if not properly managed, could lead to overheating and potential damage to the reactor core. For Small Modular Reactors (SMRs), the design of decay heat removal systems is a crucial aspect that ensures the plant's safety, particularly in emergency scenarios where active cooling mechanisms may be compromised. SMRs incorporate advanced passive decay heat removal systems to address this challenge effectively, utilizing natural physical processes to dissipate heat without the need for mechanical or electrical components.

Passive Decay Heat Exchangers: One of the primary methods SMRs use to handle decay heat is through passive heat exchangers. These systems are specifically designed to remove heat from the reactor core and transfer it to the environment using air or water as a cooling medium. The passive heat exchangers rely on natural convection to circulate air or water, drawing heat away from the reactor vessel.

In designs where water is used, these exchangers are integrated with secondary loops that dissipate heat into water bodies or cooling towers. However, in air-cooled systems, heat exchangers release heat directly into the atmosphere, making them particularly useful in areas where access to large quantities of water is limited. The key advantage of passive heat exchangers is that they do not require active pumps or external power sources to function, ensuring continuous heat removal even during power outages or system failures.

Air-Cooled Cooling Towers: For SMRs located in regions where water resources are scarce, air-cooled cooling towers serve as an efficient alternative for managing decay heat. These cooling towers utilize ambient air to absorb and dissipate heat from the reactor's heat exchangers. The design leverages the natural flow of air through the towers to cool the system, eliminating the need for large volumes of water typically required in traditional nuclear power plants.

By using air as a cooling medium, SMRs equipped with air-cooled towers can be sited in arid or remote locations where water availability is a constraint. This also reduces the environmental impact associated with water usage, making SMRs a more sustainable option for low-water regions. Additionally, air-cooled systems are less susceptible to issues such as thermal pollution, which can affect local ecosystems when large amounts of heated water are discharged into rivers or lakes.

Gravity-Fed Emergency Water Tanks: Another essential component of SMR decay heat management is the use of gravity-fed emergency water tanks. These tanks are strategically placed above the reactor core to ensure a continuous flow of water into the cooling system in the event of a primary system failure. By relying solely on gravity, these tanks can supply coolant to the reactor core without the need for pumps or external power, which is particularly beneficial during emergencies like a total loss of power (blackout scenarios).

The water from these emergency tanks flows directly into the core, absorbing the decay heat and keeping the fuel submerged, thereby preventing overheating. This gravity-driven mechanism is designed to operate autonomously for extended periods, providing a fail-safe layer of protection that requires no manual intervention or mechanical activation.

Key Advantage - Autonomous, Long-Term Cooling: The key advantage of passive decay heat removal systems in SMRs is their ability to function autonomously, ensuring that the reactor remains cool even in the absence of external power or operator intervention. By incorporating passive systems such as natural convection heat exchangers, air-cooled towers, and gravity-fed water tanks, SMRs can effectively manage decay heat over extended durations. This not only enhances the safety profile of SMRs but also simplifies their operation, reducing the reliance on complex mechanical systems that are prone to failure.

These passive cooling strategies provide multiple layers of redundancy, making SMRs inherently safer than traditional nuclear reactors. The integration of diverse, passive heat removal techniques ensures that even in worst-case scenarios—such as a complete loss of power or cooling system failure—the reactor core remains adequately cooled, thereby preventing fuel damage and minimizing the risk of radioactive release.

Core Damage Mitigation Techniques

Small Modular Reactors (SMRs) are engineered with advanced safety features to mitigate the risk of core damage and prevent severe nuclear incidents, such as core-melting scenarios. These techniques are integrated into the reactor's design to enhance its resilience, even in extreme situations. The primary objective is to contain any potential core damage and avoid the release of radioactive material, ensuring that the reactor remains safe and stable under all operating conditions. Here's an in-depth look at some of the core damage mitigation techniques utilized in SMR technology.

Core Catchers for Enhanced Containment: One of the innovative safety features incorporated into some SMR designs is the core catcher. A core catcher is a structure located beneath the reactor vessel, specifically engineered to capture and cool molten core material—known as corium—should a core meltdown occur. In the event that the reactor core begins to melt due to severe overheating, the molten material would flow downward into the core catcher, which is designed to contain and rapidly cool it.

The core catcher prevents the corium from breaching the containment structure, thereby minimizing the risk of a radioactive release. By spreading the molten material over a large surface area and using materials that can withstand extreme temperatures, the core catcher dissipates heat more efficiently, reducing the likelihood of damage to the containment vessel. This passive cooling mechanism requires no external power or active intervention, making it a robust safeguard during a severe accident.

High-Temperature Coolants: Molten Salt and Lead: Certain SMRs utilize alternative coolants, such as molten salt or lead, which have significantly higher boiling points than water. These coolants are highly effective in managing heat even at elevated temperatures, thereby reducing the risk of core overheating and meltdown. For example:

- **Molten Salt Reactors (MSRs):** Molten salt coolants can operate at temperatures exceeding 700°C without reaching their boiling point. This high thermal tolerance allows them to effectively transfer heat away from the core, even under accident conditions. If the core temperature begins to rise, the molten salt coolant can continue to absorb and dissipate heat without transitioning to a gaseous state, thus maintaining stable reactor operations.

- **Lead-Cooled Reactors:** Lead and lead-bismuth eutectic coolants are used in some SMR designs due to their excellent thermal conductivity and high boiling point (above

1,700°C). These properties allow lead-cooled reactors to remain stable even in extreme thermal conditions, reducing the risk of a core meltdown.

The use of these high-temperature coolants not only enhances the reactor's ability to withstand extreme scenarios but also simplifies the cooling system, as they can operate at atmospheric pressure. This eliminates the need for high-pressure containment structures, thereby reducing the risk of pressure-related accidents.

Self-Regulating Fuel Design: Another critical aspect of core damage mitigation in SMRs is the use of self-regulating **fuel** designs that inherently reduce reactor power output as temperatures increase. For instance:

- **Molten Salt Reactors**: In designs where the fuel is dissolved in the coolant (liquid fuel reactors), the reactor becomes self-regulating due to the thermal expansion of the fuel. As the temperature rises, the liquid fuel expands, reducing its density. This decrease in density lowers the likelihood of neutron interactions, thereby slowing down the fission reaction and reducing the reactor's power output. This negative temperature coefficient of reactivity acts as a natural safety mechanism, preventing the core from overheating.

- **Solid Fuel Designs**: SMRs using solid fuel assemblies incorporate materials that expand or change properties at higher temperatures, reducing reactivity. For example, some fuel assemblies are designed to slightly expand under heat, increasing the distance between fuel rods and thereby reducing neutron flux. This self-limiting behaviour helps prevent runaway reactions in the event of a cooling system failure.

By integrating self-regulating fuel mechanisms, SMRs can adjust their power output in response to rising temperatures without the need for operator intervention or active control systems. This adds an extra layer of protection against overheating and core damage.

Key Advantages - Comprehensive Core Damage Mitigation: The combination of core catchers, high-temperature coolants, and self-regulating fuel designs in SMRs significantly enhances their safety profile, especially in handling extreme scenarios that could lead to core damage in traditional reactors. These core damage mitigation techniques are not just reactive measures but are proactively built into the reactor's design to prevent incidents from escalating into severe accidents.

The ability to passively control temperature and contain molten materials ensures that, even in the unlikely event of a serious malfunction, the reactor core remains intact, and radioactive materials are contained. This drastically reduces the potential for environmental contamination and enhances public confidence in the safety of nuclear energy. By prioritizing safety through robust engineering solutions, SMRs represent a significant step forward in the evolution of nuclear reactor technology.

Below-Ground Containment Structures: Enhancing Safety and Security

One of the most significant design features in Small Modular Reactors (SMRs) is the use of below-ground containment structures. These structures are engineered to offer enhanced protection against a wide range of external threats while also improving the safety of the reactor in case of internal failures. By placing the reactor containment vessel underground, SMRs benefit from an additional layer of physical security that is not typically available in traditional above-ground nuclear reactors. Below-ground designs are particularly effective in shielding the reactor from natural disasters, potential accidents, and malicious attacks.

Reinforced Concrete and Steel Lining for Robust Protection: The construction of below-ground containment structures involves the use of reinforced concrete and steel linings to ensure the integrity of the containment vessel. The containment structure is designed to withstand both internal pressures generated during operational anomalies and external impacts, such as earthquakes, floods, or even aircraft strikes. The reinforced concrete acts as a physical barrier that not only prevents the release of radioactive materials but also absorbs and mitigates external shocks.

To further enhance durability, the steel lining within the containment structure acts as a secondary layer that prevents the escape of radioactive gases or liquids. This dual-layer protection system ensures that even if the outer concrete wall is damaged, the inner steel lining will continue to serve as a containment barrier. Additionally, the underground placement naturally shields the reactor from above-ground impacts, reducing the risk of structural damage.

Passive Cooling Channels for Heat Dissipation: One of the innovative features of below-ground containment structures in SMRs is the integration of passive cooling channels. These channels are designed to dissipate heat through natural convection, eliminating the need for active cooling systems that rely on electrical power or mechanical pumps. The passive cooling system is particularly crucial in scenarios such as a Loss of Coolant Accident (LOCA) or a core-melting event, where the reactor must be cooled rapidly to prevent overheating.

The cooling channels are strategically built into the containment structure, allowing ambient air or water to circulate around the reactor vessel. This natural circulation of coolant helps to remove excess heat from the containment area, reducing the risk of core damage. By relying on passive heat removal mechanisms, SMRs can continue to cool the reactor core even in the event of a total power outage, thereby significantly enhancing safety.

Key Advantages - Enhanced Physical Security and Safety: The use of below-ground containment structures offers several key advantages in ensuring the safety and security of SMRs:

1. **Enhanced Physical Security:** The underground placement makes it much harder for external threats, such as acts of terrorism or accidents, to reach the reactor core. This added layer of security reduces the vulnerability of the plant to intentional sabotage or high-impact incidents.

2. **Minimizing Radioactive Release:** In the unlikely event of a severe reactor malfunction, the below-ground containment structure acts as a robust barrier to prevent the release

of radioactive materials into the environment. This design feature significantly reduces the risk of contamination and protects nearby communities.

3. **Resilience to Natural Disasters**: The underground design inherently protects the reactor from natural disasters, such as earthquakes, hurricanes, and tsunamis, which are more likely to damage above-ground facilities. The reinforced structure is engineered to absorb seismic shocks, further enhancing the plant's resilience.

4. **Long-Term Safety Assurance**: By integrating passive cooling channels and robust containment systems, below-ground SMRs are designed to maintain safe operational conditions without relying on external power sources. This self-sufficiency ensures that, even during extended power outages or emergency scenarios, the reactor can safely cool itself.

Overall, below-ground containment structures represent a significant advancement in nuclear safety design, making SMRs a more secure and resilient option for the future of nuclear power. By combining robust physical barriers with passive safety features, these reactors are better equipped to handle both internal failures and external threats, ensuring a higher level of protection for the public and the environment.

Redundant Safety Barriers: Ensuring Multiple Layers of Protection

Small Modular Reactors (SMRs) are engineered with multiple redundant safety barriers to contain radioactive materials effectively and prevent any potential environmental contamination. These safety barriers are integral to the design philosophy of SMRs, which emphasizes robust defence-in-depth strategies. By incorporating several layers of containment, SMRs ensure that even if one barrier fails, others are in place to maintain the reactor's integrity and protect the environment.

Fuel Cladding: The First Line of Defence: The fuel cladding is the innermost barrier designed to prevent the release of radioactive materials. In SMRs, fuel rods are typically enclosed in a zirconium alloy cladding that offers high corrosion resistance and mechanical strength. This cladding acts as a seal around the ceramic uranium dioxide (UO_2) pellets, containing radioactive fission products that could otherwise escape into the reactor coolant.

The choice of zirconium alloy is crucial due to its excellent performance under high temperatures and radiation. In the event of a Loss of Coolant Accident (LOCA) or other operational anomalies, the cladding provides an essential layer of protection, maintaining its integrity long enough for emergency cooling systems to activate. By keeping the radioactive fission products isolated, the fuel cladding minimizes the risk of radioactive release at the source.

Reactor Vessel - Isolating the Core: The second layer of protection in an SMR is the reactor vessel, which encloses the entire reactor core and coolant system. Constructed from high-strength steel, the reactor vessel is designed to withstand high pressures and temperatures. This containment ensures that, even if the fuel cladding is compromised, radioactive materials remain contained within the reactor system.

The reactor vessel is engineered to resist corrosion, thermal stress, and radiation damage over the reactor's operational lifespan. It is equipped with pressure relief valves and other safety mechanisms to prevent over-pressurization, thereby reducing the risk of a catastrophic failure. By isolating the core from the surrounding environment, the reactor vessel provides a critical additional barrier against radioactive leaks.

Containment Building: The Final Barrier: The containment building serves as the outermost protective layer, designed to prevent the release of radioactive materials into the environment. Made from reinforced concrete and steel, this robust structure is engineered to withstand both internal pressures from potential accidents and external threats, such as natural disasters or impacts from aircraft.

The containment building is often integrated with passive cooling systems that utilize natural airflows to dissipate heat in case of an emergency, thus preventing the buildup of pressure inside the structure. The building's design is also optimized to resist seismic activity, ensuring that even in the event of an earthquake, the containment remains intact. This barrier is the final line of defence, providing comprehensive protection against the uncontrolled release of radioactive substances.

Key Advantages - Comprehensive Protection Through Redundancy: The use of multiple redundant safety barriers in SMRs offers several significant advantages:

1. **Robust Defence-in-Depth**: By incorporating multiple containment layers, SMRs provide a high level of safety assurance. Even if one layer fails, subsequent barriers are in place to prevent radioactive releases, significantly reducing the risk of environmental contamination.

2. **Enhanced Plant and Public Safety**: The combination of fuel cladding, reactor vessel containment, and reinforced containment buildings ensures that both plant personnel and nearby communities are protected from radiation exposure, even during severe accidents.

3. **Adaptability to Extreme Scenarios**: The multi-layered design enhances the reactor's ability to withstand extreme scenarios, such as earthquakes, floods, or accidental impacts. This makes SMRs a safer and more resilient option for nuclear power generation, especially in regions prone to natural disasters.

By implementing redundant safety barriers, SMRs achieve a level of safety that is superior to traditional nuclear reactors. This design approach not only meets stringent regulatory standards but also addresses public concerns about nuclear safety, making SMRs a viable solution for sustainable and secure energy generation.

Is an SMR Safer than a Large Nuclear Reactor?

The safety of Small Modular Reactors (SMRs) is often evaluated in terms of two key factors: the reduction in the likelihood of severe accidents and the potential for minimizing radiological consequences in the event of such incidents. The first factor, commonly quantified as Core Damage Frequency (CDF), measures the probability of a reactor experiencing core damage under extreme conditions. SMRs are generally considered to have a lower CDF compared to traditional large reactors. This is largely because SMRs leverage passive safety features more effectively, which reduces reliance on active components and human intervention. The second factor relates to the potential radiological impact if an accident does occur. This is assessed by measuring the concentration of radioactive materials at certain distances from the reactor, which directly influences the required size of the emergency planning zone (EPZ) [132].

One of the main reasons for the enhanced safety of SMRs is rooted in their fundamental physics. The reduced size of an SMR inherently lowers the thermal output, which simplifies the challenge of heat removal during both normal operations and emergencies. In nuclear reactors, safety is often compromised when heat cannot be effectively dissipated, leading to dangerous increases in temperature and pressure. This buildup of heat can jeopardize the structural integrity of the reactor. Thus, a critical aspect of nuclear safety revolves around maintaining effective heat removal. If the heat generated by the reactor can be efficiently transferred out of the system, the risk of accidents is significantly minimized [132].

The principles of heat transfer are key to understanding why SMRs are considered safer. Heat removal depends on three primary factors: the heat transfer coefficient, the temperature difference between the reactor core and the environment, and the available surface area for heat exchange. Traditional reactors rely on high heat transfer coefficients achieved through active components like pumps, which are vulnerable to mechanical failures and require electricity to function. In contrast, SMRs utilize passive cooling systems that exploit a larger surface area for heat dissipation. By leveraging natural convection and heat conduction, SMRs can maintain safe operating temperatures without the need for complex active systems. This not only simplifies the reactor design but also enhances reliability by eliminating the potential for pump or power failures [132].

To illustrate the advantages of SMRs in heat management, one can examine the relationship between reactor size, power output, and heat transfer. As the power capacity of a reactor decreases, the ratio of heat transfer to power output improves due to the larger surface area relative to the reactor's volume. For instance, a 300 MW SMR has approximately 49% more heat transfer capacity per unit of power compared to a 1,000 MW large reactor. If the SMR's power density is reduced even further, the heat transfer efficiency can increase by as much as 233% compared to traditional reactors. This efficiency gain is crucial for SMRs, enabling them to rely on passive cooling systems that do not require external power sources, thereby enhancing safety [132].

However, reducing the power density to increase safety comes with trade-offs. While it enhances the ability to dissipate heat passively, it also decreases the compactness of the reactor, which could impact the economic benefits of SMRs. A core advantage of SMRs is their small size, which reduces construction costs and allows for modular deployment. Thus, maintaining a balance between power density and safety is essential to maximize the benefits of SMR technology [132].

Beyond reducing the likelihood of severe accidents, SMRs also focus on limiting the radiological consequences if an incident were to occur. This requires an understanding of the source term, which refers to the quantity and type of radioactive materials present in the reactor core. The amount of radioactive material depends on factors such as the reactor's thermal efficiency, the type of fuel used, and its enrichment level. Additionally, the composition of the radioactive material can influence the severity of its impact, as different isotopes emit varying levels and types of radiation. Therefore, estimating the potential radiological impact of an SMR requires a comprehensive analysis of its fuel characteristics and operating conditions [132].

For SMRs to be deployed near populated areas, it is critical to demonstrate that even in the unlikely event of a severe accident, the radiological consequences would be minimal. This is essential for reducing the size of the EPZ, which determines the area where population control measures would be required in the event of a radioactive release. If an SMR can limit its radiological impact to the immediate vicinity of the reactor site, the EPZ can be reduced to a few hundred meters, significantly lowering the economic and logistical burdens associated with emergency planning [132].

The dispersion of radioactive material in the environment is a complex process influenced by factors like wind speed, atmospheric stability, and release height. A simplified Gaussian plume model can be used to estimate how radioactive materials disperse in the event of an uncontrolled release. According to this model, as the power output of a reactor decreases, the required distance for safe evacuation also decreases. For instance, a 1,000 MW reactor might require a 30 km EPZ, whereas a 300 MW SMR could reduce this zone substantially. However, achieving an EPZ limited to the plant's immediate boundary requires further reductions in the source term, which can only be accomplished through advanced containment and fuel management strategies [132].

SMRs offer a promising enhancement in nuclear safety by reducing both the probability of severe accidents and their potential radiological impact. By utilizing passive safety features and optimizing heat transfer, SMRs can operate more reliably without the complex active systems used in larger reactors. Moreover, their ability to minimize the EPZ makes them suitable for deployment closer to populated areas, opening up new possibilities for nuclear energy as a flexible and scalable power source. However, realizing these benefits requires continued innovation in reactor design and fuel management to address the challenges of scaling down nuclear technology while maintaining economic feasibility [132].

Modular Construction Techniques for SMRs

Small Modular Reactors (SMRs) are strategically designed to incorporate modular construction techniques, which simplify the building process, lower costs, and reduce the time required for project completion. Traditional large-scale nuclear reactors are typically constructed on-site, with substantial labour and resources dedicated to building each part of the reactor directly at the location. In contrast, SMRs employ prefabricated modules created in a controlled factory environment, which are then transported to the construction site for assembly. This approach allows SMRs to be constructed with greater efficiency and flexibility, making them a viable option for areas with limited infrastructure or challenging environmental conditions [133].

The modular construction of SMRs provides significant advantages, including cost reduction, faster construction timelines, improved safety, and a minimized environmental footprint. By transferring a large portion of the construction process from the site to the factory, SMRs reduce the risk of weather-related delays and allow for more consistent quality control. This factory-based approach aligns with global objectives to expand nuclear energy in a sustainable manner, supporting efforts to decarbonize the energy sector. With modular construction, SMRs offer a solution that is both economical and adaptable, catering to diverse energy demands while maintaining high standards of safety and reliability.

Modular construction is a critical concept for Small Modular Reactors (SMRs) and plays a significant role in making these reactors economically viable. Unlike traditional large-scale nuclear reactors that are primarily constructed on-site, the SMR approach involves manufacturing the majority of components in controlled factory environments. These components are then transported to the installation site in modules for final assembly. This shift in construction methodology is essential because it addresses the economic challenge posed by the "economy of scale" that traditionally benefits larger nuclear power plants. If SMR developers want to avoid the high costs associated with large-scale reactors, they must adopt fundamentally different manufacturing and construction methods. Instead of treating SMR projects like traditional chemical or energy plant construction projects, which are heavily dependent on economies of scale, they should take inspiration from industries like automobile or ship manufacturing. In these industries, the final products are largely manufactured in factories and only assembled or fine-tuned on-site [132].

The economy of scale issue is a significant challenge for SMRs. According to the widely applied "six-tenth rule" for estimating the costs of nuclear power plants, the capital cost does not decrease proportionally with the reduction in reactor size. This rule suggests that as the capacity of a power plant decreases, the cost per unit of power (per megawatt) increases. For instance, while the total construction cost of a 300 MW SMR might be about 52% lower than that of a traditional 1000 MW nuclear plant, the cost per megawatt for the SMR would actually be 62% higher [132]. This means that while smaller reactors can be cheaper overall, they are less cost-efficient on a per-MW basis if constructed using the same methodologies as large reactors. Therefore, to make SMRs cost-competitive with other energy systems, it is crucial to innovate in both design and construction. This involves developing entirely new reactor designs, transforming the manufacturing process to be more aligned with industrial mass production, and minimizing the amount of on-site construction needed [132].

The concept of modular construction fundamentally changes the business approach to nuclear energy projects. It transitions these projects from traditional, large-scale construction endeavors into something closer to a manufacturing industry where efficiency and mass production are key drivers of cost reduction. The idea is clear: by manufacturing large sections of the reactor off-site in factories, where quality can be controlled more rigorously and construction is not impacted by weather or other on-site factors, the overall costs and timelines can be significantly reduced. However, while the concept of modular construction is well understood in theory, its practical implementation in SMR projects remains challenging. Even with modular designs, there is still a substantial amount of construction that needs to take place on-site, and measuring the degree of this work quantitatively is complex. For instance, while existing large reactors claim to use modular techniques, they still require extensive on-site assembly, which limits the cost and time savings [132].

Ultimately, the potential benefits of making SMRs smaller are clear. Reducing the total construction cost can make nuclear energy more accessible and affordable, while also enhancing safety due to the simplified and inherently safer design of smaller reactors. Additionally, SMRs can be more flexible in operation, better aligning with the intermittent nature of renewable energy sources by offering improved load-following capabilities. However, the primary challenge remains overcoming the economic disadvantages of smaller scale. Modular construction is the key strategy to move away from the cost escalation associated with the loss of economy of scale. By shifting more of the construction to factory settings and treating SMR projects more like a manufacturing process than traditional plant construction, SMRs can achieve the cost efficiencies needed to make them a viable option for future energy needs [132].

Modular construction, which involves the off-site manufacturing of structural components, has been transformative across various industries, and nuclear energy is no exception. This approach is especially valuable in nuclear plant construction, as it significantly reduces the time and cost associated with on-site assembly. Modular reactors, which are smaller, standardized, and manufactured as a series of parts, can be assembled more swiftly than traditional reactors, reducing the overall cost and complexity of the process. The benefits of modular construction in nuclear projects have been reinforced by collaborative research projects, such as one undertaken by Laing O'Rourke, Arup, the Building Research Establishment (BRE), and Imperial College London. This three-year project, funded by organizations including Innovate UK and the Nuclear Decommissioning Authority, investigated the feasibility of applying Design for Manufacture and Assembly (DfMA) techniques to nuclear structures, demonstrating that modular construction can increase efficiency and quality [133].

Design for Manufacture and Assembly (DfMA) is a systematic approach that focuses on simplifying the design of components and products to make manufacturing and assembly processes more efficient, cost-effective, and reliable. DfMA integrates two complementary concepts—Design for Manufacture (DFM) and Design for Assembly (DFA). DFM emphasizes optimizing the design of individual components to streamline their production, ensuring that parts can be easily fabricated using the most efficient processes, materials, and tools. Meanwhile, DFA concentrates on designing products in a way that minimizes the complexity, time, and cost associated with assembling these components into a final product. By combining these approaches, DfMA aims to reduce the number of parts, simplify the geometry of components, and eliminate unnecessary complexities in the assembly process.

One of the fundamental principles of DfMA is the reduction of parts in a product, as fewer parts generally mean fewer assembly steps, lower costs, and reduced potential for errors during both manufacturing and assembly. By simplifying the design and reducing part count, manufacturers can achieve significant cost savings not just in production, but also in inventory management and quality control. This simplification also translates into shorter assembly times, allowing for faster project completion. For example, in industries like automotive and aerospace, where complex products require thousands of parts, DfMA can streamline the assembly process by ensuring that components fit together seamlessly and can be assembled quickly, often using automated systems.

DfMA also focuses on standardization and modularization, where components are designed to be interchangeable and reusable across different products or projects. By adopting standardized components, manufacturers can take advantage of economies of scale, reducing costs and ensuring that parts are readily available. In modular construction, for instance, standardized parts can be prefabricated in controlled factory environments and then transported to the site for quick assembly. This approach is particularly beneficial for projects like Small Modular Reactors (SMRs) or prefabricated buildings, where using DfMA techniques allows for a high degree of precision, consistent quality, and reduced on-site labor. Prefabricated modules can be assembled efficiently on-site, cutting down on construction time and minimizing the risks associated with traditional on-site construction.

Furthermore, DfMA techniques incorporate advanced technologies such as 3D modeling, computer-aided design (CAD), and digital simulations to optimize both the design and assembly processes before physical production begins. By using 3D modeling, engineers can identify and resolve potential design issues early in the process, reducing costly revisions and rework later. Digital simulations allow manufacturers to visualize the assembly process, test for potential bottlenecks, and optimize the sequence of operations to enhance efficiency. This proactive approach helps prevent delays during the actual production and assembly phases, ensuring that projects stay on schedule and within budget.

The adoption of DfMA techniques also enhances quality control by embedding checks and balances into the design and manufacturing stages. Components designed with DfMA principles are easier to inspect, handle, and assemble, which reduces the likelihood of defects or failures during the product's lifecycle. For industries with stringent safety and quality standards, such as nuclear energy, aerospace, and healthcare, DfMA ensures that every component meets the required specifications and performance criteria. By optimizing the design for ease of manufacture and assembly, companies can achieve higher reliability, reduce waste, and improve the sustainability of their operations.

Ultimately, DfMA is a powerful approach that enables organizations to deliver products more efficiently, at lower cost, and with higher quality. By focusing on the design stage to simplify manufacturing and assembly processes, DfMA not only drives operational efficiencies but also enhances product performance and customer satisfaction. The principles of DfMA are increasingly being adopted across various sectors, as companies recognize the value of reducing complexity, improving process efficiency, and achieving a faster time-to-market. As industries continue to innovate and embrace automation, the role of DfMA in enhancing productivity and competitiveness is set to grow even further.

Standardization in modular construction is particularly advantageous for repeatable structures within nuclear power stations, such as cable tunnels. For example, cable tunnels are essential in nuclear facilities, providing pathways for electrical services. Their straightforward design lends itself well to modularization. In trials conducted as part of the research project, two types of precast methods were tested for constructing cable tunnels: the solid precast method, which required minimal in-situ work, and the hybrid precast method, which utilized prefabricated reinforcement and permanent formwork. The results indicated significant on-site labour savings, with the solid precast method reducing labour by 75% and the hybrid method by almost 50%. The findings highlight the efficiency of modular construction, especially when compared to traditional methods that require extensive on-site work [133].

While some nuclear plant components, like cable tunnels, are relatively straightforward to modularize, others are more complex. For instance, fuel storage buildings within nuclear plants contain intricate design requirements, including irregular geometries, thick walls, and dense reinforcement patterns. As demonstrated by further trials, modular construction can still be applied to these complex structures by optimizing elements for factory manufacturing. In one example, traditional on-site construction of a fuel building took five days, while a modular approach reduced this to one day. The use of premanufactured pieces instead of assembling numerous individual bars and components on-site streamlines construction significantly. Such improvements underscore the potential of SMRs to decrease on-site labour and accelerate construction timelines through modularization.

The integration of DfMA in SMR construction is bolstered by advanced manufacturing techniques, which differ substantially from traditional on-site construction. In a factory, workforce productivity and conditions are highly controlled, which allows for optimization of the manufacturing process. For instance, by using straight reinforcement bars and reusable jigs, modular SMR construction eliminates complex, time-consuming tasks typically required in conventional nuclear plant construction. The result is a process that is both efficient and suited to mass production, making SMRs economically competitive with other energy sources. Robotic automation further enhances factory manufacturing, as shown in trials at the University of Sheffield's Advanced Manufacturing Research Centre, where a robotic system assembled nuclear components 75% faster than manual methods [133].

The use of 3D modelling and wireless sensor networks (WSNs) adds further efficiency and quality control to SMR construction. 3D modelling allows for accurate, real-time adjustments to be made before modules are delivered to the site, preventing misalignments and other costly on-site issues. WSNs enable continuous monitoring of environmental conditions, which is crucial in managing factors like temperature and curing strength of concrete. These technologies optimize both factory and on-site processes, ensuring that components meet rigorous safety and quality standards [133].

Modular construction has yielded impressive results, with studies showing that up to 70% of the civil engineering work for SMRs can be completed in factories, reducing on-site construction time by up to 80%. Techniques developed through research projects have refined methods for assembling both standard and complex structures off-site, demonstrating that modularization can maintain high quality and safety while also reducing costs. However, modular SMR construction holds even greater potential if designs are optimized from the outset for modular delivery. With economies of multiples rather than scale, SMRs benefit from standard designs that allow for gradual process improvements and cost reductions [133].

The broad distribution of modular SMR manufacturing could stimulate economic growth across various regions. Facilities for nuclear precasting, for example, could be located in different areas of a country, providing job opportunities and supporting regional economies. The overall success of modular construction in SMRs is not only in reducing costs and timelines but also in achieving high standards of quality, compliance, and safety. By leveraging DfMA and advanced manufacturing approaches, SMRs represent a transformative step in nuclear construction, meeting energy demands in a manner that is efficient, flexible, and sustainable [133].

Prefabrication of Components

Prefabrication of components is a central feature in the construction of Small Modular Reactors (SMRs), revolutionizing how nuclear power plants are built by significantly streamlining the process. The technique involves manufacturing critical components in controlled factory environments rather than on-site, which greatly enhances efficiency, quality, and speed of construction. In traditional nuclear power plant construction, a large portion of the work must be done on-site, often leading to delays due to weather, labour shortages, or logistical challenges. However, SMRs utilize factory-based manufacturing, where key elements like reactor vessels, heat exchangers, turbines, and safety systems are produced as standardized modules. This controlled setting ensures that components are fabricated with consistent quality, minimizing the risk of defects and ensuring that parts meet the stringent safety and performance standards required in nuclear power generation.

The adoption of assembly line production techniques in manufacturing SMR modules draws inspiration from industries like automotive and aerospace, where mass production and standardization have long been used to reduce costs and increase efficiency. By applying these methods to the production of SMR components, manufacturers can achieve significant economies of scale, which is crucial in making nuclear energy more cost-competitive. The assembly line approach allows for the rapid production of multiple, identical modules, significantly reducing lead times compared to custom-built reactors. By standardizing these components, manufacturers can also optimize inventory management and reduce waste, as parts can be easily interchanged or upgraded across different SMR projects. This scalability is particularly beneficial when multiple units are deployed to meet increasing energy demands in various regions.

Quality control is another critical advantage of prefabricating SMR components in factory environments. Unlike traditional on-site construction, where quality control can be affected by varying environmental conditions and human error, factory-based manufacturing provides a controlled setting for thorough inspections and testing. This includes non-destructive testing, pressure testing of reactor vessels, and functional testing of control systems before they are shipped to the construction site. By ensuring that all components meet rigorous quality standards prior to delivery, the likelihood of encountering issues during on-site assembly is significantly reduced. This not only accelerates the construction schedule but also enhances the overall reliability and safety of the nuclear power plant once operational.

Furthermore, prefabrication minimizes on-site labour requirements, which is particularly advantageous in locations with harsh climates or limited access to skilled labour. By producing a significant portion of the SMR off-site, the need for extensive on-site construction activities is diminished, thus reducing the project's exposure to weather-related delays and safety risks. This also leads to a more predictable construction timeline, allowing for better planning and budgeting. The reduced need for on-site labour also means fewer logistical challenges and lower construction costs, as well as a smaller environmental footprint, since less land needs to be disturbed for construction activities.

Transport and Logistics

The transport and logistics of Small Modular Reactor (SMR) components are crucial aspects of their overall construction strategy, as they leverage the modular nature of these reactors to streamline delivery and on-site assembly. One of the primary benefits of SMRs is that their prefabricated modules can be transported to the construction site using existing transport infrastructure. To facilitate this, these modules are specifically designed with dimensions that align with the size constraints of standard shipping containers, allowing them to be moved via road, rail, or barge. This design consideration enables SMR manufacturers to utilize established logistics networks, which significantly reduces transportation costs and simplifies the process of moving large, complex components over long distances.

Modular transport not only ensures efficient delivery of reactor components but also plays a vital role in enhancing the flexibility of site selection for SMR power plants. Since the modules are prefabricated in factories and shipped to the final site, SMR projects can be deployed in locations that might otherwise be challenging for traditional nuclear reactors, such as remote areas or regions with limited infrastructure. The ability to transport large reactor modules using existing roadways or waterways eliminates the need for constructing specialized infrastructure, making it possible to reach more geographically isolated regions. This capability is particularly advantageous for countries with diverse terrains or where energy needs are high in off-grid areas.

In addition to logistical efficiency, the transport of prefabricated modules helps to minimize disruptions at the construction site. By significantly reducing the amount of on-site construction work required, modular transport limits the impact on the local environment and surrounding communities. Traditional nuclear power plant construction often involves extensive excavation, heavy machinery, and prolonged on-site activities, leading to increased noise, dust, and general disruption. In contrast, with SMRs, much of the construction work is completed off-site in controlled factory environments, leaving only the final assembly to be conducted on-site. As a result, the duration of construction activities at the plant location is significantly shortened, reducing the overall environmental footprint.

The streamlined logistics of transporting SMR modules also lead to more predictable project timelines and budgets. With fewer variables to manage on-site, such as weather delays or construction challenges, project developers can better control costs and adhere to schedules. This predictability is a key factor in making SMR projects more attractive to investors and stakeholders, as it lowers the risk associated with traditional large-scale nuclear projects that often face cost overruns and delays. Additionally, the reduction in on-site labour and equipment needs translates to lower safety risks for workers, further enhancing the project's overall efficiency and reliability.

On-Site Assembly and Installation

The on-site assembly and installation of Small Modular Reactors (SMRs) represent a critical phase in the construction process, capitalizing on the advantages of modular design to achieve rapid and efficient plant deployment. Once prefabricated modules arrive at the designated site, the process of assembling them begins with precision logistics and careful planning. The modules are pre-fabricated to include all necessary components, systems, and connections, allowing for a streamlined integration process once they are on-site. Using heavy-duty cranes, large modules such as the reactor vessel, heat exchangers, and containment structures are carefully positioned in place. These modules are then securely fastened using pre-configured bolt-and-weld connections, ensuring a robust and durable setup. The use of standardized interfaces and pre-configured connections simplifies this integration process, significantly reducing the potential for errors and speeding up assembly.

One of the major benefits of modular assembly is the substantial reduction in construction time compared to traditional nuclear power plants. In conventional projects, on-site construction can take upwards of ten years due to the complexity of assembling intricate components on-site, often under unpredictable weather conditions and logistical challenges. In contrast, SMRs leverage modular construction techniques, which allow for much of the work to be completed off-site in controlled factory environments. As a result, the on-site phase is limited primarily to assembly and connection of pre-fabricated modules. This efficient approach enables SMR projects to be fully assembled within two to three years, drastically reducing construction timelines. This not only accelerates project delivery but also reduces financial risks associated with prolonged construction schedules, making SMR projects more appealing to investors and stakeholders.

Moreover, the modular approach to SMR construction supports parallel construction, allowing multiple aspects of the plant to be developed simultaneously. For example, while the main reactor unit is being assembled, work on auxiliary components such as cooling towers, electrical systems, and control rooms can proceed in parallel. This simultaneous construction of different plant sections optimizes resource utilization and reduces overall project duration. By enabling various teams to work concurrently on separate modules, the project avoids the bottlenecks and sequencing delays that are common in traditional nuclear plant construction. This method also allows for more efficient use of labour and equipment, further driving down costs and improving project efficiency.

The speed and efficiency of on-site assembly also contribute to enhanced safety. By reducing the duration of on-site work, SMRs minimize the exposure of construction workers to potential hazards and adverse environmental conditions. Additionally, the simplified assembly process, with pre-configured connections, reduces the need for complex on-site fabrication and welding, which are traditionally labour-intensive and prone to human error. As a result, the likelihood of construction-related accidents is reduced, contributing to a safer working environment.

Overall, the modular assembly and installation process of SMRs highlight the advantages of modern construction techniques in the nuclear industry. By pre-fabricating components off-site and utilizing parallel construction strategies, SMRs can be deployed more rapidly, safely, and cost-effectively than traditional reactors. This streamlined approach not only aligns with the growing need for clean energy solutions but also provides a flexible and scalable model for meeting future energy demands in diverse locations, from urban areas to remote regions.

Standardization and Scalability

Standardization and scalability are central to the design philosophy of Small Modular Reactors (SMRs), providing significant benefits in terms of efficiency, cost reduction, and flexibility. One of the key advantages of SMRs is their highly standardized design. Unlike traditional large-scale reactors, which often require custom designs tailored to specific site conditions, SMRs are developed using standardized components and construction methodologies. This uniformity allows for replication across various locations, reducing both time and costs associated with the design and regulatory approval processes. By employing standardized modules, SMR developers can streamline the permitting process, as regulators become more familiar with the technology and associated safety features. This familiarity can lead to faster approvals and lower regulatory hurdles, making it easier to deploy SMRs in different regions. Additionally, the use of standardized designs helps reduce engineering costs over time, as lessons learned from one project can be applied directly to future installations, further optimizing both design and construction.

Scalability is another critical advantage of SMRs, enabled by their modular construction. The modular approach allows utilities and energy developers to incrementally increase the plant's power capacity by adding additional reactor modules as needed. This contrasts sharply with traditional nuclear power plants, which require large upfront investments to build single, massive reactors with fixed capacities. In the case of SMRs, an initial module can be installed to meet current energy demands, with the option to add more modules as demand grows. This flexibility makes SMRs ideal for regions experiencing gradual population growth or evolving industrial needs, where it may not be practical or cost-effective to commit to a large-scale reactor from the outset. Furthermore, the ability to scale up the plant over time aligns well with market dynamics, allowing energy providers to adjust their generation capacity based on economic conditions, demand forecasts, or shifts in energy policy.

The scalability of SMRs also supports their deployment in off-grid or remote areas, where traditional large reactors may not be feasible due to infrastructure limitations. In such regions, an SMR plant can begin with a single reactor to provide stable, low-carbon power, with additional modules added over time to support community growth or industrial expansion. This incremental approach helps mitigate the financial risk associated with nuclear projects, as the capital expenditure can be spread over several years, making it easier for investors and utilities to manage cash flow and secure financing. Moreover, by deploying standardized modules that can be replicated across different sites, SMR developers can achieve economies of scale in both manufacturing and construction. This allows for mass production of reactor components, reducing unit costs as production volumes increase.

The combined benefits of standardization and scalability make SMRs particularly appealing in a world increasingly focused on clean energy transitions. As countries aim to reduce carbon emissions and transition away from fossil fuels, SMRs offer a reliable and flexible solution that can be tailored to fit a wide range of energy needs. Whether providing baseload power in urban areas or supporting off-grid mining operations, the adaptability of SMRs allows them to fit seamlessly into various energy strategies. By leveraging the principles of standardization and scalability, SMRs not only promise to lower the barriers to nuclear energy adoption but also enhance the resilience and sustainability of the global energy landscape.

Safety and Reliability Benefits

The safety and reliability benefits of Small Modular Reactors (SMRs) are among their most significant advantages, particularly in comparison to traditional large-scale nuclear reactors. One of the primary benefits of SMR construction is the reduced risk associated with on-site work. By leveraging modular construction techniques, much of the fabrication and assembly is shifted from the plant site to controlled factory environments. This reduces the need for large numbers of construction workers and heavy machinery on-site, thereby significantly lowering the chances of construction-related accidents and injuries. In traditional nuclear projects, where thousands of workers may be required on-site over several years, the risk of incidents is naturally higher due to the complex and intensive nature of the work. By contrast, the streamlined, modular approach of SMRs allows components to be prefabricated and pre-tested in factories, where safety protocols can be more easily managed. Once these modules are complete, they are transported to the site and quickly assembled, minimizing the time that workers spend in potentially hazardous conditions.

Another critical aspect of SMR safety is the integration of advanced safety systems directly into their modular design. SMRs are engineered with enhanced safety features that are often more robust than those found in older, large-scale reactors. One of the most significant design improvements is the inclusion of passive safety systems. These systems, which include passive cooling mechanisms, operate without the need for external power sources or active intervention from operators. For instance, in the event of a system failure or power outage, SMRs can rely on natural processes such as gravity, natural circulation, and convection to continue cooling the reactor core. This design greatly reduces the risk of overheating or a meltdown scenario, as the reactor can remain in a safe state without relying on mechanical pumps or human intervention. By reducing dependency on active systems, which are prone to failure, SMRs enhance overall reliability and reduce the likelihood of accidents.

Additionally, the modular design of SMRs allows for the incorporation of below-ground containment structures. These below-ground installations not only provide added protection from external threats, such as earthquakes or aircraft impacts, but also reduce the risk of radiation release in the unlikely event of a core breach. The underground containment adds an additional layer of safety by isolating the reactor core from the external environment, which is especially critical in regions that are prone to natural disasters. The design of SMRs also emphasizes the use of highly durable materials and construction methods to withstand extreme conditions, further enhancing the reliability of these plants.

By integrating these safety features during the prefabrication phase, manufacturers can ensure that each module is rigorously tested and meets strict quality standards before being transported to the site. This controlled approach not only enhances the reliability of the plant once it is operational but also reduces the need for costly retrofits or repairs after installation. As a result, SMRs are not only safer during construction and operation but also offer long-term reliability that aligns with the stringent safety requirements of the nuclear industry. These advancements make SMRs an attractive option for expanding nuclear capacity in a way that prioritizes both safety and sustainability, supporting global energy needs while reducing the risks traditionally associated with nuclear power.

Cost Efficiency

The cost efficiency of Small Modular Reactors (SMRs) is one of their most compelling advantages, especially when compared to the traditional, large-scale nuclear reactors. One of the key factors driving this efficiency is the use of modular construction techniques, which dramatically lower the initial capital costs associated with building nuclear power plants. Unlike conventional reactors that require years of intensive on-site construction and thousands of skilled labourers, SMRs leverage prefabricated modules that are constructed in controlled factory environments. These modules are then transported to the site for quick assembly. By reducing the amount of on-site work, project timelines can be significantly shortened. This reduction in construction time directly translates to lower labour costs, as fewer workers are required for shorter durations, and many of the construction risks associated with weather delays or on-site complexities are mitigated. The result is a more predictable and manageable budget, which is crucial for nuclear projects that historically have been plagued by cost overruns and delays.

The cost benefits of SMRs also extend to their impact on financing. Nuclear power projects often require substantial upfront capital, making them dependent on securing large amounts of financing before construction begins. However, with traditional nuclear plants, the long lead times and complex regulatory processes can delay revenue generation for a decade or more, increasing the financial risk for investors. In contrast, the faster construction timelines associated with SMRs mean that they can begin generating electricity and, consequently, revenue much sooner—often within two to three years of breaking ground. This quicker turnaround reduces the period during which capital is tied up without generating returns, thereby lowering the overall financing risk. Investors and utility companies are more likely to support projects where there is a clearer and faster path to profitability, making SMRs a more attractive option, especially in regions where access to capital may be limited.

Additionally, factory-based manufacturing inherent to the SMR construction model contributes to cost efficiency by optimizing economies of scale. When components are produced in factories, manufacturers can standardize processes and materials, allowing for bulk production that drives down unit costs. This contrasts with traditional reactors, where much of the work is custom-built on-site, leading to inefficiencies and higher costs. By using standardized designs and replicable components, SMRs benefit from a streamlined supply chain and reduced wastage of materials. Moreover, because factory environments offer better control over quality and testing, the likelihood of costly errors or rework during the installation phase is minimized. This focus on efficiency ensures that the budget remains predictable and controlled, which is vital for the financial success of nuclear projects.

The combination of lower upfront costs, reduced financing risks, and the ability to achieve faster returns on investment positions SMRs as a cost-efficient solution for expanding nuclear energy capacity. In an industry where cost overruns have historically been a significant barrier to nuclear adoption, SMRs offer a pragmatic alternative. By aligning the financial model with the need for quicker deployment and lower risk, SMRs provide a pathway for both developed and developing regions to incorporate nuclear energy into their energy mix. This approach not only supports energy security and sustainability goals but also aligns with economic considerations, making nuclear energy more accessible and viable in a competitive global market.

Environmental and Social Benefits

Small Modular Reactors (SMRs) offer significant environmental and social benefits, largely due to their innovative construction techniques and efficient operational design. One of the most compelling environmental advantages of SMRs is the reduction in land use and overall environmental footprint. Traditional nuclear power plants require vast amounts of land for construction, with extensive on-site work that can lead to significant land disturbance and prolonged exposure to construction-related emissions. In contrast, SMRs utilize modular construction techniques, where a substantial portion of the assembly is done off-site in controlled factory environments. This approach not only reduces the amount of land required but also minimizes the environmental disruption typically associated with large-scale construction projects. By shortening the overall construction timeline, SMRs also significantly decrease the duration during which construction-related emissions—such as dust, noise, and vehicle exhaust—impact the surrounding environment. This streamlined process helps limit the carbon footprint and aligns with global sustainability goals.

Beyond the immediate environmental benefits, SMRs offer a more socially conscious approach to infrastructure development, particularly in how they engage with local communities. The modular construction method requires fewer workers on-site and results in shorter, less disruptive construction phases. Unlike conventional nuclear plants, which can take a decade or more to build, SMRs can be constructed in just a few years. This reduction in construction time translates to less inconvenience for nearby residents, as the noise, traffic, and dust associated with traditional construction are significantly curtailed. The decreased on-site workforce also means less strain on local resources, such as housing, transportation, and public services, which can be heavily impacted during the construction of large-scale projects. This reduction in social disruption is especially important in densely populated or environmentally sensitive areas where large infrastructure projects might otherwise face resistance.

Community engagement is another area where SMRs shine. The reduced on-site construction activity not only lessens the environmental impact but also eases social concerns related to large industrial projects. With fewer workers and a shorter timeline, there is a lower risk of long-term disruption to local communities. This can make SMR projects more acceptable to residents and local stakeholders, fostering a more cooperative relationship between developers and the community. Additionally, the modular approach allows project planners to better communicate timelines and potential impacts to the public, ensuring transparency and building trust. By engaging communities early in the planning process and demonstrating the reduced impact of SMRs compared to traditional plants, developers can mitigate opposition and garner support more effectively.

Overall, the environmental and social benefits of SMRs make them a more sustainable and community-friendly option for expanding nuclear energy capacity. By limiting land use, reducing emissions during construction, and minimizing social disruption, SMRs present a modern, responsible approach to energy infrastructure. As countries around the world strive to balance the need for clean, reliable energy with environmental preservation and social acceptance, the advantages of SMRs position them as a key player in the transition to a low-carbon future.

Engineering Challenges and Solutions

The development of nuclear reactor technologies has evolved through multiple generations, each reflecting improvements in design, safety, and efficiency. The first generation, known as Generation I, focused primarily on prototypes and pilot reactors, aiming to establish the feasibility of nuclear power as a reliable energy source. These early reactors were largely experimental, serving as a testing ground for the engineering processes that would eventually lead to commercial power generation [132].

Generation II reactors form the backbone of the current global nuclear fleet. These reactors were designed for commercial-scale power generation with large capacities to provide stable baseload electricity. Countries standardized Generation II designs to reduce costs and streamline engineering processes, allowing them to produce electricity more economically. Despite their efficiency, the need for enhanced safety became evident following severe nuclear accidents, such as the incidents at Three Mile Island and Chernobyl [132].

In response, Generation III reactors were developed, integrating lessons learned from previous accidents. These reactors focused heavily on safety improvements, with both active and passive safety features to prevent core damage and mitigate risks. Most Generation III reactors are light-water reactors (LWRs), such as pressurized water reactors (PWRs) and boiling water reactors (BWRs). An advancement of this category, Generation III+ reactors, incorporates even more robust passive safety systems to enhance reliability further. Examples of these reactors include the Korean APR1400, Westinghouse's AP1000, and the European Pressurized Reactor (EPR). These designs are currently in operation or under construction globally, providing safer and more efficient power [132].

Generation IV technologies represent the next step in nuclear innovation, with a focus on sustainability, safety, and cost-effectiveness. These reactors move beyond light-water designs, utilizing coolants like liquid metal, gas, and molten salt to achieve higher operating temperatures and efficiency. By departing from traditional water-cooling methods, Generation IV reactors are designed to be more efficient and flexible, potentially enabling nuclear energy to play a larger role in a low-carbon future [132].

Interestingly, Small Modular Reactors (SMRs) do not fit neatly into the generational classification. SMRs are not a specific type of reactor but rather a new category of nuclear technology focusing on smaller, modular units that can be manufactured in a factory and assembled on-site. These reactors can draw from either Generation III or Generation IV technologies, depending on the specific design. The International Atomic Energy Agency (IAEA) defines SMRs as reactors with an electrical output of less than 300 megawatts electric (MWe). SMRs can be categorized into two main types: LWR-based designs, which utilize well-established technologies, and non-LWR designs that explore innovative cooling methods, such as liquid metal, gas, or molten salt coolants [132].

Several examples of LWR-based SMRs are currently being developed. The Korean SMART reactor was among the first SMRs to receive design certification, integrating compact design features to eliminate large coolant pipes, thus reducing the risk of major coolant loss accidents. The U.S.-based NuScale Power has developed a modular PWR design with passive safety systems that allow it to operate without pumps, reducing system complexity and increasing safety. NuScale's design is particularly notable for its ability to house up to 12 modules in a single plant, each capable of operating independently. This approach reduces operational costs by allowing a single control room to manage multiple modules [132].

China's ACP100, developed by the China National Nuclear Corporation (CNNC), exemplifies another integral-type PWR SMR. It focuses on incorporating passive safety systems to enhance reliability. The construction of the world's first commercial ACP100 demonstration reactor began in Hainan Province, marking significant progress in SMR deployment. The European Union has also joined the SMR race with the development of NUWARD by a consortium led by Électricité de France (EDF). NUWARD aims to leverage existing PWR technology to provide a flexible and scalable power solution with a focus on streamlined regulatory approval across multiple countries [132].

The i-SMR from South Korea represents a new wave of SMRs that aim to be highly efficient, cost-effective, and versatile. This design includes a boron-free core, which eliminates complex chemical systems, reducing maintenance needs and waste production. The i-SMR's modular design integrates major components into a single unit, thereby simplifying the reactor's structure and eliminating large coolant pipes that could lead to accidents. This design approach not only reduces costs but also enhances safety by integrating passive cooling systems that do not require external power sources [132].

In addition to LWR-based designs, non-LWR SMRs are being developed to explore the potential of alternative coolants. These include high-temperature gas-cooled reactors (HTGRs), molten salt reactors (MSRs), and sodium-cooled fast reactors (SFRs). For instance, TerraPower's Natrium reactor combines sodium cooling with molten salt energy storage, offering both base-load and flexible power output. The ARC-100, another sodium-cooled fast reactor, aims to recycle nuclear fuel, extending its usability and reducing waste. The Aurora reactor by Oklo is a micro-reactor designed for off-grid applications, using heat pipes and metal fuel to achieve compact, efficient power generation [132].

The use of molten salt as a coolant is being explored in designs like the Molten Chloride Fast Reactor (MCFR) by TerraPower and the KP-FHR by Kairos Power, which utilizes fluoride salt coolant to achieve high operating temperatures. These reactors offer the potential for high thermal efficiency and process heat applications, such as hydrogen production and desalination. The BREST-OD-300 in Russia focuses on lead-cooled technology, aiming to demonstrate a closed nuclear fuel cycle that reduces waste and increases sustainability [132].

Overall, the development of SMRs represents a paradigm shift in nuclear power. By focusing on modular construction, factory manufacturing, and innovative cooling technologies, SMRs offer a promising pathway to more flexible, safer, and cost-effective nuclear energy. These reactors are positioned to complement renewable energy sources, provide power in remote areas, and contribute to global decarbonization efforts. However, challenges remain in terms of regulatory approvals, construction costs, and market acceptance. As SMR technology continues to evolve, it holds the potential to revolutionize the nuclear industry by making nuclear power more adaptable, scalable, and economically viable in the 21st century [132].

The construction of Small Modular Reactors presents a unique set of engineering challenges and potential solutions that are important to the advancement of nuclear energy technology. SMRs are designed to address several issues associated with traditional large nuclear reactors, including high capital costs, lengthy construction times, and safety concerns. This synthesis explores the primary engineering challenges in SMR construction and the innovative solutions being proposed to overcome them.

One of the foremost challenges in SMR construction is the high capital cost associated with nuclear power plants. Although SMRs are designed to be less expensive than their larger counterparts, the loss of economies of scale can lead to increased specific capital costs [134]. The construction of multiple units is often necessary to achieve cost competitiveness, which can complicate financing and project management [67]. To mitigate these costs, proponents suggest a modular construction approach, where reactor components are manufactured off-site and transported to the installation location. This method not only reduces on-site construction time but also allows for standardized designs that can be replicated across different sites, thereby enhancing cost efficiency [11, 135].

Another significant challenge is the regulatory and licensing framework that has historically been tailored for large reactors. The licensing process for SMRs is often cumbersome and slow, hindering the deployment of multiple identical units [67]. To address this, there is a growing call for regulatory reform that accommodates the unique characteristics of SMRs, such as their modularity and passive safety features. Streamlining the licensing process could facilitate faster approvals and encourage investment in SMR technology [67].

Safety is a paramount concern in nuclear engineering, and while SMRs are designed with enhanced safety features, ensuring their reliability during operation remains a challenge. The inherent safety mechanisms of SMRs, such as passive cooling systems, are intended to minimize the risk of accidents [16]. However, the engineering community must continue to validate these safety systems through rigorous testing and simulation to ensure they perform as intended under various operational scenarios [16, 136]. Additionally, the integration of advanced monitoring technologies can enhance operational safety by providing real-time data on reactor performance, which is crucial for early detection of potential issues [136].

The construction timeline for SMRs is another critical factor that influences their viability. Traditional nuclear plants often face significant delays due to complex construction processes and project management challenges [137]. SMRs, by contrast, are designed to have shorter construction periods due to their modular nature and factory fabrication [135, 137]. However, achieving these timelines requires careful planning and execution, as well as a skilled workforce familiar with modular construction techniques [138]. Implementing advanced project management methodologies and leveraging digital tools for construction scheduling can help streamline the building process and reduce delays [138].

Overcoming Key Challenges for SMRs

Reducing Source Term: While SMRs inherently contain a smaller quantity of radioactive material than large nuclear reactors, further reducing the source term is crucial for minimizing the Emergency Planning Zone (EPZ) to the power plant site. This reduction is essential to ensure that SMRs can be sited closer to populated areas, making them more practical for distributed energy generation. One approach to achieve this is the adoption of advanced passive safety systems. For instance, Light Water Reactor (LWR) SMRs are being designed with robust steel containment vessels capable of withstanding significantly higher pressures than the conventional concrete structures used in large reactors. This enhances the containment's ability to prevent radioactive leaks during accidents. Non-LWR SMRs, such as those utilizing liquid metal or molten salt coolants, operate at atmospheric or low pressures, significantly reducing the likelihood of high-pressure failures and, consequently, minimizing the potential release of radioactive materials [132].

For gas-cooled SMRs, nuclear fuel designs like TRISO (Tristructural Isotropic) fuel particles are utilized, which incorporate multiple protective coatings around the nuclear fuel. These coatings act as mini containment barriers, preventing the release of radioactive isotopes even if the reactor's primary containment is compromised. By combining robust physical containment with inherently safe reactor designs, the source term can be drastically reduced. Demonstrating this reduced source term is vital for gaining regulatory approval to site SMRs in densely populated areas, thus expanding their potential applications beyond isolated regions [132].

Reducing Manufacturing and Construction Costs: A significant challenge for SMRs is avoiding the traditional "economy of scale" constraints that have historically driven the cost-effectiveness of large nuclear reactors. To make SMRs economically viable, their manufacturing and construction processes must differ fundamentally from those of large reactors. This involves minimizing the use of heavy equipment and reducing the amount of on-site construction work. Manufacturing critical components in controlled factory settings ensures consistent quality and reduces lead times. By leveraging modular construction, the need for specialized on-site labour and heavy machinery is drastically reduced, translating into significant cost savings [132].

Furthermore, SMRs must be designed with site-agnostic features to allow for widespread deployment without extensive site-specific modifications. Recent advancements in manufacturing, such as automation and robotics, can further accelerate production timelines and reduce costs. Another innovative approach to reduce capital expenses is to repurpose existing infrastructure, such as retired coal or gas power plants, which can lower SMR construction costs by 15-35%. By reusing elements like cooling towers, grid connections, and support buildings, developers can achieve faster project completion with less financial risk [132].

Reducing the Need for Operation and Maintenance Personnel: For SMRs to be truly cost-effective, they must operate with fewer personnel than large nuclear plants. Traditionally, nuclear reactors require significant staffing for operations and maintenance (O&M), but SMRs cannot afford this overhead if they are to achieve economic viability. Therefore, SMRs are being designed to rely on advanced automation, artificial intelligence (AI), and digital twin technologies to minimize human intervention. Autonomous control systems are being developed to monitor and manage reactor operations with minimal oversight. This requires both hardware and software innovations: simpler reactor designs reduce the variables that need to be controlled, while AI-driven software assists operators in managing routine tasks and responding to anomalies [132].

Digital twins—virtual replicas of the physical reactor—can simulate performance in real-time, predicting wear and tear on components to optimize maintenance schedules. This predictive maintenance approach can extend the operational lifespan of reactor parts, reduce downtime, and improve overall efficiency. By leveraging digital technologies, SMRs can reduce staffing requirements without compromising safety, thus lowering operational costs [132].

Reducing Uncertainties in Licensing: One of the biggest hurdles for the widespread adoption of SMRs is the uncertainty and complexity of the current licensing process. Traditional nuclear reactors are subject to stringent regulations that assume large-scale, high-risk operations. However, SMRs, by design, have lower risk profiles due to their smaller size and enhanced safety features. Therefore, applying the same conservative regulatory frameworks used for large reactors to SMRs can be counterproductive. To overcome this, a shift towards performance-based and risk-informed regulatory frameworks is needed. Such frameworks would focus on the actual risk posed by SMRs rather than using blanket regulations meant for larger reactors [132].

Developing a global, standardized approach to SMR licensing could streamline the approval process, reducing costs and speeding up deployment. Collaborative efforts between regulatory bodies and SMR developers are essential to establish clear guidelines that reflect the unique safety characteristics of SMRs. A standardized international licensing process would also facilitate the export of SMR technology, enabling broader adoption across multiple countries and reducing the time and cost associated with site-specific reviews [132].

Diversifying the Application Areas of SMRs: Beyond electricity generation, SMRs have the potential to decarbonize other sectors, such as industrial heat, desalination, and even transportation. Unlike traditional nuclear reactors, which are often constrained to large-scale electricity production, the flexibility of SMRs makes them ideal for various applications. For instance, high-temperature gas-cooled SMRs can provide process heat for industrial operations, reducing reliance on fossil fuels. SMRs can also be used in maritime transport, offering an alternative to diesel engines for ships, thereby contributing to decarbonization in this sector [132].

Recent collaborations between SMR developers and industries such as chemical processing and shipbuilding demonstrate the potential for expanding nuclear energy's role beyond the power sector. By utilizing nuclear reactors for district heating, hydrogen production, and desalination, SMRs can contribute to global decarbonization efforts while also enhancing their market viability. This diversification not only helps reduce greenhouse gas emissions but also strengthens the economic case for SMRs, making them more attractive to investors and policymakers [132].

To successfully deploy SMRs on a large scale, developers must overcome challenges related to safety, cost, staffing, licensing, and market diversification. By adopting advanced technologies like passive safety systems, modular construction, AI-driven automation, and digital twins, SMRs can achieve the necessary cost reductions and safety improvements. Additionally, streamlining the regulatory process and expanding the use cases for SMRs will be essential to realizing their full potential in the transition to a low-carbon future [132].

Key Components and System Integrations

At the heart of SMR power plants are several critical components and systems that are carefully integrated to ensure efficient, safe, and reliable operation. In examining the key components and system integrations of Small Modular Reactors (SMRs), it is essential to understand the intricate design and operational features that distinguish these reactors from traditional nuclear power plants. The following sections detail the primary components, their construction, and integration into the overall system.

Reactor Core and Pressure Vessel

The reactor core of an SMR is typically a compact assembly of fuel rods, often utilizing low-enriched uranium or alternative fuels such as thorium or mixed oxide (MOX). These fuel rods are encased in zirconium alloy cladding, which serves to contain radioactive fission products and enhance safety [123]. The core is housed within a robust pressure vessel constructed from high-strength steel, designed to withstand extreme operational pressures and temperatures. Many SMRs integrate the reactor core and primary coolant system within a single pressure vessel, which simplifies the design by reducing the complexity associated with external piping and connections [5].

The construction of the reactor pressure vessel is executed in a controlled factory environment, allowing for precise fabrication and rigorous quality control. This factory-based approach facilitates extensive testing before the vessel is transported to the plant site, thereby minimizing the risk of defects and enhancing the overall safety of the reactor [5]. The compact, integral design of SMRs not only reduces the likelihood of leaks but also streamlines the on-site assembly process, allowing for installation with minimal field welding [123].

Primary Coolant System and Heat Exchangers

The primary coolant system plays a critical role in transferring heat generated in the reactor core to the secondary system, where it is converted into steam for electricity generation. In water-cooled SMRs, pressurized water is typically used as the primary coolant, while advanced designs may utilize alternative coolants such as liquid sodium or helium to achieve higher operational temperatures [5]. Heat exchangers are integral to this system, facilitating efficient heat transfer to the secondary loop. These components are often prefabricated off-site, which allows for quicker installation and reduces on-site construction time [5].

In designs utilizing natural circulation, the primary coolant system can operate without pumps, relying instead on temperature-driven density differences to circulate the coolant. This passive circulation method not only simplifies the system but also enhances safety by reducing the risk of mechanical failure associated with pumps [5, 7].

Secondary Loop and Turbine-Generator System

The secondary cooling loop receives heat from the primary system via heat exchangers and converts it into steam to drive a turbine-generator set, producing electricity. The turbine-generator units in SMRs are smaller and more compact than those in traditional reactors, which contributes to a reduced overall footprint [7]. The components of the secondary loop, including steam generators and turbines, are typically prefabricated as standardized modules, allowing for efficient construction and testing before installation [7].

This modular approach not only streamlines the installation process but also ensures that all components are fully operational prior to being integrated on-site, thereby minimizing commissioning activities [7].

Containment Structure

SMRs utilize compact containment structures made from reinforced concrete and steel to provide a robust barrier against radiation release. Many designs feature below-ground containment to enhance protection against external threats such as seismic events and flooding [7]. The containment structure often incorporates passive cooling systems that can dissipate heat without relying on external power sources, further enhancing safety [7].

The prefabrication of containment modules in large sections allows for rapid assembly on-site, significantly reducing construction time compared to traditional reactors [7].

Control and Instrumentation Systems

Advanced digital control systems are integral to SMR operations, managing reactor functions and ensuring safety through real-time monitoring of critical parameters such as temperature and pressure [7]. These systems are developed and tested off-site, ensuring that they are fully operational before installation, which reduces the need for extensive commissioning activities [7]. The integration of cybersecurity measures is also critical to protect against potential digital threats, ensuring secure communication and control [7].

Passive Safety Systems and Emergency Core Cooling

A defining feature of SMRs is their reliance on passive safety systems, which can function without external power or mechanical intervention. Emergency Core Cooling Systems (ECCS) are designed to activate automatically in the event of a Loss of Coolant Accident (LOCA), utilizing gravity-fed water tanks and heat exchangers to remove decay heat from the reactor core [7]. The prefabrication of these systems enhances reliability by reducing the complexity and potential failure points associated with active systems [7].

Spent Fuel Storage and Waste Management

SMRs are designed to produce less radioactive waste due to higher fuel efficiency and longer burn-up cycles. Spent fuel is initially stored in on-site pools for cooling before being transferred to dry cask storage for long-term management [7]. The modular construction of spent fuel storage facilities allows for quick installation and future expansion as needed, ensuring safe management of radioactive waste throughout the operational life of the plant [7].

System Integration and Final Assembly

The integration of all components into a cohesive system is a hallmark of the modular construction approach. Each module is designed to interface seamlessly with others through standardized connections, facilitating a streamlined assembly process [7]. This modularity can significantly reduce the overall build time for SMRs, potentially shortening the construction period from the typical 8-10 years for large reactors to as little as 2-3 years [7].

Chapter 4
Regulatory and Licensing Framework for SMRs

Understanding International and National Nuclear Regulations

International Nuclear Regulations: Overview and Importance

International nuclear regulations are a complex and crucial framework designed to ensure the safe, secure, and peaceful use of nuclear energy. These regulations aim to protect public health, the environment, and national security while promoting the responsible use of nuclear technology for energy, medicine, and scientific research. Given the potential risks associated with nuclear reactors, radioactive materials, and nuclear proliferation, these regulations are enforced by international organizations and national regulatory bodies to standardize practices across borders.

The international nuclear regulatory framework primarily revolves around principles and guidelines established by key organizations such as the International Atomic Energy Agency (IAEA), World Nuclear Association (WNA), Nuclear Energy Agency (NEA) under the Organisation for Economic Co-operation and Development (OECD), and treaties like the Nuclear Non-Proliferation Treaty (NPT). These bodies and agreements ensure that nuclear activities are conducted in a manner that minimizes risks and enhances global cooperation.

International Atomic Energy Agency (IAEA)

The International Atomic Energy Agency (IAEA), established in 1957, is a pivotal entity in the realm of nuclear governance, focusing on the promotion of peaceful nuclear energy use, the establishment of safety standards, and the prevention of nuclear proliferation. The agency's multifaceted mandate is critical in shaping international nuclear policy and ensuring compliance among its member states.

The IAEA plays an important role in formulating safety standards that member states are encouraged to adopt. These standards encompass a wide range of areas, including radiation protection, nuclear waste management, reactor safety, and emergency preparedness. Although these guidelines are not legally binding, they serve as essential benchmarks for national regulations, significantly influencing how countries manage their nuclear programs [139]. For instance, IAEA standards are utilized in the absence of national regulations, underscoring their importance in operational management and safety protocols for nuclear reactors [139]. Furthermore, adherence to IAEA guidelines is critical for minimizing environmental and health risks associated with radioactive waste management [140, 141].

In terms of safeguards and verification, the IAEA is instrumental in ensuring compliance with the Nuclear Non-Proliferation Treaty (NPT). The agency conducts inspections to verify that nuclear materials are not diverted for military purposes, requiring member states to declare their nuclear facilities and materials [142]. This verification process is essential for maintaining international trust and security regarding nuclear capabilities. The legitimacy of the IAEA's safeguards is derived from its technical authority and the rigorous methodologies employed during inspections, which are crucial for upholding global non-proliferation norms [142, 143]. The agency's role in this area has been highlighted in various studies, emphasizing its importance in the global nuclear governance framework [144].

Figure 32: International Training Course on Developing Regulations and Associated Administrative Measures for Nuclear Security. Twenty participants from Member States attended the International Training Course on Developing Regulations and Associated Administrative Measures for Nuclear Security, held at the Agency headquarters in Vienna, Austria. 13 May 2022. IAEA Imagebank, CC BY 2.0, via Flickr.

The IAEA also facilitates technical cooperation, particularly in developing countries, to promote the safe and peaceful use of nuclear energy. This includes capacity-building initiatives that enhance the capabilities of member states to utilize nuclear technology effectively and safely [145, 146]. For example, the IAEA has been involved in supporting nuclear cardiology practices in Latin America and the Caribbean, demonstrating its commitment to addressing health needs through nuclear techniques [146]. Additionally, the agency's International Project on Innovative Nuclear Reactors and Fuel Cycles (INPRO) aims to foster collaboration among member states to explore sustainable nuclear energy solutions, further illustrating its role in promoting technological advancement and cooperation [147, 148].

The Nuclear Non-Proliferation Treaty (NPT)

The Nuclear Non-Proliferation Treaty (NPT), which entered into force in 1970, serves as a fundamental framework for global nuclear governance, aiming to prevent the spread of nuclear weapons and promote peaceful uses of nuclear energy. The treaty is built upon three essential pillars: non-proliferation, disarmament, and the peaceful use of nuclear energy.

The non-proliferation aspect of the NPT stipulates that non-nuclear weapon states commit to refraining from acquiring nuclear weapons, while nuclear-armed states agree not to transfer nuclear weapons or related technology to these states. This principle is crucial in maintaining a balance in international security and preventing the escalation of nuclear arms races [149, 150]. The International Atomic Energy Agency (IAEA) plays a pivotal role in this regard, implementing safeguards to verify compliance with the treaty. These safeguards include inspections and material accounting to ensure that nuclear materials are not diverted for weapons use [142, 151].

Disarmament is another critical pillar of the NPT, which encourages nuclear-armed states to pursue negotiations aimed at nuclear disarmament. Despite the treaty's intentions, progress in disarmament has been slow, leading to frustrations among non-nuclear states and contributing to the emergence of alternative frameworks, such as the Treaty on the Prohibition of Nuclear Weapons (TPNW) [149, 152]. The lack of significant disarmament measures has raised concerns about the treaty's effectiveness and has prompted discussions on the future of nuclear governance [149, 153].

The NPT also recognizes the right of all signatories to develop nuclear energy for peaceful purposes, provided they adhere to IAEA safeguards. This aspect is vital for promoting the use of nuclear technology in energy production while ensuring that such developments do not contribute to proliferation risks [143]. The IAEA's safeguards system is designed to monitor and verify that nuclear materials are used solely for peaceful applications, thus reinforcing the treaty's objectives [142, 151].

World Nuclear Association (WNA) and Nuclear Energy Agency (NEA)

The World Nuclear Association (WNA) and the Nuclear Energy Agency (NEA) serve pivotal roles in the nuclear energy landscape, complementing the regulatory oversight provided by the International Atomic Energy Agency (IAEA). The WNA focuses on collaboration among industry stakeholders to promote best practices and advocate for the sustainable use of nuclear energy. This includes facilitating discussions on regulatory challenges, technological advancements, and public acceptance of nuclear power, which are essential for the industry's growth and acceptance [154].

The NEA, as a specialized agency within the Organisation for Economic Co-operation and Development (OECD), assists its member countries in maintaining high safety standards and enhancing the economic and environmental performance of nuclear energy. The NEA's initiatives, such as the International Criticality Safety Benchmark Evaluation Project (ICSBEP) and the International Reactor Physics Experiment Evaluation Project (IRPhEP), exemplify its commitment to providing reliable data and benchmarks that support safety and operational efficiency in nuclear energy [154-156]. These projects have established comprehensive databases that are crucial for the validation of nuclear data and the development of advanced reactor technologies [154, 155].

Moreover, the NEA's work extends to the development of tools and systems that facilitate the verification and validation of nuclear data, which is vital for ensuring the safety and reliability of nuclear operations [155]. The collaboration between the WNA and NEA fosters an environment where best practices can be shared and implemented, ultimately contributing to the safe and sustainable advancement of nuclear energy [154].

Key International Treaties and Conventions

In addition to the NPT, several other international treaties and conventions govern nuclear activities.

Convention on Nuclear Safety (CNS): The Convention on Nuclear Safety (CNS) was adopted in 1994 in response to the growing concerns over the safety of nuclear power plants following several high-profile incidents, including the Chernobyl disaster of 1986. The CNS is designed to enhance the safety of nuclear reactors worldwide by creating a legally binding framework that obligates signatory countries to regularly review and improve their nuclear safety measures. This convention requires member states to establish and maintain robust regulatory frameworks, ensuring that nuclear power plants operate in accordance with stringent safety standards. It emphasizes transparency and cooperation by requiring countries to submit reports on their nuclear safety practices for peer review by other signatories. The CNS promotes international collaboration to share best practices, improve safety culture, and minimize the risk of accidents. This peer review process, conducted in meetings every three years, helps identify gaps in safety practices and encourages continuous improvements across the global nuclear industry.

Joint Convention on the Safety of Spent Fuel Management and the Safety of Radioactive Waste Management: Adopted in 1997, the Joint Convention on the Safety of Spent Fuel Management and the Safety of Radioactive Waste Management is the first legally binding international treaty to address the challenges of managing spent nuclear fuel and radioactive waste. This convention focuses on ensuring that radioactive waste is handled, stored, and disposed of in ways that protect human health and the environment. The convention covers all aspects of radioactive waste management, including transport, storage, treatment, and disposal. It obliges signatory countries to implement appropriate measures to prevent accidents and to minimize the release of radioactive substances during waste handling. Additionally, the convention promotes transparency by requiring member states to report on their waste management practices and undergo peer reviews. This ensures that countries adhere to high safety standards, share information, and collaborate on best practices for the safe disposal of nuclear waste. The convention is critical as it addresses long-term environmental and safety concerns associated with the growing amounts of radioactive waste generated by nuclear power plants.

Convention on Early Notification of a Nuclear Accident and Convention on Assistance in the Case of a Nuclear Accident or Radiological Emergency: In the wake of the Chernobyl disaster, the international community recognized the need for timely communication and mutual support in the event of nuclear accidents. The Convention on Early Notification of a Nuclear Accident and the Convention on Assistance in the Case of a Nuclear Accident or Radiological Emergency were both adopted in 1986 to address this need. The Early Notification Convention requires countries to promptly inform the International Atomic Energy Agency (IAEA) and neighbouring states of any nuclear incident that could result in radioactive releases beyond their borders. This early warning system helps mitigate the impact of nuclear accidents by enabling affected countries to take protective measures swiftly.

Meanwhile, the Assistance Convention establishes a framework for international cooperation in providing assistance during nuclear emergencies. Under this agreement, countries can request and offer support, including technical expertise, equipment, and resources, to manage nuclear incidents and mitigate their consequences. These conventions are vital for improving global preparedness and response capabilities, ensuring that countries can respond effectively to nuclear emergencies, thus reducing the potential harm to public health and the environment.

Comprehensive Nuclear-Test-Ban Treaty (CTBT): The Comprehensive Nuclear-Test-Ban Treaty (CTBT) is an ambitious international agreement aimed at banning all nuclear explosions, whether for military or civilian purposes. Although the treaty was opened for signature in 1996, it has yet to enter into force because several key countries, including the United States, China, and India, have not ratified it. The CTBT is significant because it seeks to curb the development of new nuclear weapons and the proliferation of existing nuclear arsenals. By banning nuclear tests, the treaty aims to halt the qualitative improvement of nuclear weapons, thus contributing to global disarmament efforts.

To verify compliance, the CTBT includes a robust verification regime with a global network of 321 monitoring stations and 16 laboratories that detect seismic, hydroacoustic, infrasound, and radionuclide signals indicative of nuclear explosions. These monitoring stations are strategically located around the world to ensure that any nuclear test can be detected with high accuracy, even if it is conducted underground or in remote areas. The treaty also allows for on-site inspections to investigate suspicious activities, thereby enhancing its enforcement capabilities. Although the CTBT is not yet legally binding, it has had a significant impact, with many countries adhering to a de facto moratorium on nuclear testing in anticipation of the treaty's full ratification. The CTBT is seen as a crucial step toward nuclear disarmament, contributing to the long-term goal of a world free of nuclear weapons.

Challenges in International Nuclear Regulation

The challenges in international nuclear regulation are multifaceted and stem from various factors that impact global nuclear safety and security. These challenges can be categorized into four main areas: varying national standards, the emergence of new technologies, nuclear proliferation risks, and public acceptance.

Varying National Standards: The International Atomic Energy Agency (IAEA) establishes baseline safety standards for nuclear operations; however, individual countries often implement additional regulations that can differ significantly from these international guidelines. This discrepancy complicates international collaboration and technology transfer, as nations may have unique regulatory frameworks that do not align with IAEA standards [157]. The IAEA has developed extensive guidelines for nations aspiring to nuclear development, emphasizing the need for adherence to high international standards to mitigate risks associated with nuclear accidents and proliferation [157]. The lack of uniformity in regulatory practices can lead to challenges in ensuring consistent safety measures across borders, thereby undermining global nuclear governance [139].

New Technologies: The rise of advanced nuclear technologies, such as small modular reactors (SMRs), fast reactors, and molten salt reactors, presents significant challenges to existing regulatory frameworks. These technologies often require updated guidelines and streamlined licensing processes to foster innovation while maintaining safety [158]. The current regulatory landscape may not adequately address the unique safety and operational characteristics of these new technologies, necessitating a re-evaluation of existing regulations to ensure they are fit for purpose in the context of modern nuclear advancements [159]. As highlighted in the literature, the precautionary principle has not been effectively integrated into nuclear safety regulation, which poses additional challenges in adapting to new technological developments [158].

Nuclear Proliferation Risks: The potential for nuclear technology to be diverted for weapons development remains a critical concern in international nuclear regulation. The emergence of advanced fuel cycles and enrichment processes heightens the need for stringent safeguards to prevent proliferation [157]. The IAEA plays a crucial role in establishing safeguards and promoting compliance among member states, but the effectiveness of these measures can be hindered by varying national interests and regulatory approaches [139]. The complexity of ensuring that nuclear materials are not misused for military purposes underscores the necessity for robust international cooperation and adherence to established non-proliferation treaties [160].

Public Acceptance: Building public trust in nuclear energy is essential for the continued development and acceptance of nuclear power. High-profile incidents, such as the Fukushima and Chernobyl disasters, have significantly influenced public perceptions and acceptance of nuclear energy [161]. Research indicates that public attitudes toward nuclear power are shaped more by value predispositions than by scientific knowledge, suggesting that regulatory bodies must engage in transparent communication and proactive safety measures to address public concerns effectively [162]. The establishment of a transparent regulatory process is vital for fostering public confidence in nuclear safety and ensuring that communities feel secure about the operations of nuclear facilities [161].

International Regulations Specific to Small Modular Reactors (SMRs)

While existing international nuclear regulations primarily address large nuclear reactors, the growing interest in Small Modular Reactors (SMRs) has prompted regulatory bodies to adapt and develop specific frameworks tailored to these advanced technologies. The unique characteristics of SMRs, such as their modular construction, enhanced safety features, and potential deployment in diverse environments, necessitate adjustments to the regulatory landscape. Below is a detailed explanation of the key international regulations and guidelines specifically relevant to SMRs:

International Atomic Energy Agency (IAEA) Guidance on SMRs: The International Atomic Energy Agency (IAEA) plays a role in developing safety standards and regulatory guidance for nuclear technologies, including SMRs. The IAEA has been actively engaged in creating specific guidelines to address the safety, licensing, and operation of SMRs, recognizing that traditional regulatory frameworks designed for large reactors may not fully apply to these smaller, modular systems.

The IAEA's Specific Safety Guides (SSG) have been expanded to incorporate SMRs, focusing on their passive safety features and novel design approaches. For instance, SSG-12, which addresses reactor siting, has been updated to account for the reduced emergency planning zones (EPZs) associated with SMRs due to their inherently safer designs. Additionally, the IAEA's TECDOC series provides guidance on areas such as design safety assessment, accident management, and licensing procedures tailored to the unique attributes of SMRs.

The IAEA also facilitates international collaboration through Coordinated Research Projects (CRPs), where member states can share experiences and best practices related to SMR safety, licensing, and deployment. These projects aim to harmonize regulatory frameworks across countries, enabling the safe and efficient deployment of SMRs globally.

Convention on Nuclear Safety (CNS): The Convention on Nuclear Safety (CNS) is a legally binding international treaty that focuses on improving the safety of nuclear power plants. While the CNS was originally designed with large reactors in mind, its principles apply to SMRs as well. The convention requires member states to establish regulatory frameworks that ensure the safety of all nuclear installations, including SMRs.

Under the CNS, countries are obligated to submit reports on their nuclear safety measures, which are then reviewed by other signatories. As SMRs are deployed, the review process will increasingly include discussions on how new reactor designs meet safety objectives, particularly regarding passive safety features and modular construction. The CNS also emphasizes transparency and peer review, which are essential for building confidence in the safety of SMR technologies.

Harmonization of Licensing and Safety Standards: One of the primary challenges in the deployment of SMRs is the lack of harmonized international licensing standards. SMRs are designed to be mass-produced and potentially deployed across multiple countries, which means that inconsistent regulatory requirements can significantly delay their deployment and increase costs. In response, the IAEA, the Nuclear Energy Agency (NEA), and other international bodies are working to harmonize safety standards for SMRs.

The IAEA has proposed adopting a technology-neutral, performance-based approach for SMR licensing, which focuses on safety outcomes rather than prescriptive design requirements. This approach allows for greater flexibility in accommodating innovative SMR designs that differ from traditional reactors. Efforts are also underway to develop a common set of safety standards that can be recognized across multiple jurisdictions, reducing the need for redundant licensing processes.

The Multinational Design Evaluation Programme (MDEP) is another initiative aimed at harmonizing safety evaluations for SMRs among participating countries. MDEP facilitates the exchange of technical information between nuclear regulators and promotes cooperation in the licensing of new reactor designs, including SMRs.

Emergency Planning and Reduced EPZs: Traditional nuclear reactors require extensive emergency planning zones (EPZs) due to the potential for significant radiological releases in the event of an accident. However, SMRs, with their enhanced safety features and passive cooling systems, are designed to reduce the likelihood of severe accidents and limit the potential release of radioactive materials.

Recognizing this, the IAEA and national regulators are considering adjustments to EPZ requirements for SMRs. For example, the NuScale Power Module, which received design certification from the U.S. Nuclear Regulatory Commission (NRC), demonstrated that its safety features could justify a significantly reduced EPZ, limited to the plant boundary. The IAEA is working to develop guidelines that balance safety with the practical benefits of reduced EPZs, which would allow SMRs to be deployed closer to populated areas, industrial sites, or remote communities in need of reliable power.

International Efforts on Standardized Design Approval: Given the modular nature of SMRs, which are designed to be manufactured in factories and assembled on-site, there is a strong emphasis on achieving standardized design approvals that can be recognized across multiple countries. The goal is to facilitate the deployment of SMRs by allowing designs that have been approved in one jurisdiction to be accepted in others, thereby reducing the time and cost associated with gaining regulatory approval in each new country.

Organizations like the IAEA and the NEA are working with national regulators to develop mutual recognition agreements and align safety assessment methodologies. This would enable SMR vendors to certify their designs more efficiently and deploy reactors internationally without facing significant regulatory delays.

Waste Management and Decommissioning: The Joint Convention on the Safety of Spent Fuel Management and the Safety of Radioactive Waste Management applies to SMRs as well as large reactors. SMRs, particularly those using advanced coolants like molten salt or liquid metal, may generate different types of waste compared to conventional reactors. Therefore, the IAEA and other international bodies are actively developing guidelines to address the management of spent fuel and radioactive waste specific to SMRs.

Additionally, SMRs are designed to have longer operational lifespans and simplified decommissioning processes compared to traditional reactors. The IAEA is working to ensure that these innovations in SMR designs are factored into international waste management and decommissioning standards, enabling countries to plan for the entire lifecycle of these reactors, from construction to decommissioning and waste disposal.

Non-Proliferation and Export Control: As with all nuclear technologies, SMRs are subject to the Non-Proliferation Treaty (NPT) and related international agreements to prevent the spread of nuclear weapons. SMRs, especially those designed for export, must comply with strict export control regulations to ensure that sensitive technologies do not fall into the hands of unauthorized entities.

The IAEA is also considering new safeguards and monitoring techniques for SMRs, especially given the possibility of floating nuclear power plants or reactors deployed in remote or offshore locations. These reactors present unique challenges in terms of ensuring that nuclear material is not diverted for unauthorized uses.

While the current international regulatory framework primarily addresses large reactors, significant progress is being made to adapt these frameworks to the unique characteristics of SMRs. The focus on harmonization, performance-based standards, and international collaboration aims to streamline the deployment of SMRs while ensuring the highest levels of safety, security, and non-proliferation.

National Regulatory Bodies and Harmonization

National regulatory bodies play a crucial role in the enforcement of nuclear regulations within their respective jurisdictions, guided by international standards set by organizations such as the International Atomic Energy Agency (IAEA). For instance, the U.S. Nuclear Regulatory Commission (NRC), the French Nuclear Safety Authority (ASN), Japan's Nuclear Regulation Authority (NRA), and the UK Office for Nuclear Regulation (ONR) are prominent examples of such agencies. These bodies not only implement regulations but also tailor them to fit local contexts, ensuring that national priorities and conditions are adequately addressed. The IAEA provides a framework of standards that these national regulators adapt to their specific needs, thereby fostering a regulatory environment that is both compliant with international norms and sensitive to local realities.

The harmonization of regulatory practices is particularly vital in the context of advancing nuclear technology, especially with the emergence of Small Modular Reactors (SMRs) and next-generation reactors. Initiatives like the Multinational Design Evaluation Programme (MDEP) and the SMR Regulators' Forum are pivotal in this regard. These initiatives aim to standardize design approvals and safety evaluations across different countries, thereby facilitating international collaboration and reducing barriers to the deployment of innovative nuclear technologies. The MDEP, for example, promotes a collaborative approach among national regulators to streamline the review processes for new reactor designs, which is essential for the timely and safe introduction of SMRs into the energy market.

Moreover, the role of national regulatory bodies extends beyond mere compliance; they are instrumental in fostering a culture of safety and continuous improvement within the nuclear sector. By engaging in international dialogues and sharing best practices, these agencies contribute to a global regulatory framework that enhances safety standards and operational efficiency. The IAEA's support for the development of national audit programs and methodologies for quality assurance in radiotherapy exemplifies how international cooperation can bolster national capabilities. Such collaborative efforts not only enhance regulatory effectiveness but also ensure that safety measures evolve in line with technological advancements and emerging challenges in the nuclear domain.

National Regulations

The construction and operation of Small Modular Reactors (SMRs) are governed by national regulations specific to each country, reflecting their unique approaches to nuclear safety, licensing, and environmental protection. These regulations must adapt to the distinct features of SMRs, which include modular construction, enhanced safety features, and reduced environmental footprints compared to traditional large reactors. Below, we explore how different countries are addressing the regulatory challenges of SMR deployment and construction.

United States

The U.S. Nuclear Regulatory Commission (NRC) is one of the most established nuclear regulatory bodies globally and has made significant strides in developing frameworks tailored to SMRs. In 2019, the NRC certified the design of the NuScale Power Module, making it the first SMR to receive design certification in the U.S. The NRC's regulatory approach focuses on performance-based and technology-neutral standards, which means that rather than mandating specific design features, the safety outcomes are emphasized. This flexibility allows SMR developers to innovate while meeting rigorous safety standards.

To further streamline the process, the NRC is revising its licensing procedures to accommodate the modular nature of SMRs. The NRC has introduced the Part 52 licensing process, which allows for early site permits, design certifications, and combined construction and operating licenses. This integrated process aims to reduce regulatory timelines and costs for SMRs. The U.S. is also exploring options to adjust emergency planning zones (EPZs) for SMRs, potentially limiting them to the site boundary due to the reactors' enhanced safety features.

Canada

Canada's nuclear regulatory authority, the Canadian Nuclear Safety Commission (CNSC), has been proactive in supporting the deployment of SMRs. The CNSC has established a Vendor Design Review (VDR) process, which allows SMR developers to obtain early feedback on their designs before applying for a construction license. This non-mandatory, pre-licensing process helps developers identify and address potential regulatory challenges early on, reducing project risks.

The CNSC has also focused on public engagement and transparency in the licensing process, especially given that SMRs may be deployed in remote or Indigenous communities. For example, Ontario Power Generation (OPG) is planning to deploy a BWRX-300 SMR at its Darlington site by 2028, and the CNSC is working closely with the developer to ensure compliance with national safety and environmental standards. Canada's approach is also exploring small EPZs to facilitate the construction of SMRs in closer proximity to populated areas.

United Kingdom

The United Kingdom has expressed strong interest in developing SMRs as part of its strategy to decarbonize its energy sector. The Office for Nuclear Regulation (ONR) oversees the licensing and regulation of nuclear facilities, including SMRs. The UK's regulatory approach emphasizes generic design assessment (GDA), which allows the ONR to evaluate SMR designs independently of a specific site. This process helps to streamline approvals for SMR projects by certifying designs before site-specific applications.

The UK government has also provided significant funding to support SMR development, notably the Rolls-Royce SMR project, which aims to have its first SMR operational by the early 2030s. The ONR is working on adapting its regulatory framework to accommodate the modular and flexible nature of SMRs, particularly in terms of factory-built modules and rapid on-site assembly.

European Union

Within the European Union (EU), nuclear power is governed by a combination of national regulations and overarching EU directives. The European Atomic Energy Community (Euratom) sets guidelines for nuclear safety, waste management, and radiation protection. However, individual EU member states retain the authority to regulate their nuclear industries. Countries like France and Finland are actively exploring the deployment of SMRs, with Électricité de France (EDF) leading the development of the NUWARD SMR.

The EU's regulatory framework is focused on harmonizing safety standards across member states, particularly for advanced reactors like SMRs. The challenge remains in aligning national regulatory bodies with the EU's stringent safety and environmental standards, especially for cross-border projects and supply chains. The EU is also exploring how reduced EPZs for SMRs can facilitate their deployment in more densely populated regions.

Russia

Russia is a leader in deploying SMRs, particularly floating nuclear power plants. The Federal Service for Environmental, Technological, and Nuclear Supervision (Rostekhnadzor) regulates the nuclear industry in Russia. The deployment of the KLT-40S SMR on the floating power plant Akademik Lomonosov (see Figure 33) marked a significant milestone in the use of SMRs for remote regions. Russia's regulatory framework focuses on both land-based and maritime applications of SMRs, with additional considerations for the unique challenges of operating reactors on water.

Figure 33: The first Russian floating nuclear power station, Akademik Lomonosov, being transported, 23 August 2019. Elena Dider, CC BY-SA 4.0, via Wikimedia Commons.

Russia's regulations emphasize passive safety systems and modular construction, allowing for rapid deployment in harsh environments such as the Arctic. The country is also developing lead-cooled fast reactors, like the BREST-OD-300, which are subject to regulatory oversight focusing on high-temperature operations and innovative safety features.

China

China is aggressively pursuing SMR technology to meet its energy and environmental goals. The National Nuclear Safety Administration (NNSA) regulates the country's nuclear sector. China is focused on developing both land-based and marine SMRs, such as the ACP100 (Linglong One), which is currently under construction. The NNSA has tailored its regulatory framework to support the rapid development of SMRs, emphasizing standardized designs and streamlined approvals.

China's approach includes accelerating the review process for SMRs to meet its ambitious timeline for nuclear expansion. The NNSA is also exploring export opportunities for Chinese-designed SMRs, particularly in regions where large reactors may not be suitable due to infrastructure limitations.

South Korea

South Korea has been actively developing SMR technology through projects like the SMART reactor and the newly proposed i-SMR. The Korea Institute of Nuclear Safety (KINS) is responsible for regulating the nuclear industry, including SMRs. South Korea's regulatory framework is focused on ensuring that SMRs meet international safety standards, especially given its plans to export these reactors to countries in the Middle East and Southeast Asia.

South Korea emphasizes passive safety systems and boron-free reactor designs in its SMRs to reduce operational complexity and enhance safety. KINS is also working on adapting existing licensing processes to accommodate the rapid deployment and modular construction of SMRs.

Japan

Japan's regulatory body, the Nuclear Regulation Authority (NRA), has been cautious in approving new nuclear projects following the Fukushima disaster. However, with a renewed focus on reducing carbon emissions, Japan is exploring the potential of SMRs. The NRA is developing new guidelines for the deployment of SMRs, with a focus on addressing seismic safety, tsunami protection, and the use of passive cooling systems.

Japan's approach emphasizes robust safety measures and public engagement to rebuild trust in nuclear energy. The NRA is also exploring how SMRs can be integrated into existing energy infrastructure, particularly in areas with limited grid capacity.

The construction and deployment of SMRs are influenced by diverse national regulations, reflecting each country's unique priorities and safety concerns. While there is a growing international consensus on the benefits of SMRs, regulatory bodies are still adapting existing frameworks to accommodate these innovative reactors. The focus on harmonizing safety standards, reducing EPZs, and streamlining licensing processes is essential to the global success of SMR technology. As SMRs gain traction, continued collaboration between national regulators, industry stakeholders, and international organizations like the IAEA will be critical in ensuring the safe and efficient deployment of these reactors worldwide.

Licensing Process for SMR Construction and Operation

The licensing process for the construction and operation of Small Modular Reactors (SMRs) is an essential regulatory pathway designed to ensure that these reactors meet stringent safety, environmental, and operational standards. Given the unique characteristics of SMRs, including their modular design, reliance on passive safety systems, and potential for deployment in various locations, the licensing process must adapt to accommodate these differences while maintaining rigorous safety protocols.

Pre-Licensing Engagement and Vendor Design Review

The pre-licensing engagement and Vendor Design Review (VDR) are critical components in the licensing process for Small Modular Reactors (SMRs). This initial phase involves consultations between the developer and the national regulatory authority, allowing for a non-binding review of proposed reactor designs, safety features, and technical specifications. The primary goal of this engagement is to identify potential regulatory issues early in the process, thereby minimizing delays during the formal licensing application phase. For instance, the Canadian Nuclear Safety Commission (CNSC) has established a structured three-phase VDR process specifically designed to assist SMR developers in addressing regulatory concerns prior to submitting formal applications [163].

During the pre-licensing phase, developers are required to submit comprehensive documentation that includes detailed descriptions of the reactor's design, safety features, and risk assessments. This documentation is then meticulously reviewed by the regulatory authority, which provides feedback aimed at refining the designs to ensure compliance with national safety standards. Such proactive engagement is particularly advantageous for SMRs, which often incorporate innovative technologies that may not align directly with existing regulatory frameworks designed for traditional reactors. By addressing potential compliance issues early, developers can avoid costly modifications later in the licensing process, thereby enhancing the overall efficiency of regulatory approvals [163].

Moreover, the VDR process serves as a platform for fostering dialogue between developers and regulators, which is essential for building a mutual understanding of safety expectations and technical requirements. This collaborative approach not only aids in aligning the reactor designs with safety standards but also helps in cultivating a regulatory environment that is responsive to the unique challenges posed by SMR technologies [163]. The CNSC's engagement strategy exemplifies how structured pre-licensing reviews can facilitate smoother transitions into formal licensing, ultimately contributing to the safe deployment of nuclear technologies [163].

Site Selection and Early Site Permits

The process of site selection and obtaining Early Site Permits (ESPs) for nuclear facilities is a critical phase in the development of nuclear power plants, particularly for Small Modular Reactors (SMRs). This phase involves a comprehensive evaluation of potential sites based on various factors, including seismic activity, proximity to water sources, population density, and environmental impact. Regulatory bodies require extensive studies to ensure that selected sites can support safe reactor operations under both normal and emergency conditions.

Site selection for nuclear facilities is a multifaceted process that necessitates adherence to a clearly defined set of criteria. The site selection process for SMRs must consider environmental pressures and the feasibility of deployment, highlighting the importance of a structured approach to site evaluation [164]. Similarly, it is necessary to gather ecological and health-related data to inform site selection, noting that the complexity of ecosystems and potential human exposure pathways complicate the assessment process [165]. This complexity is further underscored by characteristics such as local climate, geological conditions, and the availability of cooling water must be balanced with safety and environmental considerations [166].

In the United States, the Nuclear Regulatory Commission (NRC) plays a pivotal role in the ESP process, which allows developers to secure site approval prior to finalizing reactor designs. This regulatory framework is designed to expedite construction timelines for SMR developers, as they can bypass site-specific approvals later in the design process. The NRC evaluates site safety, environmental impact, and emergency preparedness as part of the ESP application, ensuring that all potential risks are assessed before construction begins [167]. The importance of these evaluations is echoed in the work of Basri et al., who review regulatory requirements for nuclear power plant site selection and emphasize the alignment of Malaysia's guidelines with international standards [167].

Moreover, the integration of human factors engineering (HFE) into the design and operation of nuclear facilities is crucial for ensuring safety and efficiency. The NRC's guidelines for HFE highlight the need for thorough evaluations of human-system interfaces to mitigate potential safety risks [168]. This aspect of site selection and permitting is vital, as it ensures that the operational environment is conducive to safe reactor management.

Design Certification and Standard Design Approval

The process of Design Certification and Standard Design Approval is pivotal in the licensing of nuclear reactors, particularly for Small Modular Reactors (SMRs). This phase entails a comprehensive evaluation by regulatory authorities, such as the U.S. Nuclear Regulatory Commission (NRC), to ensure that reactor designs adhere to stringent safety and performance standards. The assessment focuses on critical components such as the reactor core design, passive safety systems, containment structures, and emergency cooling systems, which are essential for mitigating risks associated with nuclear operations [169, 170].

For SMRs, the design certification process can be more efficient compared to traditional large reactors. This efficiency arises from the modular nature of SMRs, which allows for standardized designs that can be replicated across multiple sites without the need for exhaustive reviews for each new installation. The NRC's certification of the NuScale SMR, the first of its kind in the U.S., exemplifies this streamlined approach. Once certified, developers can deploy the same design in various locations, significantly reducing regulatory costs and expediting the overall deployment timeline [169, 170]. This standardization not only enhances the economic viability of SMRs but also promotes safety through proven design features that have undergone rigorous scrutiny [170].

Moreover, the incorporation of passive safety systems in SMR designs is a critical factor that influences the certification process. These systems are engineered to function without external power or operator intervention, thereby enhancing the reactor's safety profile during emergency situations. For instance, recent advancements in nuclear technology have led to designs that can effectively manage decay heat for extended periods, which is a significant improvement over older reactor designs [5, 169]. The emphasis on passive safety mechanisms aligns with the regulatory focus on minimizing potential hazards and ensuring public safety, which is paramount in the nuclear industry [170].

Combined Construction and Operating License (COL)

The Combined Construction and Operating License (COL) process is a critical regulatory framework in the United States for Small Modular Reactor (SMR) developers. This process allows developers to obtain both construction and operational permissions simultaneously, thereby streamlining the approval process. The COL encompasses comprehensive assessments of the reactor's construction plans, safety measures, quality assurance programs, and operational protocols, which are essential for ensuring the safety and reliability of nuclear facilities.

A significant aspect of the COL process is the requirement for developers to demonstrate the reactor's ability to operate safely under various scenarios, including potential accidents and natural disasters. This necessitates extensive documentation, including probabilistic risk assessments (PRA), safety analyses, and emergency response plans. The PRA methodology is particularly relevant as it systematically estimates the risks associated with complex sociotechnical systems, providing a framework for evaluating the likelihood and consequences of potential accidents [171, 172]. The importance measure analysis within PRA helps in ranking risk-contributing factors, which is crucial for decision-making in the licensing process [171].

Moreover, the COL process emphasizes the need for robust safety assessments that incorporate both deterministic and probabilistic performance objectives. For instance, the assessment of leakage risks in nuclear containment structures under accident conditions is an area where probabilistic safety assessment methodologies are applied [173]. These methodologies not only enhance the understanding of potential risks but also guide the development of effective safety measures and emergency response strategies [172].

The benefits of the COL process are particularly pronounced for SMR projects, which aim for shorter construction timelines. By reducing the time and financial resources required for separate licensing stages, the COL facilitates a more efficient pathway to bringing new nuclear technologies online. This efficiency is critical in the context of increasing energy demands and the need for sustainable energy solutions [174].

The initial step in the COL process requires developers to submit a comprehensive application. This application must include detailed plans covering various aspects such as construction, safety measures, quality assurance programs, and operational protocols. The thoroughness of this documentation is critical for the NRC to assess the proposed reactor design's compliance with regulatory standards [170]. The NRC's stringent requirements ensure that all potential risks are identified and mitigated, which is particularly important for the innovative designs characteristic of SMRs [67].

Safety assessments are a cornerstone of the COL process. Developers are mandated to demonstrate that their reactors can operate safely under a range of scenarios, including potential accidents and natural disasters. This involves conducting probabilistic risk assessments (PRAs) and safety analyses to evaluate the likelihood and consequences of various incidents [170]. Such rigorous evaluations are essential for ensuring that the reactors are resilient to events like earthquakes and extreme weather, thereby enhancing the overall safety of nuclear power generation [67]. The emphasis on safety not only protects plant personnel and the surrounding community but also fosters public confidence in nuclear energy as a safe alternative to fossil fuels [175].

Another critical component of the COL process is the requirement for comprehensive emergency response plans. These plans must outline the procedures and protocols to be followed in the event of an incident, ensuring that the safety of plant personnel, the surrounding community, and the environment is prioritized [170]. The thoroughness of these plans is vital for effective crisis management and reflects the NRC's commitment to maintaining high safety standards in nuclear operations [67].

The COL process offers several advantages that are particularly beneficial for SMR projects. One significant benefit is time efficiency; the simultaneous granting of construction and operational licenses can significantly reduce the overall timeline for bringing a new reactor online (Asuega et al., 2023). This is crucial for SMRs, which are designed for quicker deployment compared to traditional large-scale reactors. Additionally, the streamlined nature of the COL process can lead to substantial cost savings for developers, minimizing the financial resources required for regulatory compliance, which is particularly advantageous given the capital-intensive nature of nuclear projects [170]. Furthermore, the rigorous safety assessments required by the COL process enhance public confidence in nuclear energy, addressing concerns about safety and regulatory compliance [175].

Despite its benefits, the COL process presents challenges. The regulatory complexity involved can be daunting for new developers, necessitating significant expertise and resources to navigate the specific requirements and ensure compliance with NRC regulations [170]. Public perception of nuclear energy remains a significant hurdle, and developers must engage in effective communication and outreach to address public concerns and foster acceptance of SMR technology [67]. Additionally, the evolving nature of SMR technology introduces uncertainties related to design and operational performance, which developers must address through robust safety analyses and risk assessments [175].

Environmental Impact Assessment (EIA)

Environmental Impact Assessment (EIA) is a critical process mandated for nuclear projects, including Small Modular Reactors (SMRs). The EIA evaluates the potential environmental effects associated with the construction and operation of SMRs, focusing on various environmental components such as air, water, soil, and local ecosystems. This assessment is particularly crucial given the unique challenges posed by nuclear energy, including the management of radioactive waste, cooling water discharge, and the potential for accidents [14, 16].

The smaller footprint and modular nature of SMRs generally result in a reduced environmental impact compared to traditional large-scale nuclear reactors. This is due to their design, which allows for more efficient land use and potentially lower emissions during operation [128, 176]. However, despite these advantages, developers are still required to demonstrate compliance with both national and international environmental protection standards. This compliance is essential to ensure that the environmental impacts are adequately mitigated and that the local communities are protected [5, 14].

Moreover, the results of the EIA are typically made available for public review, allowing local communities to engage in the decision-making process. This transparency is vital for addressing public concerns and fostering community acceptance of nuclear projects [19, 56]. The public review process not only enhances the legitimacy of the EIA but also ensures that the voices of the affected communities are heard before any construction begins [176, 177].

Public Consultation and Stakeholder Engagement

Public consultation and stakeholder engagement are essential components of the licensing process for Small Modular Reactors (SMRs), particularly when these reactors are located near populated areas. Regulatory authorities mandate that developers engage with local communities, government agencies, and various stakeholders to address concerns regarding safety, environmental impacts, and emergency preparedness. This engagement often takes the form of public hearings and information sessions, which serve to explain the project, answer questions, and foster trust within the community. The importance of these processes is underscored by the need to build public confidence in the safety and benefits of SMRs, especially as they may be deployed in regions unsuitable for traditional large reactors. Gaining regulatory and social acceptance hinges on effective public engagement strategies.

Research indicates that stakeholder engagement significantly influences the success of energy projects, including SMRs. For instance, Geysmans et al. [178] highlight the necessity of broadening stakeholder engagement in emergency preparedness, emphasizing that both formal and informal engagement can shape power dynamics and influence project outcomes. Similarly, Zhu's study [179] illustrates that investment in renewable energy projects correlates strongly with stakeholder engagement, as local community support can mitigate risks and enhance project viability. This relationship is particularly relevant for SMRs, where community acceptance is critical for project advancement.

Moreover, the role of public hearings in urban planning and development is well documented. Ivanova discusses the importance of public hearings in understanding citizens' needs and attitudes toward proposed projects, suggesting that these forums are vital for fostering community involvement and trust [180]. Perko et al. [181] further emphasize that transparency in stakeholder engagement is crucial for effective communication and public trust, particularly in the context of nuclear or radiological emergency management. This transparency not only facilitates informed decision-making but also promotes a participatory democratic process that is essential for the acceptance of high-risk technologies like SMRs.

The integration of stakeholder engagement into the planning and implementation of SMRs is also supported by empirical studies. For example, Waris et al. [182] identify critical constructs of stakeholder engagement that can enhance the performance of renewable energy projects, indicating that a structured approach to engagement can lead to better project outcomes. Additionally, Lindner et al. [183] discuss the importance of stakeholder involvement in resilience projects, highlighting how collaborative efforts can lead to innovative solutions and improved project success. This collaborative environment is particularly beneficial for SMRs, as it allows for the exchange of knowledge and expertise among stakeholders, thereby addressing safety concerns and enhancing community trust.

Construction and Pre-Operational Testing

The construction and pre-operational testing phases of nuclear power plants (NPPs) are critical to ensuring safety and compliance with regulatory standards. Once all necessary licenses are obtained, the construction phase commences, during which regulatory authorities conduct periodic inspections to ensure adherence to approved designs and safety standards. This involves quality assurance checks, material testing, and verification of safety systems, which are essential for maintaining the integrity of the construction process [89, 184].

The modular construction process, particularly for Small Modular Reactors (SMRs), allows for many components to be prefabricated in factories, significantly reducing on-site construction time and minimizing associated risks [185]. During the construction phase, effective risk management is paramount. Studies have shown that identifying risk factors in each work type is crucial for the secure implementation of NPP construction projects [185]. The complexity of constructing a nuclear power plant necessitates meticulous planning and execution, as it typically involves multiple phases, including ground-breaking and post-tensioning, which can take several months to complete [185, 186]. Moreover, the economic implications of construction are significant; thus, advanced scheduling and cost management techniques, such as the linear scheduling method (LSM), are employed to optimize project timelines and resource allocation [187].

Before a nuclear plant can commence operations, developers must conduct extensive pre-operational testing to verify that all systems function as designed. This includes rigorous testing of the reactor core, cooling systems, control mechanisms, and emergency shutdown procedures [188]. Regulatory bodies often require detailed reports and inspections prior to granting approval to load nuclear fuel, ensuring that all safety protocols are met [184, 188]. The emphasis on safety has intensified following past nuclear incidents, prompting a re-evaluation of safety features in modern reactor designs to enhance reliability and minimize the need for emergency planning [189, 190].

Operational License and Ongoing Oversight

The operational licensing and ongoing oversight of Small Modular Reactors (SMRs) are critical components in ensuring the safety and reliability of nuclear energy generation. After successful pre-operational testing, regulatory authorities grant operational licenses, enabling SMRs to commence commercial operations. However, this initial licensing is merely the beginning of a comprehensive regulatory framework that mandates continuous monitoring and periodic inspections throughout the operational lifespan of the reactors. This ongoing oversight is essential to ensure that SMRs consistently meet stringent safety and environmental standards, allowing for prompt responses to any emerging issues that may arise during operation [15, 123].

The regulatory landscape for SMRs is particularly complex due to their innovative features, such as autonomous control systems and advanced digital monitoring technologies. These advancements necessitate additional scrutiny from regulators to ensure that these systems function reliably and safely. For instance, the integration of digital twins and AI-based monitoring systems is becoming increasingly relevant in the regulatory framework for SMRs, as these technologies can provide real-time data and predictive analytics that enhance operational safety and efficiency [70, 191]. The use of such technologies is aligned with the principles of Integrated Safety, Operations, Security, and Safeguards (ISOSS), which advocates for incorporating regulatory considerations into the design phase of SMR facilities to minimize the need for costly retrofitting and ensure compliance with safety standards from the outset [70].

Moreover, the licensing process for SMRs has evolved to incorporate more sophisticated methodologies, such as the Best Estimate Plus Uncertainty (BEPU) approach. This method allows for a more nuanced understanding of operational risks and uncertainties, moving away from conservative analysis towards a framework that better reflects the operational realities of modern nuclear reactors [170]. The emphasis on continuous improvement and adaptation in regulatory practices is crucial, especially in light of the lessons learned from past nuclear incidents, which underscore the importance of robust safety systems and proactive regulatory oversight [16].

The licensing process for SMRs involves multiple stages designed to ensure that these advanced reactors are built and operated safely. From pre-licensing consultations to combined construction and operational licenses, the process is evolving to accommodate the unique characteristics of SMRs. The goal is to streamline approvals without compromising safety, thereby enabling faster deployment of SMRs to meet global energy needs. As countries strive to achieve their decarbonization targets, refining the regulatory framework for SMRs will be crucial in ensuring that they play a significant role in the future energy landscape.

Compliance with Safety and Environmental Standards

The construction, operation, and decommissioning of Small Modular Reactors (SMRs) are governed by a range of safety and environmental standards. These standards are critical to ensuring that SMRs operate safely, minimize environmental impact, and adhere to regulatory requirements in different jurisdictions. The unique design and operational characteristics of SMRs, such as their modular construction, passive safety systems, and ability to be deployed in diverse environments, require specific adaptations of existing nuclear safety frameworks.

Standards Applicable to Small Modular Reactor (SMR) Power Stations

The construction, operation, and decommissioning of Small Modular Reactors (SMRs) are governed by a complex set of standards and guidelines to ensure safety, security, and environmental protection. These standards are developed and maintained by various international and national organizations, such as the International Atomic Energy Agency (IAEA), the International Organization for Standardization (ISO), the Nuclear Regulatory Commission (NRC) in the United States, and other regional and national regulatory authorities.

Below is a list of some of the relevant standards that are commonly applicable to SMR power stations:

1. International Safety Standards

A. IAEA Safety Standards

- **General Safety Requirements**
    - IAEA GSR Part 1: **Governmental, Legal and Regulatory Framework for Safety**
    - IAEA GSR Part 2: **Leadership and Management for Safety**
    - IAEA GSR Part 3: **Radiation Protection and Safety of Radiation Sources**

- o   IAEA GSR Part 4: **Safety Assessment for Facilities and Activities**

- o   IAEA GSR Part 5: **Predisposal Management of Radioactive Waste**

- o   IAEA GSR Part 6: **Decommissioning of Facilities**

- **Specific Safety Requirements**

  - o   SSR-2/1 (Rev. 1): **Safety of Nuclear Power Plants: Design**

  - o   SSR-2/2 (Rev. 1): **Safety of Nuclear Power Plants: Operation**

  - o   SSR-4: **Safety of Nuclear Fuel Cycle Facilities**

- **Safety Guides**

  - o   SSG-2: **Deterministic Safety Analysis for Nuclear Power Plants**

  - o   SSG-3: **Development and Application of Level 1 Probabilistic Safety Assessment**

  - o   SSG-9: **Severe Accident Management Programmes for Nuclear Power Plants**

  - o   SSG-12: **Licensing Process for Nuclear Installations**

  - o   SSG-30: **Safety Classification of Structures, Systems, and Components in Nuclear Power Plants**

B. Nuclear Non-Proliferation and Security

- IAEA INFCIRC/153: **Comprehensive Safeguards Agreement**

- IAEA INFCIRC/225/Rev.5: **Physical Protection of Nuclear Material and Nuclear Facilities**

- IAEA NSS-13: **Nuclear Security Recommendations on Physical Protection**

- IAEA Nuclear Security Series No. 8: **Preventive and Protective Measures Against Insider Threats**

2. Radiation Protection Standards

- **International Commission on Radiological Protection (ICRP)**

  - o   ICRP Publication 103: **The 2007 Recommendations of the ICRP**

  - o   ICRP Publication 111: **Protection Against Radon in Workplaces**

- **ISO Standards for Radiation Protection**

- ○ ISO 2919: **Sealed Radioactive Sources - Classification**

- ○ ISO 4037: **Radiation Protection - X and Gamma Reference Radiation for Calibrating Dosimeters**

- ○ ISO 10276: **Radiation Protection - Sealed Radioactive Sources - Leak Testing Methods**

3. Environmental Protection Standards

- **IAEA Environmental Standards**

  - ○ IAEA GSG-9: **Environmental Impact Assessment for Nuclear Installations**

  - ○ IAEA SSG-17: **Control of Orphan Sources and Other Radioactive Material in the Metal Recycling and Production Industries**

- **ISO Environmental Standards**

  - ○ ISO 14001: **Environmental Management Systems**

  - ○ ISO 14040: **Life Cycle Assessment**

  - ○ ISO 14064: **Greenhouse Gases - Carbon Footprint Measurement and Management**

- **Waste Management**

  - ○ Joint Convention on the Safety of Spent Fuel Management and on the Safety of Radioactive Waste Management (IAEA)

4. Design and Engineering Standards

- **ASME Standards**

  - ○ ASME BPVC Section III: **Nuclear Facility Components**

  - ○ ASME NQA-1: **Quality Assurance Requirements for Nuclear Facility Applications**

  - ○ ASME OM Code: **Operation and Maintenance of Nuclear Power Plants**

- **ISO Standards for Engineering and Manufacturing**

  - ○ ISO 9001: **Quality Management Systems**

  - ○ ISO 55000: **Asset Management**

- o ISO 19443: **Quality Management Systems for Nuclear Supply Chain**

- **IEEE Standards for Electrical Systems**

  - o IEEE 603: **Standard Criteria for Safety Systems for Nuclear Power Generating Stations**

  - o IEEE 379: **Application of the Single-Failure Criterion to Nuclear Power Generating Station Safety Systems**

  - o IEEE 497: **Instrumentation for Monitoring Radiation Conditions in Nuclear Power Generating Stations**

5. Cybersecurity and Information Protection

- **IAEA Nuclear Security Series**

  - o NSS-17: **Computer Security at Nuclear Facilities**

  - o NSS-33-T: **Technical Guidance on Computer Security Incident Response at Nuclear Facilities**

- **ISO/IEC Standards**

  - o ISO/IEC 27001: **Information Security Management Systems**

  - o ISO/IEC 27032: **Guidelines for Cybersecurity**

6. Licensing and Regulatory Guidelines

- **Country-Specific Regulations**

  - o **U.S. Nuclear Regulatory Commission (NRC)**

    - ▪ 10 CFR Part 50: **Domestic Licensing of Production and Utilization Facilities**

    - ▪ 10 CFR Part 52: **Licenses, Certifications, and Approvals for Nuclear Power Plants**

    - ▪ NRC Regulatory Guide 1.70: **Standard Format and Content of Safety Analysis Reports for Nuclear Power Plants**

  - o **Canadian Nuclear Safety Commission (CNSC)**

    - ▪ REGDOC-2.5.2: **Design of Reactor Facilities: Nuclear Power Plants**

    - ▪ REGDOC-2.3.2: **Accident Management: Nuclear Power Plants**

- o **French Nuclear Safety Authority (ASN)**

    - ▪ ASN Guide No. 22: **Safety Analysis Reports for Nuclear Facilities**

- **International Licensing Harmonization**

    - o WNA (World Nuclear Association): **Harmonization of Reactor Design Evaluation and Licensing**

    - o Multinational Design Evaluation Programme (MDEP)

7. Operational and Maintenance Standards

- **IAEA Safety Standards**

    - o NS-G-2.11: **A System for the Feedback of Experience from Events in Nuclear Installations**

    - o SSG-25: **Periodic Safety Review for Nuclear Power Plants**

- **ASME and IEEE Maintenance Standards**

    - o ASME OM Code: **Operation and Maintenance of Nuclear Power Plants**

    - o IEEE 1205: **Guide for Assessing, Monitoring, and Mitigating Aging Effects on Electrical Equipment Used in Nuclear Power Generating Stations**

- **ISO Standards for Asset Management**

    - o ISO 55001: **Asset Management Systems**

    - o ISO 45001: **Occupational Health and Safety Management Systems**

8. Decommissioning and Decontamination

- **IAEA Standards for Decommissioning**

    - o WS-G-2.1: **Decommissioning of Nuclear Power Plants and Research Reactors**

    - o GSR Part 6: **Decommissioning of Facilities**

- **ISO Standards for Waste Management**

    - o ISO 17873: **Waste Management Systems for Radioactive Waste**

- **Guidelines for Decontamination**

- o   IAEA TRS-467: **Management of Disused Sealed Radioactive Sources**

- o   IAEA SSG-47: **Decommissioning of Nuclear Power Plants**

9. Emergency Preparedness and Response

- **IAEA Safety Standards**

  - o   GSR Part 7: **Preparedness and Response for a Nuclear or Radiological Emergency**

  - o   EPR-NPP: **Preparedness and Response for a Nuclear Power Plant Emergency**

- **International Conventions**

  - o   Convention on Early Notification of a Nuclear Accident

  - o   Convention on Assistance in the Case of a Nuclear Accident or Radiological Emergency

Safety Standards

The safety standards for Small Modular Reactors (SMRs) encompass a comprehensive framework that addresses reactor design and construction safety, radiation protection, operational safety, and security measures. These standards are significant for ensuring the safe operation of nuclear reactors and are guided by international protocols and best practices.

Reactor Design and Construction Safety

SMRs are required to comply with the International Atomic Energy Agency (IAEA) Safety Standards, which delineate the essential safety requirements for nuclear reactors. A significant focus is placed on the incorporation of passive safety systems, which are designed to function without human intervention or external power sources. This approach enhances the reliability of the reactor during emergencies, as these systems utilize natural forces such as gravity and natural circulation to maintain safe conditions [188, 192, 193]. The design of these systems is critical, as they can effectively manage decay heat for extended periods, thereby preventing core damage [5].

Furthermore, redundancy and diversity in critical systems are emphasized to ensure functionality in the event of component failures. This is complemented by a robust containment strategy that maintains integrity under extreme conditions, including seismic events and flooding [194]. The defence-in-depth strategy mandates multiple layers of protection, which include robust containment structures, core damage mitigation measures, and backup cooling systems, thereby minimizing the likelihood of radioactive material release [5, 195]. The assessment of safety is quantitatively supported by metrics such as Core Damage Frequency (CDF) and Large Release Frequency (LRF), which are essential for evaluating the probability of severe accidents [190].

Radiation Protection

Radiation protection standards are governed by the guidelines set forth by the International Commission on Radiological Protection (ICRP). These guidelines advocate for the ALARA (As Low As Reasonably Achievable) principle, which aims to minimize radiation exposure to both workers and the public during normal operations and emergencies [196]. The establishment of exclusion zones and emergency planning zones (EPZs) is a critical aspect of radiation protection, as these zones are designed to control public exposure in the event of an accident [196].

Operational Safety

Operational safety standards for SMRs necessitate adherence to Human Factors Engineering (HFE) principles to optimize control room design for operator performance. This includes the development of Emergency Operating Procedures (EOPs) and Severe Accident Management Guidelines (SAMGs) to effectively manage unexpected situations [197]. Regular probabilistic safety assessments (PSA) and deterministic safety analyses are conducted to evaluate the safety performance of the plant, ensuring that operational protocols are continually refined and updated [197]. Additionally, periodic safety drills and inspection programs are essential for maintaining operational readiness and ensuring compliance with safety protocols [197].

Security Measures

Security measures for SMRs are critical in safeguarding against potential threats, including sabotage and cyber-attacks. Compliance with physical security standards is mandatory to protect against unauthorized access [198]. The development of cybersecurity protocols is also essential to secure digital control systems from malicious interference [198]. Furthermore, the implementation of safeguards for the non-proliferation of nuclear materials aligns with the Nuclear Non-Proliferation Treaty (NPT) and IAEA safeguards, ensuring that nuclear materials are not diverted for illicit purposes [198].

Environmental Standards

Environmental standards play an important role in ensuring that projects, particularly those involving Small Modular Reactors, adhere to regulations that protect the environment. This response synthesizes relevant literature on Environmental Impact Assessments (EIA), waste management, decommissioning, and climate resilience, providing a comprehensive overview of the standards and practices necessary for sustainable nuclear energy development.

Environmental Impact Assessments (EIA)

Before the construction of SMR projects, rigorous Environmental Impact Assessments (EIA) are mandated to identify potential environmental effects. EIAs serve as essential tools for evaluating the environmental implications of proposed projects, including their impacts on air, water, soil quality, and biodiversity. For instance, Mwanga emphasizes that EIAs inform decision-makers about the environmental consequences of development projects and help in formulating mitigation strategies to minimize adverse impacts [199]. Similarly, Karimi et al. highlight that the EIA process involves predicting and assessing influences on both the biophysical and social environments, ensuring that development aligns with environmental limits [200]. Moreover, assessments must also evaluate thermal discharges from cooling systems, which can significantly affect nearby aquatic ecosystems, as noted in various studies addressing the thermal impacts of energy projects [201].

Waste Management and Spent Fuel Handling

Compliance with international frameworks, such as the Joint Convention on the Safety of Spent Fuel Management and Radioactive Waste Management, is crucial for the safe handling and disposal of spent nuclear fuel and radioactive waste. Effective waste management strategies, including dry cask storage and geological disposal, are essential to minimize environmental contamination. Merk et al. [202] discuss the importance of developing long-term strategies for managing nuclear waste, emphasizing that while the mass of waste remains constant, the activity of fission products increases with energy production, necessitating innovative waste management approaches. Liu and Dai [141] further stress the significance of comprehensive management practices to mitigate the risks associated with radioactive waste, underscoring the need for effective treatment and disposal methods. Additionally, the integration of advanced materials for waste immobilization is critical for ensuring long-term safety [203].

Planning for the decommissioning phase during the design stage of nuclear facilities is vital for ensuring the safe removal of radioactive materials. Compliance with national and international guidelines, including those set by the International Atomic Energy Agency (IAEA), is necessary for effective decommissioning and site restoration. Solomon et al. [204] note that high-level nuclear waste management has been a longstanding political challenge, necessitating adherence to rigorous standards to address societal fears surrounding radiation. Furthermore, the restoration of sites to a safe condition for future use is crucial, as it minimizes residual radiation levels and ensures compliance with environmental standards [205]. The integration of these practices is essential for maintaining public trust and ensuring environmental safety.

The design of SMRs must incorporate resilience against extreme weather events, such as floods and storms, which are increasingly prevalent due to climate change. This necessitates the incorporation of climate adaptation strategies in site selection and design processes. Research indicates that nuclear facilities must be designed to withstand long-term environmental changes, ensuring operational continuity and safety [37]. The need for such resilience is underscored by the increasing frequency of extreme weather events, which pose significant risks to energy infrastructure [65]. By integrating climate change considerations into the planning and design phases, SMRs can contribute to a more sustainable energy future while minimizing environmental impacts.

Compliance and Licensing Standards

The regulatory landscape for Small Modular Reactors (SMRs) is characterized by a complex interplay of national and international standards, compliance requirements, and licensing processes. Each country has its own national nuclear regulatory authority, such as the U.S. Nuclear Regulatory Commission (NRC), the French Nuclear Safety Authority (ASN), and the Canadian Nuclear Safety Commission (CNSC), which are responsible for overseeing the safety and environmental compliance of nuclear facilities, including SMRs. These authorities establish specific regulatory frameworks that govern the design, construction, and operation of nuclear reactors, ensuring that they meet stringent safety and environmental standards [13].

Internationally, organizations like the International Atomic Energy Agency (IAEA) play a crucial role in standardizing safety requirements for nuclear reactors across different jurisdictions. The IAEA's guidelines aim to harmonize safety practices and reduce regulatory hurdles, facilitating the global deployment of nuclear technologies, including SMRs. This standardization is essential for promoting safety and efficiency in nuclear energy production, as it allows for the sharing of best practices and lessons learned from various countries [139].

Licensing processes for SMRs typically encompass several critical components, including design certification and pre-construction approvals. These processes ensure that the reactor designs meet safety standards before construction begins. Additionally, periodic safety reviews are mandated during the operational phase to assess compliance with safety regulations and to implement necessary upgrades or modifications based on evolving safety knowledge and technology [13]. Furthermore, compliance with country-specific environmental regulations is a fundamental aspect of the licensing process, as it addresses the potential environmental impacts of nuclear operations and ensures that they align with national environmental policies [13].

Case Study: Licensing an SMR in Various Jurisdictions

Licensing Process for Small Modular Reactors (SMRs) in the United States

The licensing and regulatory approval process for Small Modular Reactors (SMRs) in the United States is overseen by the U.S. Nuclear Regulatory Commission (NRC). The NRC has established a rigorous framework designed to ensure that SMRs meet all necessary safety, environmental, and security requirements before they can be constructed and operated. While this process is similar to that for traditional large reactors, there are specific provisions and adaptations tailored to the unique characteristics of SMRs.

1. Regulatory Framework and Governing Laws

The primary legal framework governing the licensing of SMRs in the United States is outlined in the Code of Federal Regulations (CFR), particularly 10 CFR Part 50 and 10 CFR Part 52. These regulations detail the requirements for obtaining construction permits, operating licenses, and design certifications for nuclear power plants. The NRC also issues guidance documents, known as Regulatory Guides (RGs), and reviews applications based on standards set forth in the Nuclear Energy Innovation and Modernization Act (NEIMA) of 2019, which aims to streamline the licensing process for advanced reactors, including SMRs.

2. Pathways for Licensing SMRs

The NRC offers three primary pathways for licensing nuclear reactors, including SMRs:

- **10 CFR Part 50**: This traditional two-step process involves obtaining a Construction Permit (CP) followed by an Operating License (OL).

- **10 CFR Part 52**: This process allows for a one-step Combined License (COL) that covers both construction and operation, contingent upon meeting safety and design requirements.

- **Design Certification (DC)**: An applicant can obtain a pre-approved reactor design certification, which simplifies the subsequent licensing process by eliminating the need for repeated design reviews.

A. Construction Permit and Operating License (10 CFR Part 50)

In the Part 50 process, applicants first obtain a Construction Permit (CP) based on the preliminary design and safety analysis. Once construction is completed, the applicant must then apply for an Operating License (OL), which requires a more comprehensive safety review. While this approach allows for flexibility during construction, it also involves multiple stages of regulatory scrutiny, potentially lengthening the overall timeline.

B. Combined License (COL) under 10 CFR Part 52

The Part 52 process streamlines the licensing procedure by granting a Combined License (COL) that authorizes both construction and operation. This one-step process is contingent upon the applicant's ability to demonstrate compliance with all safety, security, and environmental standards. A COL is generally preferred for SMRs because it reduces regulatory uncertainty and allows for an accelerated timeline. Once issued, a COL can only be utilized if the licensee completes a series of Inspections, Tests, Analyses, and Acceptance Criteria (ITAAC) to confirm that the plant is built according to approved specifications.

C. Design Certification (DC) Process

A Design Certification (DC) is another option available under Part 52, allowing vendors to obtain pre-approval of their reactor designs before applying for a license. The DC process is technology-neutral and valid for 15 years, which can be renewed. Once certified, a reactor design can be referenced by utilities applying for a COL, significantly reducing the need for repetitive design reviews.

3. Application Process and Requirements

The NRC licensing process involves multiple stages of application, review, and public participation:

- **Pre-Application Engagement**: The NRC encourages early engagement between SMR developers and the regulatory body to identify potential challenges, clarify requirements, and streamline the review process.

- **Formal Application Submission**: Applicants submit a comprehensive application, including safety analysis reports, environmental impact assessments, and security plans. The application must demonstrate compliance with NRC regulations and include technical specifications, plant design, operational procedures, and emergency preparedness plans.

- **Safety Review and Analysis**: The NRC conducts a thorough technical review of the safety analysis, focusing on aspects such as reactor core design, cooling systems, containment structures, and passive safety features. For SMRs, there is particular emphasis on passive safety mechanisms and smaller emergency planning zones (EPZs) due to their reduced risk profile.

- **Environmental Review**: Under the National Environmental Policy Act (NEPA), the NRC requires an Environmental Impact Statement (EIS) to assess potential environmental impacts. This includes evaluating the site's geology, hydrology, ecology, and the impact on local communities.

- **Public Hearings and Stakeholder Engagement**: The NRC holds public hearings to gather input from stakeholders, including local communities, environmental groups, and industry experts. These hearings provide transparency and ensure that the public has an opportunity to participate in the decision-making process.

4. Special Considerations for SMRs

The NRC has made adjustments to its traditional regulatory framework to accommodate the unique features of SMRs:

- **Reduced Emergency Planning Zones (EPZs)**: Due to their smaller size, lower power density, and enhanced safety features, SMRs may be eligible for reduced EPZs compared to large reactors. This adjustment can facilitate siting SMRs closer to population centres, reducing transmission costs.

- **Modular Construction**: SMRs are designed with modular construction in mind, allowing components to be prefabricated off-site and assembled on-site. The NRC reviews the modular construction process to ensure that quality and safety standards are met throughout fabrication, transportation, and assembly.

- **Innovative Safety Features**: SMRs often rely on passive safety systems that do not require active controls or human intervention. The NRC evaluates these features to ensure they provide adequate protection in case of an accident. The emphasis is on assessing how passive cooling systems, inherent safety features, and simplified designs contribute to minimizing risk.

5. Inspections, Tests, Analyses, and Acceptance Criteria (ITAAC)

Once a license is issued, the licensee must demonstrate compliance with the Inspections, Tests, Analyses, and Acceptance Criteria (ITAAC) as outlined in the COL. These criteria are critical to confirm that the SMR has been constructed according to the approved design specifications. The NRC conducts periodic inspections during construction and commissioning to verify compliance.

- **Inspections**: NRC inspectors visit the site to verify that construction activities meet safety and quality standards.

- **Tests**: Licensees must perform specific tests on critical systems and components to ensure they function correctly.

- **Analyses**: The NRC reviews analytical models and simulations to confirm the plant's safety performance under various conditions.

- **Acceptance Criteria**: The licensee must meet all specified acceptance criteria before the plant can begin commercial operation.

6. Post-Licensing Activities

Even after receiving an operating license, SMR operators are subject to ongoing regulatory oversight:

- **Operational Safety Inspections**: The NRC conducts routine inspections to monitor plant operations, maintenance practices, and safety management.

- **License Amendments**: If modifications to the plant design or operational procedures are necessary, the licensee must apply for a license amendment, which is subject to NRC review.

- **Decommissioning Requirements**: At the end of an SMR's operational life, the licensee must submit a decommissioning plan to safely dismantle the plant, manage radioactive waste, and restore the site.

7. Challenges and Future Outlook

The NRC's licensing process for SMRs is evolving to adapt to the needs of new technologies while maintaining its commitment to safety and public health. One of the key challenges is the harmonization of regulations to accommodate emerging SMR designs, including those using non-traditional coolants like liquid metals or molten salts. Additionally, the development of a risk-informed and performance-based regulatory framework is critical to supporting the deployment of advanced reactors while ensuring rigorous safety standards are upheld.

In summary, the licensing process for SMRs in the United States is extensive, involving multiple layers of review to ensure safety, security, and environmental protection. While the existing framework provides pathways for SMR approval, ongoing efforts to modernize the process are essential to accelerate the deployment of these innovative technologies and help meet future energy demands in a sustainable manner.

Licensing Process for Small Modular Reactors (SMRs) in Canada

The licensing and regulatory framework for Small Modular Reactors (SMRs) in Canada is managed by the Canadian Nuclear Safety Commission (CNSC). The CNSC is responsible for ensuring that all nuclear activities within the country, including the construction and operation of SMRs, adhere to stringent safety, security, and environmental standards. Canada has been actively working to streamline its licensing process for SMRs to encourage innovation and meet the country's goals for clean energy.

1. Regulatory Framework Governing SMR Licensing

Canada's nuclear regulatory framework is primarily governed by the Nuclear Safety and Control Act (NSCA), along with its associated regulations. These include:

- Class I Nuclear Facilities Regulations

- Radiation Protection Regulations

- Nuclear Security Regulations

- Nuclear Non-Proliferation Import and Export Control Regulations

The CNSC has also developed several Regulatory Documents (REGDOCs), which provide guidance on compliance with the regulatory requirements. In addition, the CNSC is aligned with the International Atomic Energy Agency (IAEA) standards to ensure that Canadian regulations meet global best practices.

2. Phased Licensing Approach for SMRs

The CNSC has adopted a phased approach to licensing SMRs, recognizing that these reactors differ significantly from traditional large-scale reactors in terms of design, safety features, and operational models. This phased approach helps to reduce regulatory uncertainty and allows for flexibility, especially when dealing with new and innovative reactor designs. The licensing process generally consists of three main phases:

- Phase 1: Pre-Licensing Vendor Design Review (VDR)

- Phase 2: Site Preparation and Construction Licensing

- Phase 3: Operating License and Commissioning

Each phase is designed to progressively address the safety, environmental, and security concerns of SMRs while allowing for stakeholder engagement.

A. Pre-Licensing Vendor Design Review (VDR)

The Vendor Design Review (VDR) is an optional, yet highly recommended, pre-licensing process that allows SMR developers to engage with the CNSC early in the development phase. The VDR process helps identify potential regulatory challenges, ensuring that the design aligns with Canadian requirements before a formal license application is submitted. The VDR is conducted in three stages:

1. **Stage 1: Basic Design Review** - Focuses on the general design and safety concepts, ensuring that they align with CNSC requirements.

2. **Stage 2: Detailed Design Review** - Evaluates the specific safety features, such as containment systems, reactor cooling, and passive safety mechanisms.

3. **Stage 3: System Review** - An in-depth analysis of the safety analysis report, including assessments of severe accident scenarios, emergency preparedness, and radiological protection.

The VDR does not result in a license but provides detailed feedback that can help SMR developers address regulatory concerns early in the process. This step is particularly beneficial for new reactor technologies that may not fit into traditional regulatory frameworks.

B. Site Preparation and Construction Licensing

Once the VDR is completed (if undertaken), the formal licensing process begins with the Site Preparation License and Construction License applications. This phase involves several key steps:

- **Site Evaluation:** The applicant must submit detailed reports on site selection, including geological, hydrological, and environmental assessments. The CNSC reviews the site's suitability for hosting an SMR, focusing on factors like seismic stability, potential environmental impacts, and proximity to populated areas.

- **Environmental Impact Assessment (EIA)**: Under the Canadian Environmental Assessment Act (CEAA), the project must undergo an Environmental Impact Assessment (EIA). This involves public consultations and stakeholder engagement to evaluate the potential environmental effects of the SMR project.

- **Public Hearings and Stakeholder Engagement**: The CNSC conducts public hearings where community members, environmental groups, and other stakeholders can provide input. The public consultation process ensures transparency and addresses concerns related to safety, environmental protection, and socio-economic impacts.

- **Construction License Application**: After obtaining site preparation approval, the developer applies for a Construction License. This application includes detailed engineering designs, safety analysis, emergency response plans, and quality assurance protocols. The CNSC evaluates these submissions to ensure compliance with safety and regulatory standards.

The **Construction License** allows the applicant to proceed with building the reactor and associated infrastructure. The CNSC conducts ongoing inspections during construction to ensure that all activities meet the approved safety and quality standards.

C. Operating License and Commissioning

Once construction is complete, the applicant must apply for an Operating License. This phase involves a comprehensive review of the plant's readiness for operation:

- **Pre-Operational Testing and Commissioning**: Before the reactor can begin commercial operation, it must undergo a series of tests to verify that all systems are functioning as designed. This includes tests on reactor controls, cooling systems, containment structures, and radiation protection measures.

- **Final Safety Analysis Report (FSAR)**: The applicant must submit a Final Safety Analysis Report (FSAR) that details the operational safety procedures, accident management strategies, and radiation protection protocols. The CNSC reviews this report to ensure that the plant meets all safety requirements.

- **Emergency Preparedness Plan**: SMR operators must demonstrate that they have robust emergency preparedness plans in place, including coordination with local authorities and emergency response teams.

- **Public Hearings and Final Approval**: The CNSC may hold additional public hearings before granting the Operating License. These hearings provide one last opportunity for public input and ensure that all concerns have been adequately addressed.

Once the Operating License is granted, the SMR can begin commercial operation. However, the CNSC continues to conduct regular inspections and audits to ensure ongoing compliance with safety, environmental, and security standards.

3. Key Considerations for SMR Licensing in Canada

Canada's regulatory framework for SMRs includes specific considerations to address the unique features of these reactors:

- **Passive Safety Features**: Many SMRs incorporate passive safety systems that do not require active intervention or external power sources to remain safe. The CNSC assesses these features to determine their reliability and effectiveness in preventing accidents.

- **Reduced Emergency Planning Zones (EPZs)**: Given the smaller size and lower power density of SMRs, developers may be able to reduce the size of the Emergency Planning Zone (EPZ) compared to traditional reactors. This adjustment can facilitate the siting of SMRs closer to urban areas.

- **Modular Construction**: SMRs are designed for modular construction, where components are fabricated off-site and assembled on-site. The CNSC reviews the quality assurance processes for modular construction to ensure safety standards are maintained throughout fabrication, transport, and assembly.

- **Remote and Off-Grid Applications**: SMRs are particularly suited for remote locations, such as mining sites or isolated communities. The CNSC evaluates the feasibility of deploying SMRs in these areas, focusing on logistical challenges, infrastructure requirements, and emergency response capabilities.

4. International Collaboration and Harmonization

The CNSC is actively involved in international collaborations to harmonize SMR regulations, especially with the United States and other countries exploring SMR deployment. The CNSC participates in the International Atomic Energy Agency (IAEA) forums and bilateral agreements to share best practices, reduce regulatory duplication, and streamline the approval process for SMRs intended for export.

The CNSC's proactive approach to SMR licensing aims to foster innovation while maintaining stringent safety standards. By adopting a risk-informed and performance-based regulatory framework, Canada is positioning itself as a leader in the development and deployment of advanced nuclear technologies, including SMRs. This approach not only supports Canada's clean energy goals but also contributes to the global effort to combat climate change by reducing reliance on fossil fuels.

Licensing Process for Small Modular Reactors (SMRs) in Europe

The licensing and regulatory landscape for Small Modular Reactors (SMRs) in Europe is evolving to accommodate the deployment of these innovative nuclear technologies. The European Union (EU) is characterized by a patchwork of national regulatory frameworks, as each member state is responsible for its nuclear safety regulations. However, there is a significant emphasis on harmonizing standards and approaches across the continent to ensure safety, promote technological innovation, and facilitate cross-border collaboration.

1. European Union Regulatory Framework for Nuclear Safety

Although each EU country has its own nuclear regulatory authority, the European Atomic Energy Community (EURATOM) treaty plays a crucial role in establishing a cohesive approach to nuclear safety. EURATOM has developed a comprehensive set of directives that member states must incorporate into their national laws, particularly concerning nuclear safety, radiation protection, and waste management. Key directives relevant to SMR licensing include:

- **Nuclear Safety Directive (2009/71/Euratom)**: Sets the overarching framework for nuclear safety across the EU, requiring member states to establish independent regulatory authorities and ensure high safety standards.

- **Radioactive Waste Management Directive (2011/70/Euratom)**: Mandates that all EU countries establish plans for the safe disposal of spent fuel and radioactive waste.

- **Basic Safety Standards Directive (2013/59/Euratom)**: Establishes minimum standards for radiation protection, which must be adhered to during the operation of any nuclear facility, including SMRs.

These directives require member states to adopt stringent safety measures, which form the basis for licensing nuclear facilities, including SMRs.

2. Harmonization Efforts in Europe

One of the major challenges in the European licensing process for SMRs is the lack of a unified regulatory framework across member states. However, there is a concerted effort to harmonize standards, particularly for new nuclear technologies like SMRs. To achieve this, several initiatives have been launched:

- **Western European Nuclear Regulators Association (WENRA)**: WENRA provides recommendations and reference levels for nuclear safety to ensure consistent standards among its members. These guidelines are increasingly being used by national regulators as benchmarks for evaluating SMR designs.

- **European Nuclear Safety Regulators Group (ENSREG)**: ENSREG works to align nuclear safety regulations across EU member states and facilitate the sharing of best practices. It plays a key role in promoting a common approach to safety assessments for SMRs.

- **EURATOM Research and Training Programme**: This initiative funds research projects to develop advanced nuclear technologies, including SMRs, while also supporting efforts to harmonize safety assessments and regulatory processes across the EU.

Harmonization is crucial for SMRs, as it simplifies the process for vendors seeking to deploy their technologies in multiple EU countries, reducing the time and cost associated with obtaining multiple national licenses.

3. National Licensing Processes in European Countries

While the EU strives for harmonization, individual countries retain their sovereign authority to license nuclear reactors. The licensing process for SMRs typically follows a set of national procedures that align with EURATOM directives but are adapted to fit local regulatory requirements. Below are examples of how some European countries approach SMR licensing:

A. United Kingdom

The United Kingdom, although no longer an EU member, remains a key player in the European nuclear sector. The UK's nuclear regulatory authority, the Office for Nuclear Regulation (ONR), is known for its robust, independent safety assessment process. For SMRs, the UK follows a Generic Design Assessment (GDA) process:

- **Step 1**: Preliminary Safety Evaluation, where the fundamental safety features of the SMR design are assessed.

- **Step 2**: Detailed Assessment, focusing on reactor safety systems, security, environmental protection, and emergency preparedness.

- **Step 3**: Final Assessment, which includes public consultations and results in a design acceptance confirmation if successful.

The UK's GDA process allows SMR vendors to obtain design approval before selecting a specific site, accelerating the deployment of SMRs.

B. France

France, with its strong nuclear industry, is also exploring the deployment of SMRs to diversify its energy mix. The Autorité de Sûreté Nucléaire (ASN) is the national regulatory body responsible for nuclear safety. France has a rigorous, phased approach to licensing nuclear facilities, which includes:

- **Pre-Licensing Engagement**: Vendors can engage with ASN during the design phase to clarify regulatory expectations.

- **Construction and Operating License Application**: A comprehensive safety analysis report, environmental impact assessment, and public consultations are required before licenses are granted.

- **Periodic Safety Reviews**: Once operational, SMRs must undergo periodic reviews to ensure ongoing compliance with safety standards.

France is working on aligning its regulatory processes with EU directives while maintaining its high national safety standards.

C. Finland

Finland has a reputation for stringent nuclear safety regulations and is open to new nuclear technologies like SMRs. The Radiation and Nuclear Safety Authority (STUK) oversees the licensing process, which includes:

- **Preliminary Safety Assessment**: Focuses on the feasibility of the SMR design and its alignment with Finnish safety regulations.

- **Construction License**: Requires a detailed safety case, including site suitability, environmental impact, and emergency response planning.

- **Operating License**: Issued after successful construction and commissioning, ensuring the plant meets all operational safety standards.

Finland's approach emphasizes early regulatory engagement, allowing for the identification of potential safety issues during the design phase.

4. Key Steps in the Licensing Process

Despite differences among national regulators, the general licensing process for SMRs in Europe can be summarized as follows:

- **Pre-Licensing Engagement**: This phase includes discussions between SMR vendors and regulators to clarify safety requirements and identify potential challenges. Early engagement helps align the design with national safety standards.

- **Design Review**: A detailed review of the reactor design, including safety systems, security measures, and environmental impact assessments. This phase may include a vendor design review process similar to the UK's GDA or Finland's preliminary safety assessment.

- **Site Licensing**: Once a design is approved, vendors must apply for a site-specific license. This involves assessments of geological, hydrological, and environmental factors, as well as public consultations to gauge community acceptance.

- **Construction and Commissioning**: After receiving a construction license, the reactor is built under the supervision of the national regulator. During commissioning, the plant undergoes extensive testing to verify that it meets all safety standards before it begins commercial operations.

- **Operating License**: Once construction and testing are complete, the operator must obtain an operating license. This involves submitting a final safety analysis report (FSAR) that details the reactor's operational safety procedures, emergency preparedness plans, and radiation protection measures.

- **Ongoing Compliance and Periodic Safety Reviews**: Even after the reactor is operational, it must undergo periodic safety reviews to ensure ongoing compliance with regulatory standards. This includes inspections, safety audits, and reviews of incident reports.

5. Challenges and Opportunities in Licensing SMRs in Europe

While SMRs hold promise for expanding clean energy capacity in Europe, the regulatory landscape presents challenges. The lack of a unified licensing framework means that vendors must navigate a complex web of national regulations, which can slow deployment. However, the push for harmonization by organizations like WENRA and ENSREG offers opportunities to streamline the process.

Europe's strong focus on safety, public engagement, and environmental protection means that SMR developers must demonstrate not only the safety and reliability of their technologies but also their environmental and social benefits. As European countries seek to decarbonize their energy sectors, SMRs could play a crucial role, especially in regions looking to phase out coal and gas plants. However, achieving this will require continued collaboration between regulators, vendors, and policymakers to overcome regulatory barriers and expedite the deployment of SMRs.

Licensing Process for Small Modular Reactors (SMRs) in India

The deployment of Small Modular Reactors (SMRs) in India involves navigating a complex and stringent regulatory environment. As a country with significant nuclear ambitions, India is actively exploring innovative nuclear technologies like SMRs to meet its growing energy needs while also reducing carbon emissions. However, given the safety, security, and environmental concerns associated with nuclear energy, the licensing and regulatory process for SMRs in India is comprehensive and involves multiple layers of approvals.

1. Regulatory Framework and Authorities

The regulation of nuclear power in India is primarily overseen by the Atomic Energy Regulatory Board (AERB). Established in 1983, the AERB is responsible for ensuring the safety and regulatory compliance of nuclear facilities across the country. It operates under the jurisdiction of the Atomic Energy Act of 1962 and functions as an independent body under the Department of Atomic Energy (DAE). The DAE itself oversees the entire nuclear power program in India, including research, development, and deployment.

In addition to the AERB, other regulatory bodies and institutions play crucial roles in the nuclear sector:

- **Department of Atomic Energy (DAE)**: Oversees the overall management of nuclear energy, research reactors, and the nuclear fuel cycle.

- **Nuclear Power Corporation of India Limited (NPCIL)**: The government-owned entity responsible for the construction, operation, and maintenance of nuclear power plants.

- **Bhabha Atomic Research Centre (BARC)**: Focuses on nuclear research and the development of advanced reactor technologies, including SMRs.

2. Legal and Regulatory Framework

India's nuclear regulatory framework is primarily guided by several key laws:

- **Atomic Energy Act of 1962**: Provides the DAE with the authority to develop and regulate nuclear energy for peaceful purposes, as well as to oversee the licensing of nuclear facilities.

- **Atomic Energy (Radiation Protection) Rules of 2004**: Specifies safety and radiation protection standards for nuclear facilities.

- **Environmental Protection Act of 1986**: Requires environmental impact assessments (EIAs) for projects that may have significant environmental implications, including nuclear power plants.

The existing regulatory framework focuses primarily on large nuclear reactors. For SMRs to be deployed, it may require amendments to the existing laws and guidelines to accommodate the unique safety and operational characteristics of SMRs.

3. Phased Licensing Process for SMRs

The licensing process for constructing and operating SMRs in India is likely to follow a multi-phase approach, similar to that for large nuclear power plants. However, it would also need to address the specific challenges and benefits associated with SMRs, such as their modular design, passive safety features, and scalability. The following sections outline the phased process:

A. Pre-Licensing Phase

Before initiating a formal licensing application, developers must engage with the AERB and other relevant regulatory bodies to discuss the project's feasibility, safety, and compliance requirements. This phase includes:

- Conducting feasibility studies to assess the technical and economic viability of SMR deployment in specific regions.

- Engaging in preliminary discussions with the AERB to clarify regulatory expectations, particularly around safety standards, site selection, and environmental protection.

- Preparing an initial conceptual design and safety case for the proposed SMR technology.

B. Site Selection and Environmental Assessment

SMRs, like any other nuclear power project, are subject to stringent site selection criteria. Developers must ensure compliance with the guidelines set forth by the AERB and the Ministry of Environment, Forest, and Climate Change (MoEFCC). Key steps include:

- Conducting a comprehensive Environmental Impact Assessment (EIA) to evaluate the potential environmental and social impacts of the proposed SMR site.

- Addressing concerns related to population density, natural disasters, and proximity to water sources, as these factors are critical in site approval.

- Engaging in public consultations to obtain feedback from local communities and stakeholders. Public acceptance is crucial, especially given India's sensitivity to nuclear projects after historical accidents like the Bhopal disaster.

C. Design Certification and Safety Review

Once the site is approved, developers must obtain design certification from the AERB. This phase involves a detailed review of the reactor design, including:

- Submitting a comprehensive Preliminary Safety Analysis Report (PSAR), which outlines the reactor's design, safety features, and passive cooling systems.

- Demonstrating compliance with India's stringent safety standards, particularly around passive safety systems, emergency preparedness, and radiation protection.

- The AERB will assess the design to ensure it meets international best practices, as recommended by the International Atomic Energy Agency (IAEA).

D. Construction Licensing

Upon receiving design certification, developers can proceed to apply for a construction license. This phase includes:

- Submitting detailed construction plans, including specifications for safety systems, containment structures, and auxiliary systems.

- Conducting site inspections and audits to ensure compliance with AERB guidelines.

- The construction license is granted only after the AERB is satisfied that the project meets all safety and environmental requirements.

E. Operating License

Following the construction phase, developers must apply for an operating license. The AERB requires:

- A Final Safety Analysis Report (FSAR), which includes detailed assessments of reactor safety, emergency response plans, and radiation protection measures.

- Conducting comprehensive pre-commissioning tests to verify that all systems are functioning correctly and that safety protocols are in place.

- Only after successful inspections and validation of safety measures can the operating license be issued, allowing the SMR to begin commercial operation.

4. Challenges and Opportunities for Licensing SMRs in India

Deploying SMRs in India presents both challenges and opportunities:

- **Regulatory Reform**: India's current regulatory framework is designed for large reactors, so there is a need to adapt the licensing process to accommodate the unique characteristics of SMRs. This may require new regulations and amendments to existing laws.

- **Public Perception and Acceptance**: Nuclear energy projects in India often face public resistance due to safety concerns and past incidents. Developers will need to focus on transparent communication and community engagement to gain public trust.

- **Infrastructure and Supply Chain**: India has a well-established nuclear industry, but the adoption of SMRs will require new manufacturing capabilities and supply chains tailored to modular reactors.

- **International Collaboration**: Given the novelty of SMR technology, India could benefit from **international cooperation** with countries that are already advancing SMR deployment, such as the United States, Canada, and Russia.

5. The Path Forward

India has shown interest in SMRs as a means to diversify its energy mix, reduce carbon emissions, and achieve energy security. However, for SMRs to become a reality, the country needs to address regulatory challenges, build public confidence, and develop the necessary infrastructure. The government, through the DAE, has already initiated discussions on exploring SMR technology for remote areas and industrial applications. Moving forward, a combination of regulatory reform, international collaboration, and public engagement will be critical for the successful deployment of SMRs in India. By leveraging its existing expertise in nuclear technology and focusing on streamlined licensing processes, India has the potential to become a leader in SMR deployment in the Asia-Pacific region.

Licensing Process for Small Modular Reactors (SMRs) in China

China has emerged as a global leader in nuclear energy development, particularly through its active pursuit of Small Modular Reactors (SMRs). This strategic initiative is part of China's broader effort to diversify its energy portfolio and achieve its ambitious carbon neutrality goals by 2060. The deployment of SMRs is seen as a crucial step in transitioning to a low-carbon energy system, as these reactors offer enhanced safety features, reduced construction times, and the ability to be deployed in a variety of settings, including remote areas where large reactors may not be feasible [11, 16].

The licensing and regulatory process for SMRs in China is complex and involves multiple layers of approvals from various national regulatory authorities. This framework has been evolving to accommodate the unique characteristics of SMRs while ensuring compliance with stringent safety and environmental standards. The legal development surrounding commercial nuclear energy in China has progressed through three phases: Beginning (1985–2002), Growth (2003–2015), and Maturity (2016–present), reflecting the increasing sophistication of regulatory frameworks to support nuclear energy expansion [206]. The Chinese government has placed significant emphasis on nuclear safety licensing, which is critical given the lessons learned from past nuclear incidents, such as the Fukushima disaster [207].

Moreover, the development of SMRs in China is supported by a robust national policy framework that aims to integrate nuclear power into the country's energy mix effectively. The Medium and Long-term Nuclear Power Development Plan outlines specific targets for nuclear capacity, which includes the deployment of SMRs as a key component of this strategy [208, 209]. The regulatory landscape is continuously being refined to facilitate the approval and deployment of these advanced reactor designs, which are recognized for their potential to enhance energy security and contribute to environmental sustainability [67].

1. Regulatory Framework and Authorities

The regulation of nuclear energy in China is overseen by several key organizations:

- **National Nuclear Safety Administration (NNSA)**: The primary regulatory authority responsible for the safety of nuclear installations, including the licensing and oversight of nuclear power plants and reactors. It operates under the Ministry of Ecology and Environment (MEE) and is responsible for ensuring compliance with nuclear safety standards, conducting inspections, and enforcing regulations.

- **National Energy Administration (NEA)**: Oversees the overall development and management of China's energy sector, including nuclear energy. The NEA is involved in policy formulation, planning, and regulation of the energy industry.

- **China Atomic Energy Authority (CAEA)**: Focuses on the peaceful use of nuclear energy, including international cooperation, technology development, and the overall strategic direction of nuclear projects.

These regulatory bodies collaborate to ensure that SMRs meet national safety requirements, align with international best practices, and are deployed in a manner that supports China's energy security and environmental goals.

2. Legal and Regulatory Framework

China's nuclear regulatory environment is governed by a series of laws, regulations, and guidelines:

- **Nuclear Safety Law (2018)**: This is the cornerstone of China's nuclear regulatory framework, establishing the legal basis for nuclear safety, emergency management, and radiation protection.

- **Regulations on the Safety of Civilian Nuclear Installations (1986, revised 2015):** Outlines the safety standards for the construction, operation, and decommissioning of nuclear facilities.

- **Environmental Protection Law (2015):** Requires comprehensive Environmental Impact Assessments (EIAs) for all major infrastructure projects, including nuclear power plants.

These laws and regulations are designed to ensure that nuclear projects, including SMRs, adhere to stringent safety, security, and environmental standards.

3. Phased Licensing Process for SMRs

The licensing process for SMRs in China is similar to that for large-scale nuclear reactors but tailored to accommodate the unique design and operational features of SMRs. It follows a phased approach that includes site approval, construction licensing, commissioning, and operating licensing. The key phases are as follows:

A. Pre-Licensing and Feasibility Assessment

The process begins with feasibility studies and early-stage assessments. During this phase, SMR developers engage with the NNSA and other relevant authorities to clarify regulatory requirements and align their projects with national policies. Key steps include:

- Conducting pre-feasibility studies to assess the technical, economic, and environmental viability of the SMR project.

- Preparing initial design concepts and engaging in pre-licensing consultations with the NNSA to discuss safety standards, siting criteria, and potential regulatory hurdles.

- Submitting a preliminary project proposal to the National Development and Reform Commission (NDRC) for strategic approval.

B. Site Selection and Environmental Impact Assessment (EIA)

Once the project receives initial approval, the focus shifts to site selection and environmental assessments. This phase includes:

- Conducting a comprehensive Environmental Impact Assessment (EIA) to identify potential risks to the environment and public health. The EIA report must be approved by the Ministry of Ecology and Environment (MEE).

- Assessing the seismic stability, hydrology, and population density of the proposed site to ensure compliance with China's stringent siting criteria.

- Engaging with local stakeholders and communities to obtain feedback and address concerns related to the proposed project.

C. Design Certification and Safety Review

Design certification is a critical phase where the reactor design undergoes rigorous evaluation by the NNSA. This involves:

- Submitting a Preliminary Safety Analysis Report (PSAR), which includes details on the reactor's safety features, passive cooling systems, and emergency preparedness plans.

- Demonstrating compliance with Chinese nuclear safety standards as well as guidelines set by the International Atomic Energy Agency (IAEA).

- Conducting technical assessments to evaluate the robustness of the SMR design, including its ability to withstand external hazards like earthquakes and floods.

The NNSA may require additional independent safety assessments or third-party audits to verify the reactor's safety features.

D. Construction Licensing

Upon receiving design approval, the developer can apply for a construction license. This phase includes:

- Providing detailed engineering blueprints, construction plans, and safety protocols to the NNSA for review.

- Conducting on-site inspections and quality assurance audits to ensure that construction activities adhere to regulatory standards.

- Receiving a formal construction permit, which authorizes the start of construction activities.

E. Commissioning and Operating License

After the construction is completed, the SMR must undergo a series of tests and inspections before it can become operational. Key steps include:

- Conducting pre-commissioning tests to verify the functionality of all safety systems, containment structures, and reactor components.

- Submitting a Final Safety Analysis Report (FSAR), which includes detailed assessments of operational safety, emergency response plans, and radiation protection measures.

- Obtaining an operating license from the NNSA, which allows the SMR to begin commercial operation.

The operating license is granted only after the NNSA is satisfied that the reactor meets all safety, security, and environmental requirements.

4. Challenges and Opportunities for SMR Licensing in China

Challenges:

- **Regulatory Adaptation**: The current regulatory framework in China is primarily designed for large reactors. Adapting it to accommodate the unique features of SMRs requires extensive revisions to existing laws and guidelines.

- **Public Acceptance**: As with any nuclear project, public perception is a significant challenge. China's nuclear authorities must engage in transparent communication with local communities to build trust and gain public support.

- **Supply Chain and Infrastructure**: Developing the necessary supply chain for modular construction and assembly of SMRs presents logistical challenges, especially in remote areas where China plans to deploy SMRs.

Opportunities:

- **Carbon Neutrality Goals**: China's commitment to achieving carbon neutrality by 2060 provides a strong impetus for the deployment of SMRs, particularly for decarbonizing remote regions and industrial processes.

- **Export Potential**: China aims to become a leader in exporting SMR technology, especially to developing countries that lack the infrastructure for large nuclear reactors. Streamlining the licensing process will enhance China's competitive edge in the global nuclear market.

- **Technological Innovation**: The adoption of advanced technologies, such as digital twins and AI-based monitoring systems, can optimize SMR operations and enhance safety, making these reactors more appealing for both domestic and international markets.

5. Conclusion

China's regulatory framework for SMRs is evolving to balance the need for innovation in nuclear technology with the imperative of maintaining high safety and environmental standards. The phased licensing process, overseen by the NNSA and supported by other regulatory bodies, ensures that SMRs can be deployed safely and efficiently. As China continues to refine its regulatory approach, it is well-positioned to leverage SMR technology to meet its energy and environmental goals, both domestically and as a key player in the international nuclear market.

Licensing Process for Small Modular Reactors (SMRs) in Australia

The process for licensing Small Modular Reactors (SMRs) in Australia is both complex and evolving, as the country currently does not have an established nuclear power industry, as at November 2024. Australia has historically maintained a policy against nuclear energy, largely due to public and political concerns. However, the growing focus on reducing carbon emissions, achieving net-zero targets, and exploring sustainable energy alternatives has revived interest in nuclear power, especially with the emergence of SMRs. These compact reactors are seen as a potentially safer and more flexible solution to Australia's energy needs. Nevertheless, the country faces significant regulatory, legal, and social hurdles in establishing a framework for the deployment of SMRs.

1. Regulatory Authorities and Framework

The regulation of nuclear activities in Australia is primarily governed by the Australian Radiation Protection and Nuclear Safety Agency (ARPANSA), which oversees radiation protection, nuclear safety, and licensing. Additionally, the Australian Nuclear Science and Technology Organisation (ANSTO) is involved in nuclear research and the use of radioactive materials, although it does not have direct authority over nuclear power generation.

The legislative framework for nuclear regulation in Australia includes:

- **Australian Radiation Protection and Nuclear Safety Act 1998**: This act empowers ARPANSA to regulate all Commonwealth entities involved in activities related to radiation and nuclear energy.

- **Environment Protection and Biodiversity Conservation Act 1999 (EPBC Act)**: Any significant project, including a nuclear power plant, would require an environmental assessment under the EPBC Act to evaluate its impact on biodiversity, ecosystems, and communities.

- **Nuclear Non-Proliferation (Safeguards) Act 1987**: This act ensures that Australia's nuclear activities comply with international safeguards and treaties, particularly those related to the Non-Proliferation of Nuclear Weapons (NPT).

2. Legal Restrictions on Nuclear Power

Australia currently has legal restrictions on the construction of nuclear power plants. The Environment Protection and Biodiversity Conservation Act 1999 explicitly prohibits the approval of nuclear power plants, and the Australian Radiation Protection and Nuclear Safety Act 1998 limits the use of nuclear materials to research and medical purposes. Additionally, several state laws, such as the Nuclear Activities (Prohibitions) Act in Victoria, specifically ban the construction of nuclear facilities.

For SMRs to be deployed in Australia, these legal barriers would need to be addressed through legislative amendments at both federal and state levels. This would require significant political will, public consultation, and possibly a shift in public perception towards nuclear energy.

3. Potential Pathway for Licensing SMRs

Despite the current prohibitions, there is growing interest in exploring SMRs as a low-carbon energy source. If Australia were to move forward with SMRs, the licensing process would likely involve a multi-phase approach, similar to the processes used in countries with established nuclear power sectors, such as the United States, Canada, or the United Kingdom. Below is an outline of how the licensing process for SMRs might be structured in Australia:

A. Pre-Licensing Engagement and Feasibility Studies

Before a formal licensing application can be submitted, prospective SMR developers would likely engage with ARPANSA, the Department of Industry, Science, and Resources, and other relevant agencies to discuss regulatory expectations and project feasibility. This phase would involve:

- Conducting feasibility studies to assess the economic viability and potential locations for SMR deployment.

- Engaging in preliminary discussions with regulators to understand safety, environmental, and security requirements.

- Preparing a high-level conceptual design and safety case to demonstrate the safety features of the SMR technology.

B. Environmental Assessment and Community Consultation

Given Australia's stringent environmental protection laws, any proposal to build an SMR would require a comprehensive environmental assessment under the EPBC Act. This would include:

- Assessing the environmental impact of the proposed SMR site, including effects on local ecosystems, water resources, and potential radiation exposure.

- Conducting public consultations to gauge community sentiment and address concerns. Public opposition has historically been a major barrier to nuclear energy projects in Australia, so gaining community support would be crucial.

- Ensuring compliance with Indigenous land rights and consulting with Indigenous communities if the project is proposed on or near their lands.

C. Design Certification and Safety Review

After completing the environmental assessments and receiving initial regulatory feedback, SMR developers would need to obtain design certification from ARPANSA. This phase would include:

- Submitting a detailed safety analysis report, which includes assessments of reactor design, passive safety features, radiation protection, and emergency response plans.

- Demonstrating that the SMR meets Australia's stringent safety and security standards, as well as international best practices, such as those outlined by the International Atomic Energy Agency (IAEA).

- Ensuring compliance with Australia's non-proliferation commitments under the Nuclear Non-Proliferation (Safeguards) Act.

D. Construction and Operating Licenses

If the design certification is approved, developers would then apply for a construction license. This process would involve:

- Submitting detailed construction plans, including the design of critical safety systems, infrastructure, and site security measures.

- Conducting site inspections and audits to ensure compliance with safety and environmental regulations.

- Once construction is complete, a final safety assessment would be required before an operating license is granted. This would include testing the reactor systems, emergency preparedness drills, and verifying radiation protection measures.

4. Challenges and Considerations for Licensing SMRs in Australia

Deploying SMRs in Australia presents several unique challenges:

- **Regulatory and Legislative Reform**: The current legal framework prohibits nuclear power generation, so substantial legal reforms would be required to enable SMR deployment. This would involve amending federal and state laws and developing new regulations specific to SMRs.

- **Public Perception**: There is significant public resistance to nuclear power in Australia due to historical concerns about safety, waste management, and environmental impact. To gain public support, extensive educational campaigns and transparent stakeholder engagement would be necessary.

- **Supply Chain and Infrastructure**: Australia lacks the existing nuclear infrastructure that countries like the United States or Canada have. Developing a domestic supply chain and training a skilled workforce would be essential to support the construction and operation of SMRs.

- **Site Selection and Environmental Protection**: Australia's stringent environmental regulations and sensitivity to Indigenous land rights mean that site selection would need to be approached with great care. Proposals would have to demonstrate minimal environmental impact and include robust plans for waste management and decommissioning.

5. The Path Forward

While SMRs present a promising solution for Australia's energy challenges, particularly in achieving decarbonization goals, the path to deployment is filled with regulatory, legal, and social complexities. Moving forward, Australia would need to take a phased approach, starting with feasibility studies, legislative reform, and pilot projects to demonstrate the safety and viability of SMR technologies. By addressing these challenges, Australia could potentially leverage SMRs as a cornerstone of its future energy strategy, providing a low-carbon, reliable energy source to complement renewables and reduce the country's reliance on fossil fuels.

Future Trends in Nuclear Regulatory Policies

The future trends in nuclear regulatory policies are shaped by a confluence of safety concerns, technological advancements, environmental considerations, and public perception. As nations grapple with energy security and climate change, the regulatory frameworks governing nuclear energy are evolving to address these multifaceted challenges.

One significant trend is the strengthening of safety regulations in response to past nuclear accidents. The Fukushima Dai-ichi disaster in 2011 prompted many countries to reassess their nuclear policies and enhance safety protocols for existing reactors and new constructions [210]. This shift reflects a broader global trend towards prioritizing safety in nuclear energy, which is crucial for gaining public trust and acceptance. For instance, in South Korea, regulatory bodies have intensified safety standards and considered decommissioning older plants to mitigate risks associated with radiation exposure [211]. Similarly, the establishment of independent regulatory agencies, such as India's Nuclear Safety Regulatory Authority, illustrates a move towards more robust governance structures that can adapt to emerging safety concerns [212].

Another trend is the increasing integration of environmental considerations into nuclear regulatory policies. As countries strive to meet Sustainable Development Goals (SDGs) and combat climate change, nuclear energy is being recognized as a low-carbon alternative to fossil fuels [213]. This recognition is leading to a re-evaluation of nuclear energy's role within national energy strategies, particularly in countries with high energy demands. For example, nations like France and Slovakia have successfully integrated nuclear power into their energy mix, achieving over 50% of their energy needs through nuclear sources [214]. This trend suggests that future regulatory frameworks will likely emphasize the environmental benefits of nuclear energy while addressing waste management and sustainability issues [215].

Technological advancements, particularly in reactor design, are also influencing regulatory policies. The emergence of small modular reactors (SMRs) presents new opportunities for safer and more flexible nuclear energy deployment. These reactors are designed to be more economically viable and have lower risks of accidents, which could lead to more favourable regulatory environments [60]. However, the successful integration of SMRs into existing regulatory frameworks will require careful consideration of safety, economic, and environmental factors [216]. Furthermore, the development of thorium-based nuclear technologies is gaining attention as a potential future fuel source, which may necessitate new regulatory approaches to accommodate its unique characteristics [217].

Public perception remains a critical factor in shaping nuclear regulatory policies. Studies indicate that public attitudes towards nuclear energy vary significantly across regions, influenced by historical accidents and current energy needs [218]. In OECD countries, scepticism towards nuclear energy is prevalent, while non-OECD countries often exhibit more favourable views [219]. As such, regulatory bodies must engage with the public to foster understanding and acceptance of nuclear energy, which is essential for the successful implementation of future nuclear projects.

Chapter 5
Construction Process of SMR-Based Power Stations

Planning and Timeline for SMR Construction

Design Phases for SMR Development

The design and development of Small Modular Reactors (SMRs) are complex processes that involve multiple phases to ensure the safety, efficiency, and compliance of the reactor systems. Globally, reactor vendors, operators, and regulators may use different terminologies to describe the stages of reactor design, which can lead to misunderstandings, especially when a reactor design is being licensed across multiple jurisdictions with well-established nuclear regulations. To overcome these inconsistencies, it is crucial to establish a clear, universally accepted framework that categorizes the stages of design progression. This report breaks down the design process into four distinct phases: Conceptual Design, Plant-Level Engineering Design, System-Level Engineering Design, and Component-Level Engineering Design. These phases represent increasing levels of detail and are tailored to address the specific safety and environmental assessments required at each stage. The description of each phase below is based on the requirements for a first-of-a-kind (FOAK) SMR, with a reduced scope for nth-of-a-kind (NOAK) reactors, especially if they are licensed and constructed in the same country as the initial FOAK unit [220].

Phase 1: Conceptual Design

The first stage, known as the Conceptual Design Phase, involves selecting and enhancing design options. This phase is dedicated to addressing critical questions, developing preliminary solutions, identifying major risks, and formulating mitigation strategies. The main objective here is to establish a strong foundation for the overall reactor design by exploring various concepts and aligning them with the desired safety principles [220].

The outcome of this phase is typically a comprehensive set of documents outlining the initial design approach, key safety principles, and the rationale behind critical decisions. These documents serve as a blueprint for future design developments. Ideally, the majority of this work is completed before formal engagement with regulatory authorities, though some jurisdictions may allow limited pre-licensing discussions to address initial regulatory concerns [220].

Phase 2: Plant-Level Engineering Design

In the Plant-Level Engineering Design Phase, the focus shifts to defining the essential systems, structures, and components (SSCs) of the reactor. During this stage, key elements such as process flow diagrams, initial instrumentation and control (I&C) architecture, preliminary layout designs, and the first drafts of 3D models are developed. This phase also involves defining safety specifications and establishing preliminary design parameters for critical SSCs [220].

At this point, the design is advanced enough to allow initial assessments of the reactor's compliance with regulatory safety standards, environmental impact, and security requirements. The detailed documentation produced during this phase serves as a foundation for regulatory review and sets the stage for more detailed engineering work in subsequent phases [220].

Phase 3: System-Level Engineering Design

The System-Level Engineering Design Phase is characterized by further refinement of the SSCs and the development of additional systems necessary for reactor functionality. This phase expands on the groundwork laid in Phase 2, incorporating detailed piping and instrumentation diagrams, a comprehensive plant and item list, and updated architectural plans for the control systems. It also includes a more sophisticated 3D model and detailed specifications for critical components.

During this phase, the design team typically expands, often involving external contractors or partners to handle specific subsystems. Key assessments are conducted, such as design basis conditions, probabilistic safety assessments (PSA Levels 1 and 2), human factors engineering, and preliminary internal and external hazard evaluations. The design at this stage is sufficiently detailed to support the creation of a Preliminary Safety Analysis Report (PSAR), which is essential for securing regulatory approval before moving to the construction phase [220].

Phase 4: Component-Level Engineering Design

The final phase, known as the Component-Level Engineering Design Phase, focuses on the detailed engineering required to manufacture all SSCs for the entire plant. This is a critical phase for SMRs due to their modular construction approach, which requires many systems to be fully designed and fabricated before on-site assembly can begin. Unlike traditional large nuclear reactors, where certain non-safety-critical systems can be designed during construction, SMRs necessitate that most components be finalized early in the process [220].

The primary objective of this phase is to produce the final detailed engineering designs, including manufacturing specifications, component layouts, and complete 3D models. The output also includes finalized documentation for safety assessments, covering deterministic analysis, probabilistic safety assessments, and comprehensive evaluations of human factors, internal and external hazards, decommissioning, and environmental impact. This phase culminates in the production of a Final Safety Analysis Report (FSAR), which is crucial for obtaining the necessary licenses to proceed with reactor construction and operation [220].

Summary of the Design Phase Integration

The four-phase approach to SMR design provides a structured methodology that enhances clarity and ensures consistency across international regulatory frameworks. By progressively increasing the level of detail and focusing on safety and environmental assessments at each stage, this methodology facilitates a smoother regulatory approval process, particularly for FOAK projects. For NOAK reactors, the established documentation from earlier phases can significantly reduce the time and cost of licensing, especially if the design remains consistent with the approved FOAK reactor [220].

Overall, this phased approach not only improves regulatory compliance but also optimizes the development timeline, allowing for more efficient deployment of SMRs to meet global energy needs. The structured progression from conceptual design to detailed component engineering ensures that each reactor is built to the highest safety standards, while also accommodating the unique advantages of modular construction inherent to SMR technology [220].

Planning for an SMR Power Plant Project

The development and construction of a Small Modular Reactor (SMR) power plant is a complex and multi-phase process that requires meticulous planning to ensure regulatory compliance, safety, cost efficiency, and timely project execution. The planning process involves a detailed understanding of design phases, regulatory requirements, site assessments, and coordination between various stakeholders, including reactor vendors, regulatory authorities, and engineering contractors. The planning and execution of SMR projects must be aligned with the specific design phases outlined below to ensure the successful deployment of a first-of-a-kind (FOAK) or nth-of-a-kind (NOAK) SMR reactor. This structured approach not only enhances the project's feasibility but also streamlines regulatory approvals and optimizes the overall project timeline.

Phase 1: Conceptual Design

The initial planning phase, known as the Conceptual Design phase, involves establishing the foundational elements of the SMR project. This stage focuses on selecting optimal design options, identifying critical risks, and developing preliminary solutions to address potential challenges. During this phase, reactor vendors and project stakeholders engage in high-level assessments to explore various reactor concepts, evaluate technological innovations, and align the project with regulatory expectations.

A comprehensive set of documents is produced during this phase, which outlines the safety principles, design rationale, and key decisions taken. These documents serve as blueprints for subsequent stages of development. While regulatory engagement is often limited during this phase, some jurisdictions allow early pre-licensing consultations to address initial regulatory concerns and establish a clear pathway for the licensing process. The output from the Conceptual Design phase provides a strategic direction for the project, setting the stage for detailed engineering and safety assessments in subsequent phases.

Phase 2: Plant-Level Engineering Design

Once the conceptual framework is established, the project moves into the Plant-Level Engineering Design phase. At this stage, the focus is on defining the essential systems, structures, and components (SSCs) that will form the backbone of the SMR plant. Detailed process flow diagrams, initial instrumentation and control (I&C) architecture, preliminary layout designs, and 3D models are developed to provide a clearer picture of the reactor's physical and operational structure.

During this phase, project teams also begin to define safety specifications and establish preliminary design parameters for critical systems. This is crucial for conducting initial assessments of regulatory compliance, environmental impact, and security measures. The detailed documentation produced at this stage forms the basis for regulatory review, enabling stakeholders to demonstrate adherence to safety standards and environmental protection requirements. Additionally, this phase provides the foundation for more advanced engineering work, setting the project up for system-level integration in the next phase.

Phase 3: System-Level Engineering Design

The System-Level Engineering Design phase is where the project transitions from high-level planning to detailed engineering. During this phase, the definitions of SSCs are refined, and additional systems necessary for the reactor's functionality are integrated into the design. This includes the development of detailed piping and instrumentation diagrams, a comprehensive plant and equipment list, and enhanced architectural plans for control systems.

At this stage, the design team expands, often incorporating external contractors with specialized expertise in subsystems like reactor chemistry, radiological protection, and radioactive waste management. Critical assessments are carried out, such as design basis conditions, probabilistic safety assessments (PSA Levels 1 and 2), and preliminary hazard evaluations. The output of this phase includes a Preliminary Safety Analysis Report (PSAR), which is a key requirement for obtaining regulatory approval to proceed with construction. This report encompasses all safety-related design details, human factors engineering, and environmental impact assessments, ensuring that the reactor meets the stringent requirements for operational safety and environmental protection.

Phase 4: Component-Level Engineering Design

The final phase in the planning process is the Component-Level Engineering Design phase, where the focus shifts to the detailed engineering required for manufacturing and construction. This phase is critical for SMRs due to their modular construction approach, which requires most systems to be fully designed, fabricated, and tested before they are transported to the site for assembly.

In this phase, the project team produces the finalized design documentation, including manufacturing specifications, component layouts, and complete 3D models of the reactor systems. The aim is to minimize design modifications once construction begins, thereby reducing costs and project timelines. The output also includes the Final Safety Analysis Report (FSAR), which covers deterministic safety analysis, probabilistic safety assessments, human factors, internal and external hazards, and decommissioning plans. This comprehensive documentation is crucial for obtaining the final operating license, enabling the reactor to move into the construction and commissioning stages.

Integrating Regulatory Compliance and Site Planning

Throughout the planning process, compliance with national and international regulatory standards is paramount. In countries with well-established nuclear regulatory frameworks, SMR projects must align with safety guidelines, environmental protection laws, and licensing requirements. This necessitates close collaboration with regulatory authorities such as the National Nuclear Safety Administration (NNSA) in China, the Nuclear Regulatory Commission (NRC) in the United States, or the Canadian Nuclear Safety Commission (CNSC) in Canada.

Site selection is another critical component of the planning process. Factors such as seismic stability, proximity to water sources, population density, and environmental impact assessments play a significant role in determining the feasibility of an SMR project. Comprehensive Environmental Impact Assessments (EIAs) are conducted to identify potential risks to the environment and public health, ensuring that the project aligns with sustainable development goals.

Project Execution: From Design to Deployment

Once the design phases are completed and the necessary regulatory approvals are secured, the focus shifts to the construction and commissioning of the SMR power plant. Given the modular nature of SMRs, much of the construction occurs off-site in controlled factory environments, allowing for faster assembly and reduced costs. Modular construction also minimizes the environmental footprint and reduces the time required for on-site activities, which is especially beneficial for projects located in remote or environmentally sensitive areas.

The commissioning phase involves rigorous testing and validation of all systems to ensure they meet safety and performance standards. This includes cold and hot functional tests, reactor start-up tests, and full-load operation trials. Once the reactor passes these tests, it receives its operating license, allowing it to begin commercial operations.

Including Building Codes, Regulations and Standards in the Planning Process

The construction of a Small Modular Reactor (SMR) power plant is a highly regulated and complex process that must adhere to a range of building codes, construction regulations, and standards. These requirements are essential to ensure safety, environmental protection, and compliance with both national and international guidelines. The following provides an overview of key building codes, construction regulations, and standards that need to be accommodated when planning for an SMR power plant construction project.

When planning the construction of a Small Modular Reactor (SMR) power plant, compliance with various building codes, construction regulations, and nuclear-specific standards is essential to meet the safety, environmental, and operational requirements of the host country. Each country has its own set of regulations and standards, as well as international guidelines that influence their frameworks. Below is an example of some of the key applicable codes and standards for SMR projects in the United States, Europe, China, India, the United Kingdom, Canada, France, Russia and Australia, noting that this is not an exhaustive list.

1. United States

When planning the construction of a Small Modular Reactor (SMR) power plant in the United States, a variety of building codes, construction regulations, and nuclear-specific standards must be adhered to. The United States has a highly regulated framework for nuclear power projects, and SMRs must comply with both nuclear regulatory requirements and general construction standards. Below is a detailed explanation of the codes, standards, and regulations that apply to the construction of SMR power plants in the U.S.

1. Nuclear Regulatory Commission (NRC) Regulations

The primary regulatory authority for nuclear power in the U.S. is the Nuclear Regulatory Commission (NRC). The NRC has established comprehensive regulations for the design, construction, and operation of nuclear power plants, including SMRs. The following parts of the Code of Federal Regulations (CFR) are particularly relevant:

- **10 CFR Part 50**: This regulation covers the licensing of production and utilization facilities, including traditional nuclear reactors. SMR developers can apply for construction permits and operating licenses separately under this part.

- **10 CFR Part 52**: For SMRs, it is often more efficient to apply under Part 52, which covers combined construction and operating licenses (COLs). This part simplifies the licensing process by integrating safety reviews, construction permits, and operating licenses into a single process.

- **10 CFR Part 72**: Covers the licensing requirements for the independent storage of spent nuclear fuel, which is essential for SMRs that generate radioactive waste.

- **10 CFR Part 73**: Establishes physical security requirements to protect nuclear facilities from sabotage or unauthorized access.

- **10 CFR Part 100**: Specifies site criteria for nuclear power plants, including geological and seismological evaluations.

These NRC regulations ensure that the safety, security, and environmental protection requirements for nuclear power plants are met throughout the design, construction, and operational phases.

2. American Society of Mechanical Engineers (ASME) Boiler and Pressure Vessel Code

The **ASME Boiler and Pressure Vessel Code (BPVC)** is critical for the design and construction of nuclear reactor components:

- **ASME Section III**: Focuses specifically on the construction of nuclear power plant components, such as reactor vessels, piping, and steam generators.

- **ASME Section XI**: Provides guidelines for the in-service inspection of nuclear power plant components to ensure ongoing safety and integrity.

The ASME BPVC ensures that all pressure-containing components are constructed to withstand operational stresses and maintain safety throughout the plant's life cycle.

3. Institute of Electrical and Electronics Engineers (IEEE) Standards

Electrical systems in SMRs must comply with IEEE standards, which are essential for ensuring the reliability and safety of electrical and control systems in nuclear facilities:

- **IEEE 603**: Establishes criteria for nuclear power plant protection and control systems.

- **IEEE 7-4.3.2**: Focuses on software requirements for safety systems in nuclear power plants, which is particularly relevant for the advanced control systems used in SMRs.

These standards ensure the reliability of instrumentation, control systems, and software critical to reactor safety.

4. National Fire Protection Association (NFPA) Codes

Fire protection is crucial for the safety of nuclear facilities:

- **NFPA 804**: Provides guidelines for fire protection in nuclear power plants.

- **NFPA 805**: Allows a risk-informed, performance-based approach to fire protection, providing flexibility in how fire safety is achieved in SMRs.

The application of these NFPA codes is necessary to minimize the risk of fire-related incidents within nuclear facilities.

5. International Building Code (IBC)

The **International Building Code (IBC)** is the primary code used for general construction in the U.S. and applies to the structural design of SMR buildings:

- **IBC 2018/2021**: Provides guidelines on structural integrity, seismic design, wind resistance, and fire safety. Nuclear power plants, including SMRs, must comply with these requirements to ensure the safety of buildings housing reactor systems.

- **ASCE 7-16 (Minimum Design Loads for Buildings and Other Structures)**: Used in conjunction with the IBC to determine the structural loads due to wind, seismic activity, snow, and other factors.

Compliance with IBC ensures that the construction of SMR facilities meets the highest standards for structural integrity and safety.

6. Environmental Protection Agency (EPA) Regulations

Environmental protection is a significant aspect of SMR projects, requiring compliance with EPA regulations:

- **National Environmental Policy Act (NEPA)**: Requires an Environmental Impact Statement (EIS) to assess the potential environmental impacts of SMR construction and operation.

- **Clean Air Act (CAA)** and **Clean Water Act (CWA)**: Ensure that emissions and discharges from nuclear facilities meet federal environmental standards.

These regulations protect air and water quality, ensuring that SMRs have minimal impact on the surrounding environment.

7. Occupational Safety and Health Administration (OSHA) Standards

Worker safety is governed by **OSHA standards**, which apply to the construction, operation, and maintenance of nuclear facilities:

- **29 CFR Part 1910**: Covers general industry standards, including electrical safety, hazardous materials handling, and personal protective equipment (PPE).

- **29 CFR Part 1926**: Provides safety standards specifically for construction projects, ensuring that SMR construction sites adhere to best practices in worker safety.

Compliance with OSHA standards is essential to protect workers during the construction and operation of SMR plants.

8. Seismic and Structural Design Codes

Given that SMRs may be deployed in various locations, including those with seismic activity, compliance with **seismic design standards** is crucial:

- **ASCE 43-19 (Seismic Design Criteria for Structures, Systems, and Components in Nuclear Facilities)**: Specifies design criteria to ensure that SMR structures can withstand seismic events.

- **IBC Chapter 16**: Covers the structural design requirements for buildings in seismically active regions.

These standards help ensure the resilience of SMR facilities in the event of an earthquake.

9. Other Relevant Standards

- **ANSI/ANS (American Nuclear Society) Standards**: These cover various aspects of nuclear facility design, such as ANS-2.2 (Earthquake Instrumentation Criteria) and ANS-58.14 (Safety and Pressure Relief Systems).

- **Nuclear Quality Assurance (NQA-1)**: Establishes quality assurance requirements for nuclear facility construction, ensuring that components and systems meet the highest standards of quality and reliability.

2. Europe (General Framework)

Europe is a region with diverse regulatory frameworks due to its member countries' varying national requirements, alongside overarching European Union (EU) directives. Below is an in-depth explanation of the key standards and regulations that must be taken into account for SMR power plant construction in Europe.

1. European Nuclear Safety Framework

In Europe, the construction and operation of nuclear facilities, including SMRs, are governed by a combination of national regulations and European Union (EU) directives. These regulations ensure that nuclear power plants meet stringent safety, environmental, and construction standards. Key EU directives include:

- **Council Directive 2009/71/Euratom (amended by Directive 2014/87/Euratom): Establishes a Community framework for nuclear safety, ensuring that member states** implement high standards of safety for nuclear installations. It requires regular safety assessments, peer reviews, and continuous improvement.

- **Council Directive 2011/70/Euratom**: Governs the responsible management of spent fuel and radioactive waste. It mandates that EU member states have national frameworks for the safe disposal of radioactive waste, including provisions for spent fuel from SMRs.

- **Council Directive 2013/59/Euratom**: Sets basic safety standards for radiation protection, covering the protection of workers, the public, and the environment from ionizing radiation. This directive is relevant for the operation and maintenance of SMRs.

These EU directives establish a common legal framework for nuclear safety, waste management, and radiation protection across Europe, which must be integrated into the planning and construction phases of SMR projects.

2. National Regulatory Authorities and Licensing

Each European country has its own **national nuclear regulatory authority** responsible for overseeing the licensing and construction of nuclear power plants, including SMRs. These authorities are responsible for ensuring that nuclear projects comply with both national laws and EU directives. Examples include:

- **France**: The **Autorité de Sûreté Nucléaire (ASN)** is responsible for nuclear safety, licensing, and regulatory oversight.

- **Germany**: The **Bundesamt für die Sicherheit der nuklearen Entsorgung (BASE)** and the **Federal Ministry for the Environment, Nature Conservation, Nuclear Safety, and Consumer Protection (BMUV)** oversee nuclear safety.

- **United Kingdom**: The **Office for Nuclear Regulation (ONR)** manages the regulation of nuclear safety and licensing, even after Brexit.

- **Finland**: The **Radiation and Nuclear Safety Authority (STUK)** handles nuclear safety regulations.

Each country may have additional national regulations that must be followed during the planning, construction, and operation phases, so it is crucial for SMR developers to understand the specific requirements of the host country.

3. International Atomic Energy Agency (IAEA) Guidelines

Although not specific to the EU, the International Atomic Energy Agency (IAEA) provides internationally recognized guidelines that are often adopted by European countries:

- **IAEA Safety Standards**: These cover aspects such as site selection, design safety, operational safety, and decommissioning. European regulators typically align their requirements with IAEA standards to ensure global best practices.

- **IAEA GSR Part 4**: Applies to safety assessments during the design, construction, and operation of nuclear reactors, including SMRs.

These guidelines provide a comprehensive framework for ensuring the safety and reliability of SMR projects and are typically used alongside national regulations.

4. European Building and Construction Codes

The construction of an SMR power plant in Europe must also comply with general European building and construction standards, including:

- **Eurocodes**: The Eurocodes are a set of harmonized structural design standards used across Europe. Key Eurocodes relevant to SMR construction include:

  - **Eurocode 2 (EN 1992)** for the design of concrete structures.

  - **Eurocode 3 (EN 1993)** for the design of steel structures.

  - **Eurocode 7 (EN 1997)** for geotechnical design, crucial for assessing soil stability and foundation requirements for SMRs.

  - **Eurocode 8 (EN 1998)** for seismic design, ensuring that nuclear facilities can withstand earthquakes.

- **EN 1090**: Specifies requirements for the execution of steel structures and aluminium structures, ensuring the integrity of construction materials.

- **EN 61513**: Covers **instrumentation and control systems** specific to nuclear reactors, ensuring reliability and safety in operations.

These codes are essential for ensuring the structural integrity and resilience of SMR power plants, particularly in seismically active regions or areas with challenging environmental conditions.

5. *Environmental and Impact Assessments*

Environmental protection is a critical aspect of SMR construction in Europe. Developers must adhere to the following directives:

- **Environmental Impact Assessment (EIA) Directive (2014/52/EU)**: Requires an EIA for projects likely to have significant environmental impacts, including nuclear power plants. This involves assessing the potential effects of the SMR on local ecosystems, water sources, and air quality.

- **Habitats Directive (92/43/EEC)** and **Birds Directive (2009/147/EC)**: Ensure that SMR projects do not negatively impact protected natural habitats and bird species.

- **Water Framework Directive (2000/60/EC)**: Regulates the impact of nuclear facilities on water bodies, ensuring that the construction and operation of SMRs do not contaminate water resources.

These environmental regulations ensure that the construction and operation of SMRs align with Europe's commitment to sustainability and environmental protection.

6. *Fire Protection and Safety Standards*

Fire safety is critical in nuclear facilities, including SMRs:

- **EN 1366**: Sets standards for fire resistance of service installations, such as ventilation and electrical systems.

- **EN 54**: Governs fire detection and alarm systems to ensure rapid response to potential fire hazards.

- **NFPA Standards**: Although primarily American, certain National Fire Protection Association (NFPA) standards are also referenced in Europe to ensure comprehensive fire protection strategies.

7. Electrical and Instrumentation Standards

The electrical and control systems in SMRs must comply with standards such as:

- **IEC 61513**: Covers the instrumentation and control systems for nuclear power plants, ensuring the safety and reliability of control systems.

- **IEC 61226**: Classifies the safety functions of instrumentation and control systems, which is critical for ensuring the integrity of SMR operations.

These standards ensure that the electrical systems supporting SMR reactors are designed for maximum safety and efficiency.

The construction of an SMR power plant in Europe requires compliance with a comprehensive set of international, regional, and national regulations. This includes nuclear-specific safety standards, general construction codes, environmental protection laws, and fire safety measures. By adhering to these diverse standards, SMR developers can ensure that their projects meet the rigorous safety, environmental, and construction quality requirements necessary for successful deployment in Europe.

Planning for an SMR in Europe involves not only addressing regulatory and technical challenges but also navigating the complexities of the multi-layered regulatory environment. Developers must engage early with national regulatory authorities, comply with EU directives, and align with international guidelines to achieve the necessary approvals for construction and operation.

3. China

China has a well-established regulatory framework for nuclear power plants, which emphasizes both safety and efficient project execution. Below is a detailed explanation of the key standards and regulations that must be accounted for when planning the construction of an SMR power plant in China.

1. China's Nuclear Regulatory Framework

China's regulatory oversight of nuclear facilities, including SMRs, is primarily managed by the National Nuclear Safety Administration (NNSA), which operates under the Ministry of Ecology and Environment (MEE). The NNSA sets the standards for nuclear safety, reactor design, construction, operation, and decommissioning. Key regulations and standards enforced by the NNSA include:

- **Regulations on the Safety of Civil Nuclear Facilities (HAF001)**: Establishes the overall safety framework for nuclear facilities in China, covering design, construction, commissioning, operation, and decommissioning.

- **Nuclear Safety Law of the People's Republic of China (2017)**: This law sets forth the legal foundation for nuclear safety, establishing the responsibilities of nuclear operators, regulatory authorities, and third parties involved in nuclear projects. It includes requirements for safety assessments, inspections, and licensing procedures.

- **HAF102 and HAF103**: Provide detailed safety guidelines for the siting, design, and construction of nuclear power plants, ensuring that all projects meet rigorous safety standards to protect public health and the environment.

These regulations serve as the legal foundation for the safe development and operation of SMRs in China, ensuring that all nuclear projects are conducted under strict regulatory oversight.

2. Design and Safety Standards for SMRs

China has adopted and adapted various international nuclear safety standards, primarily those recommended by the International Atomic Energy Agency (IAEA), to align its own regulations with global best practices. However, China has also developed its own standards tailored to local conditions and specific nuclear technologies. Relevant standards for SMR construction include:

- **GB/T 50318-2017**: This national standard covers the safety requirements for nuclear power plant design, including seismic design criteria, structural integrity, and fire protection measures.

- **GB/T 50739-2012**: Addresses the seismic safety evaluation of nuclear facilities, which is crucial given China's diverse seismic activity zones.

- **HJ 660-2013**: Sets forth environmental protection standards specifically for nuclear power plants, covering radioactive emissions, waste management, and radiation protection for the surrounding environment.

China's safety standards ensure that SMRs are designed to withstand natural disasters, including earthquakes and extreme weather events, while also minimizing environmental impacts.

3. Building Codes and Construction Regulations

In addition to nuclear-specific standards, SMR projects in China must comply with general construction codes that apply to all large-scale infrastructure projects. These include:

- **GB 50229-2006**: Covers the code for the design of nuclear power plants, with a focus on structural design, material specifications, and construction techniques to ensure safety and durability.

- **GB 50009-2012**: The building load code, which defines the structural loads that must be considered during the design phase, such as wind loads, snow loads, and seismic forces.

- **GB 50204-2015**: Outlines the construction quality standards for concrete structures, ensuring the integrity of reactor containment buildings and other critical infrastructure.

- **GB 50191-2012**: Provides specifications for the seismic design of buildings and structures, particularly important for nuclear facilities in seismically active regions.

- **JGJ 18-2012**: The standard for fire safety in construction, ensuring that SMRs are built with adequate fire protection systems, including fire-resistant materials and detection systems.

These codes are crucial for ensuring that the SMR is constructed to withstand not only operational stresses but also external hazards, such as earthquakes and fires.

4. Environmental Protection and Impact Assessment

The Ministry of Ecology and Environment (MEE) in China oversees environmental protection for all infrastructure projects, including nuclear power plants. Key regulations include:

- **Environmental Impact Assessment (EIA) Law (2003, revised 2018)**: Requires a comprehensive EIA before any construction begins. This assessment evaluates the potential impacts of the SMR on air quality, water resources, and local ecosystems.

- **HJ 2.1-2011**: Guidelines for the preparation of EIA reports for nuclear facilities, detailing the methodologies for assessing radiological impacts, waste management, and emergency response plans.

- **Regulations on the Prevention and Control of Environmental Pollution by Solid Waste (2020)**: Governs the disposal of radioactive waste, ensuring that SMRs manage waste in an environmentally safe manner.

These environmental regulations ensure that SMR projects minimize their impact on the surrounding environment and comply with China's stringent pollution control standards.

5. Fire Protection and Safety Standards

Fire safety is a critical aspect of nuclear power plant construction in China:

- **GB 50261-2005**: The national code for fire protection in nuclear facilities, covering fire-resistant materials, emergency response systems, and fire detection and suppression systems.

- **NFPA Standards**: While China primarily uses its own national standards, some facilities also reference National Fire Protection Association (NFPA) standards to ensure comprehensive fire safety.

6. Instrumentation and Control Standards

The control and safety systems in SMRs must comply with rigorous standards to ensure operational safety:

- **GB/T 19671-2011**: Specifies requirements for the instrumentation and control systems of nuclear power plants, covering control room design, automation, and system redundancies.

- **GB/T 16813-2018**: Focuses on the classification of safety-related I&C systems, ensuring that critical systems are designed to prevent failures and mitigate potential risks.

These standards are crucial for maintaining the reliability of the control systems that govern reactor operations and emergency shutdown procedures.

Planning for the construction of an SMR power plant in China requires careful adherence to a comprehensive set of regulations, building codes, and safety standards. This includes compliance with both national and international nuclear safety guidelines, structural and seismic codes, environmental impact assessments, and fire safety measures.

The integration of these standards ensures that SMR projects in China are not only safe and efficient but also environmentally sustainable. Developers must work closely with the NNSA and MEE to navigate the complex regulatory landscape, ensuring that their projects meet all legal and technical requirements for construction, operation, and decommissioning. By adhering to these stringent standards, SMR projects can contribute to China's ambitious goals for clean energy and carbon neutrality.

4. India

The construction of Small Modular Reactor (SMR) power plants in India involves navigating a rigorous framework of building codes, construction regulations, nuclear safety standards, and environmental protection guidelines. India has a well-established nuclear power industry, governed by national standards as well as international guidelines to ensure the safety, efficiency, and sustainability of nuclear projects. Below is an in-depth description of the key regulations and standards that must be considered when planning for the construction of an SMR in India.

1. India's Nuclear Regulatory Framework

India's nuclear power sector is primarily regulated by the Atomic Energy Regulatory Board (AERB), which falls under the jurisdiction of the Department of Atomic Energy (DAE). The AERB is responsible for ensuring that nuclear power plants, including SMRs, comply with all safety regulations throughout the entire lifecycle of the plant, from design to decommissioning. The key regulatory documents that impact SMR construction include:

- **Atomic Energy Act, 1962**: This act provides the legal framework for the regulation of all nuclear activities in India, including the construction, operation, and decommissioning of nuclear power plants.

- **AERB Safety Code for Nuclear Power Plants (AERB/NPP/SC)**: Establishes the safety requirements for the design, construction, commissioning, and operation of nuclear power plants, ensuring compliance with both national and international safety standards.

- **AERB Safety Guide on Siting of Nuclear Power Plants (AERB/SG/S-1)**: Provides guidelines for selecting suitable sites for nuclear power plants, including the evaluation of geological, seismic, and environmental factors.

These regulations ensure that all nuclear facilities, including SMRs, are designed and operated with the highest safety standards to protect public health and the environment.

2. Design and Safety Standards for SMRs

The design and safety assessment of SMRs in India must comply with both national and international standards, especially those recommended by the International Atomic Energy Agency (IAEA). The AERB has adopted several IAEA guidelines and tailored them to Indian conditions. Key safety standards relevant to SMRs include:

- **AERB/NPP/SG/D-3**: This safety guide focuses on the design safety of nuclear power plants, including seismic safety, structural integrity, fire protection, and radiation shielding.

- **IS 875 (Part 3): 2015**: This Indian Standard covers the design loads for buildings and structures, specifically wind loads, which is crucial for the structural design of SMRs to withstand extreme weather conditions.

- **IS 1893 (Part 1): 2016**: Specifies the criteria for earthquake-resistant design of structures, particularly important for ensuring the safety of nuclear facilities located in seismically active zones.

India's adherence to both national standards and IAEA guidelines ensures that SMRs are designed to handle various natural and man-made hazards.

3. Building Codes and Construction Regulations

SMR projects in India must also comply with general construction codes and standards that apply to all infrastructure projects, particularly those related to building safety, structural integrity, and fire protection:

- **National Building Code of India (NBC), 2016**: The NBC is the overarching guideline for construction practices in India. It covers all aspects of building design and construction, including fire safety, structural design, and safety during construction.

- **IS 456: 2000**: This standard covers the general requirements for the design and construction of reinforced concrete structures, which is critical for the construction of SMR containment buildings and other concrete structures.

- **IS 3370 (Part 2): 2009**: Provides guidelines for the design of concrete structures for the containment of hazardous materials, ensuring that SMRs are built with materials that can handle radioactive content securely.

These construction codes ensure that SMR projects in India are built to withstand structural stresses while maintaining high safety standards.

4. Environmental Protection and Impact Assessment

The **Ministry of Environment, Forest and Climate Change (MoEFCC)** regulates the environmental aspects of SMR construction projects in India. Nuclear projects must undergo a rigorous environmental assessment to ensure that they comply with the country's environmental protection laws:

- **Environment (Protection) Act, 1986**: Establishes the legal framework for environmental protection in India and mandates the assessment of potential environmental impacts for all major infrastructure projects.

- **Environmental Impact Assessment (EIA) Notification, 2006**: Requires a comprehensive EIA report for nuclear power projects, covering areas such as air and water quality, radiation protection, waste management, and the social impact on nearby communities.

- **AERB/SG/G-8**: Provides specific guidelines for the management of radioactive waste to minimize environmental contamination and ensure the safe disposal of spent fuel.

These environmental regulations are designed to protect ecosystems, water resources, and communities surrounding nuclear power plants.

5. Fire Safety and Protection Standards

Fire safety is a critical concern in nuclear power plants, including SMRs, to prevent accidents that could lead to catastrophic consequences:

- **AERB/SG/D-4**: Establishes fire protection requirements for nuclear facilities, covering fire detection, suppression systems, and emergency response protocols.

- **NBC Fire and Life Safety Provisions (Part 4)**: Specifies fire safety standards for industrial facilities, ensuring that SMRs are equipped with adequate fire suppression and emergency evacuation systems.

6. Instrumentation and Control Systems

The control systems for SMRs in India must comply with stringent standards to ensure safe operation and effective emergency responses:

- **AERB/SG/D-9**: Covers the safety classification and design requirements for instrumentation and control systems in nuclear power plants.

- **IS 3043: 2018**: Provides guidelines for earthing in electrical installations, which is crucial for ensuring the reliability of control systems in SMRs.

- **IS 732: 2019**: Specifies the code of practice for electrical wiring installations, which must be followed to prevent electrical faults that could compromise the safety of SMRs.

Planning for the construction of an SMR power plant in India requires comprehensive adherence to a diverse set of regulations, building codes, and safety standards. The integration of these guidelines ensures that SMR projects are both safe and efficient, while also minimizing their environmental footprint.

5. United Kingdom

The regulatory landscape in the UK is robust, aiming to ensure that all nuclear projects, including SMRs, are developed safely, efficiently, and sustainably. Below is a detailed breakdown of the key regulations and standards that must be considered when planning the construction of an SMR in the UK.

1. UK Nuclear Regulatory Framework

In the UK, the regulation of nuclear power plants is primarily overseen by the Office for Nuclear Regulation (ONR). The ONR is responsible for ensuring that all nuclear activities comply with safety, security, and environmental requirements. The ONR's regulatory approach focuses on risk management, safety assessments, and compliance with both national and international standards.

- **Nuclear Installations Act, 1965**: This act provides the legal foundation for the regulation of nuclear installations, including the licensing and oversight of nuclear power plants.

- **Health and Safety at Work Act, 1974**: Establishes the general framework for health and safety in workplaces, including construction sites for nuclear power projects.

- **Nuclear Industries Security Regulations, 2003**: Focuses on securing nuclear materials and protecting nuclear sites from potential threats, including cybersecurity concerns.

The ONR uses a **goal-setting regulatory approach**, which allows flexibility for developers to propose their own methods to achieve safety objectives, provided they can demonstrate compliance with the established safety principles.

2. Design Safety Standards for SMRs

The design and safety assessment of SMRs must comply with rigorous safety standards, including both UK-specific guidelines and international best practices established by the **International Atomic Energy Agency (IAEA)**. The key standards relevant to SMRs include:

- **ONR Safety Assessment Principles (SAPs)**: Provides detailed guidance on the expectations for nuclear safety, covering areas such as reactor core design, containment, and safety systems.

- **Western European Nuclear Regulators Association (WENRA) Safety Reference Levels**: Provides a harmonized approach to safety standards across Europe, which the UK continues to align with post-Brexit.

- **BS EN 61513**: This standard focuses on the design and management of instrumentation and control (I&C) systems for nuclear reactors, ensuring high levels of reliability and safety.

These safety standards ensure that SMRs are designed to mitigate risks associated with nuclear power generation, especially in areas like accident prevention, emergency response, and radiation protection.

3. Building Codes and Construction Regulations

SMR projects in the UK must comply with a comprehensive set of building codes and construction regulations to ensure structural integrity, fire safety, and overall compliance with national construction standards:

- **Building Regulations 2010**: These regulations cover all aspects of building design and construction, including fire safety, structural stability, and energy efficiency. Compliance with these regulations is essential for all new construction projects in the UK.

- **BS EN 1993 (Eurocode 3)**: Provides standards for the design of steel structures, which is particularly relevant for the construction of reactor buildings and containment structures.

- **BS EN 1992 (Eurocode 2)**: Applies to the design of concrete structures, ensuring that SMR construction meets high standards for durability and structural integrity.

- **Construction (Design and Management) Regulations, 2015**: These regulations focus on health and safety during the planning, design, and construction phases, emphasizing the need to minimize risks to construction workers.

These construction standards ensure that SMR projects are built to withstand various environmental and structural challenges, maintaining the highest levels of safety and quality.

4. Environmental and Planning Permissions

Environmental protection is a critical concern for nuclear projects in the UK. The Environment Agency (EA) and the Scottish Environment Protection Agency (SEPA) regulate the environmental aspects of nuclear power plants, including SMRs. Key regulations include:

- **Environmental Permitting Regulations, 2016**: This framework requires nuclear power projects to obtain environmental permits, covering aspects like radioactive waste management, emissions control, and water usage.

- **Planning Act, 2008**: Requires major infrastructure projects, including nuclear power plants, to undergo a thorough planning application process, which includes public consultations and environmental impact assessments (EIA).

- **Environmental Impact Assessment (EIA) Directive**: Although originally an EU regulation, the UK has retained this directive, requiring a comprehensive EIA for all large-scale construction projects to assess potential impacts on local ecosystems, air quality, and water resources.

5. Fire Safety and Protection Standards

Fire safety is a critical concern for SMRs to prevent accidents that could have catastrophic consequences. Key standards include:

- **BS 9999: 2017**: Focuses on fire safety in the design, management, and use of buildings, providing guidance on fire detection, suppression, and emergency evacuation procedures.

- **BS EN 13501-1**: Establishes the classification of building materials based on fire resistance, ensuring that materials used in SMR construction meet fire safety requirements.

These fire safety standards ensure that SMRs are equipped with adequate fire protection systems, reducing the risk of fire-related accidents.

6. Instrumentation and Control Systems

The instrumentation and control systems for SMRs in the UK must comply with stringent standards to ensure safe and efficient reactor operation:

- **BS EN 61508**: This standard focuses on the functional safety of electrical, electronic, and programmable electronic safety-related systems, ensuring reliability and reducing the risk of failure.

- **BS EN 61513**: Provides specific guidance on the requirements for instrumentation and control systems in nuclear power plants, emphasizing safety and redundancy.

7. Decommissioning and Waste Management

SMRs must also adhere to strict regulations regarding decommissioning and radioactive waste management to ensure minimal environmental impact:

- **Radioactive Waste Management Code of Practice**: Sets out the requirements for the safe storage, treatment, and disposal of radioactive waste generated by nuclear power plants.

- **Decommissioning Strategy (UK Government)**: The UK has specific guidelines for the safe decommissioning of nuclear facilities, including financial provisions and safety protocols to protect workers and the environment.

The construction and operation of an SMR power plant in the UK involve navigating a complex regulatory landscape that covers safety, environmental protection, structural integrity, fire safety, and waste management. By adhering to these comprehensive regulations, SMR developers can ensure that their projects meet the stringent safety standards expected in the UK, reducing risks to both the public and the environment.

Planning for an SMR project in the UK requires early engagement with regulatory bodies like the ONR, Environment Agency, and local planning authorities to ensure compliance and secure the necessary permits. By integrating the UK's regulatory requirements into the project planning process, SMR developers can optimize project timelines and facilitate smoother regulatory approvals.

6. Canada

The Canadian Nuclear Safety Commission (CNSC) is the primary regulatory authority that oversees all nuclear-related activities, including SMR projects. Below is an in-depth explanation of the key regulations, standards, and codes that must be adhered to when planning for an SMR construction project in Canada.

1. Regulatory Framework and Licensing

Canada has a well-defined regulatory framework for nuclear power projects, including SMRs. The Canadian Nuclear Safety Commission (CNSC) oversees all nuclear-related activities in the country to ensure safety, security, and environmental protection.

- **Nuclear Safety and Control Act (NSCA), 1997**: This is the foundational legislation that establishes the CNSC's authority. It provides the framework for licensing nuclear facilities, including SMRs, and sets out provisions for radiation protection, nuclear security, and environmental assessments.

- **CNSC Regulations**: Under the NSCA, there are several regulations relevant to SMR projects, including:

  o **General Nuclear Safety and Control Regulations**

- ○ **Class I Nuclear Facilities Regulations**
- ○ **Radiation Protection Regulations**
- ○ **Nuclear Security Regulations**

The CNSC requires that SMR developers undergo a multi-phase licensing process that includes site preparation, construction, operation, and decommissioning. Each phase involves rigorous safety assessments, public consultations, and environmental reviews.

2. Design and Safety Standards for SMRs

SMRs in Canada must comply with detailed design and safety standards that align with both national and international best practices. These standards ensure that SMRs are developed with robust safety measures and advanced technological features.

- **CNSC Regulatory Documents (REGDOCs)**: The CNSC provides comprehensive guidance through its REGDOC series, which includes:
  - ○ **REGDOC-2.5.2**: Covers design requirements for nuclear power plants, emphasizing safety systems, accident prevention, and risk management.
  - ○ **REGDOC-2.4.4**: Provides guidance on probabilistic safety assessments, which are crucial for evaluating the safety of SMRs.
  - ○ **REGDOC-2.9.1**: Addresses environmental protection policies for nuclear facilities, ensuring compliance with Canadian environmental laws.
- **International Atomic Energy Agency (IAEA) Standards**: Canada often aligns its nuclear safety standards with the IAEA's recommendations to ensure global best practices are followed.
- **CSA Standards (Canadian Standards Association)**: The CSA develops and maintains nuclear standards that are frequently referenced by the CNSC, including:
  - ○ **CSA N286**: Focuses on quality management in nuclear facilities.
  - ○ **CSA N290.1**: Covers requirements for nuclear facility construction.
  - ○ **CSA N293**: Addresses fire protection for nuclear power plants.
  - ○ **CSA N288.6**: Provides guidelines for environmental risk assessments.

3. Building Codes and Construction Regulations

Canada's construction regulations for SMRs ensure that nuclear facilities are built to the highest standards of safety, durability, and sustainability.

- **National Building Code of Canada (NBC)**: The NBC is the primary building code that sets the standards for the construction and design of buildings across the country. For nuclear projects, additional considerations are made for structural integrity, fire safety, and resilience to natural disasters.

- **National Fire Code of Canada (NFC)**: Ensures that fire safety measures are in place during the construction and operation of SMRs, including fire suppression systems, emergency evacuation protocols, and fire-resistant materials.

- **CSA N291**: Specifies requirements for the construction of nuclear facility structures, focusing on aspects such as earthquake resistance and structural robustness.

4. Environmental Assessments and Permitting

Environmental protection is a critical aspect of SMR development in Canada. The CNSC and other federal agencies ensure that SMR projects comply with stringent environmental regulations.

- **Impact Assessment Act (IAA), 2019**: Under the IAA, SMR projects may require an environmental impact assessment (EIA) to evaluate potential impacts on air quality, water resources, wildlife, and local communities. This assessment also includes public consultations to address concerns of local stakeholders.

- **Canadian Environmental Protection Act (CEPA), 1999**: This act ensures that SMR projects comply with environmental standards for emissions, waste disposal, and chemical management.

- **Fisheries Act**: Protects Canadian waterways and aquatic life, which is particularly relevant for SMRs located near water bodies.

- **CNSC REGDOC-2.9.1**: Provides guidelines for environmental protection during the planning, construction, and operation phases of nuclear facilities.

5. Fire Safety and Protection Standards

Fire safety is critical for nuclear facilities, including SMRs, to prevent accidents and ensure the safety of personnel and surrounding communities.

- **CSA N293**: Focuses on fire protection for nuclear power plants, outlining requirements for fire detection, suppression systems, and emergency response plans.

- **National Fire Code (NFC)**: Sets out standards for fire safety measures, including fire-resistant materials, fire exits, and emergency protocols.

6. Instrumentation and Control Systems

SMRs require sophisticated instrumentation and control (I&C) systems to ensure safe and efficient reactor operation.

- **CSA N290.6**: Provides standards for nuclear instrumentation, ensuring that SMRs have reliable control systems for monitoring and safety.

- **IEC 61513**: Covers the design of instrumentation and control systems for nuclear reactors, ensuring compliance with international safety standards.

7. Decommissioning and Waste Management

Canada has stringent regulations for the decommissioning of nuclear facilities and the management of radioactive waste.

- **REGDOC-2.11.2**: This CNSC regulatory document provides guidance on the management of radioactive waste, ensuring that SMR operators have comprehensive waste management strategies.

- **CSA N294**: Focuses on the decommissioning of nuclear facilities, outlining requirements for site cleanup, waste disposal, and radiation protection.

8. Security and Cybersecurity

Ensuring the security of nuclear facilities is a high priority in Canada, especially for SMRs, which may be deployed in remote or decentralized locations.

- **Nuclear Security Regulations**: CNSC requires SMR projects to implement stringent security measures, including physical security, cybersecurity, and protection against sabotage.

- **CSA N290.7**: Provides standards for cybersecurity in nuclear facilities, ensuring that digital control systems are protected from cyber threats.

The planning, design, and construction of an SMR power plant in Canada involve navigating a comprehensive set of building codes, construction regulations, and safety standards. By adhering to the requirements set by the CNSC, CSA standards, and other regulatory frameworks, SMR developers can ensure that their projects meet the highest standards of safety, security, and environmental protection.

Successful project planning for SMRs in Canada requires early engagement with regulatory bodies, thorough environmental assessments, and adherence to rigorous design and safety protocols. These efforts are essential for gaining public trust, securing regulatory approval, and ensuring the safe deployment of SMR technology in Canada.

7. France

The construction and operation of Small Modular Reactors (SMRs) in France are governed by a robust set of regulations, codes, and standards. France has a well-established nuclear regulatory framework that ensures the highest levels of safety, security, and environmental protection for nuclear facilities, given its strong reliance on nuclear energy. Below is a detailed breakdown of the building codes, construction regulations, and standards that need to be considered when planning an SMR power plant in France.

1. Regulatory Framework and Licensing Authority

The Autorité de Sûreté Nucléaire (ASN) is the primary regulatory authority overseeing nuclear activities in France. The ASN is responsible for licensing, regulatory compliance, and ensuring that nuclear facilities, including SMRs, meet stringent safety and environmental standards.

- **French Nuclear Safety Law (Loi TSN, 2006):** The law establishes the framework for nuclear safety and radiation protection, giving the ASN the authority to regulate nuclear facilities and enforce compliance.

- **Decree No. 2007-1557:** Governs nuclear facility construction, operation, and decommissioning, detailing the procedures for licensing and regulatory oversight.

- **INB Order (Arrêté INB):** Defines the safety rules applicable to Basic Nuclear Installations (Installations Nucléaires de Base, INB), including SMRs. The INB regulations cover safety assessments, design requirements, and operational procedures.

The licensing process for SMRs in France involves multiple stages, including site selection, safety assessments, construction authorization, and operating permits. This process is closely monitored by the ASN to ensure that all safety and environmental requirements are met.

2. Design and Safety Standards for SMRs

France has stringent standards for the design and safety of nuclear reactors, ensuring that SMRs are developed with advanced safety measures and cutting-edge technology.

- **RCC-M Code:** The **Règles de Conception et de Construction des Matériels Mécaniques des Ilots Nucléaires (RCC-M)** code sets the standards for the design and construction of mechanical equipment used in nuclear facilities, including pressure vessels, piping, and heat exchangers.

- **RCC-E Code:** Focuses on the electrical systems used in nuclear power plants, ensuring that electrical components meet safety and reliability requirements.

- **RCC-CW Code:** Applies to the civil engineering works of nuclear facilities, specifying requirements for the structural integrity and durability of buildings housing nuclear reactors.

- **IAEA Safety Standards:** France aligns many of its nuclear safety regulations with the **International Atomic Energy Agency (IAEA)** guidelines, ensuring that SMRs comply with global best practices.

- **EDF Standards:** As France's primary nuclear operator, Électricité de France (EDF) has its own set of internal safety and engineering standards that supplement the national regulatory framework.

3. Building and Construction Codes

The construction of SMRs in France must adhere to a comprehensive set of building and construction codes to ensure safety, durability, and environmental sustainability.

- **Code de la Construction et de l'Habitation (CCH)**: The French Building Code governs construction practices, including structural integrity, fire safety, and energy efficiency. Although primarily focused on residential and commercial buildings, certain provisions apply to nuclear facilities.

- **Eurocodes**: As a member of the European Union, France adheres to the Eurocodes, which are a set of European standards for the structural design of buildings and civil engineering works. These include:

    o **Eurocode 2**: Design of concrete structures.

    o **Eurocode 3**: Design of steel structures.

    o **Eurocode 8**: Design of structures for earthquake resistance, which is critical for the seismic safety of nuclear installations.

- **NF Standards**: The **Norme Française (NF)** standards are developed by the French Association for Standardization (AFNOR) and cover various aspects of construction, materials, and safety.

4. Fire Safety and Environmental Protection

Fire safety and environmental protection are crucial aspects of SMR projects to prevent accidents and ensure compliance with stringent environmental regulations.

- **Arrêté INB**: The INB Order mandates fire protection systems, fire-resistant materials, and emergency response plans for nuclear facilities.

- **Environmental Code (Code de l'Environnement)**: This code governs the protection of natural resources, waste management, and pollution control. SMR projects must undergo thorough environmental impact assessments to identify and mitigate potential environmental risks.

- **Seveso III Directive Compliance**: Nuclear facilities in France must comply with the **Seveso III Directive**, which aims to prevent major industrial accidents involving hazardous substances and minimize their impact on human health and the environment.

5. Instrumentation, Control Systems, and Cybersecurity

The reliability of instrumentation, control (I&C) systems, and cybersecurity is vital to the safe operation of SMRs in France.

- **RCC-E Code**: Provides standards for the design and construction of I&C systems in nuclear facilities, ensuring that they are reliable and capable of managing reactor operations safely.

- **ISO/IEC 27001**: Addresses information security management systems, which are crucial for protecting digital infrastructure in nuclear power plants from cyber threats.

- **Cybersecurity Requirements for Critical Infrastructure**: France has specific regulations under the Loi de Programmation Militaire (LPM) that require critical infrastructure, including nuclear facilities, to implement robust cybersecurity measures.

6. Decommissioning and Waste Management

France places a strong emphasis on the safe decommissioning of nuclear facilities and the management of radioactive waste.

- **Plan National de Gestion des Matières et Déchets Radioactifs (PNGMDR)**: The national plan outlines strategies for managing radioactive waste, including waste generated by SMRs.

- **RCC-Waste**: This code provides guidelines for the storage, treatment, and disposal of radioactive waste from nuclear reactors.

- **ASN Guidelines**: The ASN requires comprehensive decommissioning plans to be submitted before the construction of an SMR to ensure that facilities can be safely dismantled at the end of their operational life.

7. Quality Assurance and Manufacturing Standards

SMR projects must adhere to stringent quality assurance standards to ensure the safety and reliability of all components.

- **ISO 9001**: Quality management systems for ensuring that all processes, from design to construction, meet the highest standards.

- **RCC-M**: Covers the quality requirements for the manufacturing of mechanical components in nuclear facilities.

- **CE Marking**: Ensures that components used in SMR construction comply with EU regulations, particularly for pressure equipment, electrical systems, and construction materials.

Planning and constructing an SMR power plant in France requires adherence to a comprehensive set of building codes, construction regulations, and safety standards. Compliance with ASN regulations, RCC codes, Eurocodes, and environmental laws ensures that SMRs are developed and operated safely, efficiently, and sustainably.

8. Russia

The construction and operation of Small Modular Reactors (SMRs) in Russia are subject to a comprehensive set of regulatory frameworks and standards, driven by the country's extensive experience in nuclear energy development. Russia has established itself as a global leader in the nuclear industry, with significant expertise in the design, construction, and operation of various types of reactors, including SMRs. Below, the building codes, construction regulations, and standards that must be considered when planning an SMR power plant project in Russia are outlined.

1. Regulatory Framework and Licensing Authority

In Russia, the nuclear industry is tightly regulated by several government bodies to ensure safety, compliance, and environmental protection:

- **Rostekhnadzor (Federal Environmental, Industrial and Nuclear Supervision Service):** The primary regulatory authority responsible for overseeing nuclear safety, construction, licensing, and operation of nuclear facilities, including SMRs.

- **Rosatom State Atomic Energy Corporation**: Oversees the development and operation of nuclear power plants in Russia. Rosatom plays a significant role in promoting the deployment of SMRs domestically and internationally.

- **Federal Law No. 170-FZ (On the Use of Atomic Energy):** This law establishes the legal framework for the use of nuclear energy in Russia, covering licensing, safety regulations, and nuclear liability.

The licensing process for SMRs in Russia involves several stages, including design approval, construction permits, commissioning, and operational licenses. Rostekhnadzor ensures that each stage complies with national safety and environmental standards.

2. Design and Safety Standards for SMRs

Russia has developed stringent standards for the design, safety, and construction of nuclear reactors, particularly for SMRs, which leverage modular construction techniques and passive safety features.

- **NP-001-15 (General Safety Provisions for Nuclear Power Plants):** This regulation outlines the general safety requirements for the design, construction, and operation of nuclear power plants, including SMRs. It emphasizes the use of passive safety systems and advanced reactor technology.

- **NP-082-07 (Requirements for the Design of Nuclear Power Plants with Small Reactors):** Focuses specifically on the safety requirements for SMRs, covering reactor core design, cooling systems, and containment structures.

- **GOST R Standards**: Russia's national standards for engineering, materials, and construction, which include provisions for nuclear facilities.

- **RB Series (Regulatory Guidelines)**: These guidelines cover specific technical requirements related to reactor physics, thermal-hydraulics, and radiation protection.

Russian SMR designs, such as the KLT-40S used in floating nuclear power plants, emphasize robust safety features, including passive cooling systems and advanced containment structures, to minimize the risk of accidents.

3. Building and Construction Codes

The construction of SMRs in Russia must adhere to a range of national building codes to ensure structural integrity, durability, and safety:

- **SNIP (Construction Norms and Rules)**: The Russian construction code that sets requirements for structural engineering, civil construction, and infrastructure projects. These codes cover everything from soil analysis to fire safety and are crucial for the construction of nuclear facilities.

- **SP (Set of Rules)**: These are supplementary guidelines to the SNIP standards, providing more detailed technical requirements for specific construction activities, including those related to nuclear installations.

- **GOST 27751-2014**: Establishes the principles of structural reliability for buildings and structures, focusing on safety, durability, and resistance to environmental impacts.

- **Seismic Standards**: For regions with seismic activity, additional requirements outlined in **SP 14.13330.2018** (Seismic Building Standards) must be adhered to, ensuring that SMR facilities can withstand potential earthquakes.

4. Fire Safety and Environmental Protection

Fire safety and environmental protection are critical components of SMR project planning in Russia:

- **Federal Law No. 123-FZ (Technical Regulations on Fire Safety Requirements)**: This law mandates the installation of fire prevention systems, emergency response protocols, and fire-resistant materials in nuclear facilities.

- **NP-096-15 (Safety Requirements for Handling Radioactive Waste)**: Regulates the safe management of radioactive waste, ensuring minimal environmental impact.

- **Federal Law No. 7-FZ (On Environmental Protection)**: Requires comprehensive environmental impact assessments (EIAs) to be conducted before construction begins. This includes evaluating the impact on local ecosystems and water bodies, which is particularly important for floating SMRs.

5. Instrumentation, Control Systems, and Cybersecurity

The integration of modern instrumentation, control systems (I&C), and cybersecurity measures is essential to ensure the safe and reliable operation of SMRs in Russia:

- **NP-011-99 (Requirements for Nuclear Power Plant Control Systems)**: Defines standards for the design, implementation, and maintenance of I&C systems to ensure reactor safety.

- **ISO/IEC 27001**: While not specific to Russia, international cybersecurity standards like ISO/IEC 27001 are increasingly being adopted to protect the digital infrastructure of nuclear power plants.

- **Federal Security Service (FSB)**: Provides guidelines and regulations for cybersecurity, especially for critical infrastructure like nuclear power plants.

6. Decommissioning and Waste Management

Russia places a strong emphasis on the decommissioning of nuclear facilities and the management of radioactive waste:

- **NP-020-01 (Requirements for the Decommissioning of Nuclear Facilities)**: Provides detailed guidelines for the safe decommissioning of nuclear reactors, including SMRs.

- **Federal Target Program for Radioactive Waste Management**: Focuses on developing infrastructure for the safe disposal and recycling of radioactive materials, aligning with international best practices.

7. Quality Assurance and Manufacturing Standards

Quality assurance is a critical component of the construction and operation of SMRs to ensure safety and reliability:

- **ISO 9001**: Although an international standard, ISO 9001 is widely used in Russia to establish quality management systems across various industries, including nuclear energy.

- **GOST Standards for Nuclear Equipment**: Cover the manufacturing and inspection of nuclear components to ensure compliance with safety and performance requirements.

Planning and constructing an SMR power plant in Russia requires adherence to a comprehensive set of building codes, construction regulations, and safety standards. The regulatory framework is robust, with Rostekhnadzor playing a central role in ensuring compliance with stringent safety protocols. Russia's experience with nuclear technology, combined with its strong regulatory oversight, makes it a leader in the deployment of SMR technologies, both domestically and internationally.

9. Australia

The planning, construction, and operation of Small Modular Reactors (SMRs) in Australia involve a comprehensive framework of building codes, construction regulations, and nuclear safety standards. Given that Australia is considering SMRs as part of its future energy strategy, adhering to stringent safety and environmental standards is crucial to gaining public trust and regulatory approval.

1. Regulatory Framework and Licensing Authorities

In Australia, nuclear energy activities are heavily regulated, primarily by federal laws and overseen by dedicated regulatory bodies:

- **Australian Radiation Protection and Nuclear Safety Agency (ARPANSA)**: ARPANSA is the primary federal agency responsible for regulating nuclear safety and radiation protection. It sets safety standards, issues licenses, and monitors compliance for nuclear facilities, including SMRs.

- **Australian Nuclear Science and Technology Organisation (ANSTO)**: While primarily focused on research, ANSTO provides technical expertise and support for nuclear-related projects.

- **State and Territory Authorities**: Each state and territory has its own regulations concerning building codes, environmental protection, and land use. These must be aligned with federal regulations for any nuclear project.

- **Environmental Protection Agencies (EPAs)**: These agencies oversee environmental assessments, approvals, and monitoring of potential environmental impacts related to SMR construction and operation.

2. Relevant Federal Laws and Safety Standards

Australia currently does not have commercial nuclear reactors; however, existing regulatory frameworks can accommodate SMRs under stringent conditions:

- **Australian Radiation Protection and Nuclear Safety Act 1998 (ARPANS Act)**: This is the foundational legislation governing the use of nuclear energy and radiation protection in Australia. It requires nuclear facilities to undergo rigorous safety assessments and licensing processes.

- **Australian Radiation Protection and Nuclear Safety Regulations 2018**: These regulations provide detailed requirements on radiation protection, safety management, and the safe handling of radioactive materials.

- **Environment Protection and Biodiversity Conservation Act 1999 (EPBC Act)**: This act mandates a comprehensive environmental impact assessment (EIA) for nuclear facilities, including SMRs, to evaluate their impact on biodiversity, water resources, and ecosystems.

3. Design and Safety Standards for SMRs

SMRs must adhere to national and international safety standards to ensure safe and efficient operations. Australia's approach to SMR development would likely align with international best practices:

- **International Atomic Energy Agency (IAEA) Safety Standards**: Australia is a member of the IAEA and would incorporate these safety standards into its regulatory framework for SMRs. These include requirements for reactor design, safety analysis, and emergency preparedness.

- **ISO Standards (ISO 9001, ISO 14001, ISO 45001)**: These international standards cover quality management, environmental management, and occupational health and safety. Compliance with these standards ensures that SMR projects meet global safety benchmarks.

- **IAEA Safety Requirements (SSR-2/1, SSR-2/2)**: These documents provide safety requirements for the design and operation of nuclear power plants, including passive safety systems and containment measures, which are essential for SMRs.

4. Building Codes and Construction Regulations

The construction of SMRs in Australia would need to align with national building codes to ensure the structural integrity, safety, and durability of facilities:

- **National Construction Code (NCC)**: The NCC sets out the minimum requirements for building design and construction throughout Australia. It includes provisions for fire safety, structural integrity, and energy efficiency, which are essential for SMR facilities.

- **Australian Standards (AS/NZS)**: These standards cover a wide range of construction-related activities. Relevant standards include:

  - **AS 1170.4 (Earthquake Actions)**: Ensures that nuclear facilities are designed to withstand seismic events, which is critical for the safety of SMRs.

  - **AS/NZS 3000 (Electrical Installations - Wiring Rules)**: Governs the electrical systems of SMR facilities, ensuring compliance with safety requirements.

  - **AS 2885 (Pipelines - Gas and Liquid Petroleum)**: Relevant if the SMR is integrated with other infrastructure for process heat applications or cogeneration.

5. Fire Safety and Emergency Preparedness

Given the nature of nuclear facilities, fire safety and emergency response are crucial:

- **AS 3745 (Planning for Emergencies in Facilities)**: Provides guidelines for emergency planning, including the development of an emergency response plan for nuclear facilities.

- **Fire and Rescue NSW and other state agencies**: These bodies may impose additional requirements for fire safety, especially in the case of an SMR located in remote areas or near populated regions.

6. Environmental Protection and Waste Management

Environmental protection is a critical aspect of SMR project planning in Australia:

- **EPBC Act Environmental Impact Assessments**: SMRs would require a detailed EIA to assess the potential impact on air quality, water resources, and local biodiversity. This is especially important given Australia's sensitive ecosystems.

- **Radioactive Waste Management**: The management of radioactive waste would need to comply with ARPANSA's Radioactive Waste Management Policy and the IAEA Joint Convention on the Safety of Spent Fuel Management and on the Safety of Radioactive Waste Management.

- **ISO 14001 (Environmental Management Systems)**: Ensures that the SMR project minimizes its environmental footprint during construction and operation.

7. Instrumentation, Control Systems, and Cybersecurity

SMRs involve sophisticated instrumentation and control systems that must be robust and secure:

- **AS/NZS IEC 61508 (Functional Safety of Electrical/Electronic Systems)**: Ensures that control systems for SMRs are designed to meet high safety integrity levels.

- **Cybersecurity Guidelines (Australian Cyber Security Centre - ACSC)**: Nuclear facilities must be equipped with strong cybersecurity measures to protect critical infrastructure from potential threats.

8. Quality Assurance and Manufacturing Standards

Ensuring quality during the construction and operation of SMRs is essential:

- **ISO 9001 (Quality Management Systems)**: This standard ensures that all processes, from design to construction and operation, meet the highest quality standards.

- **AS/NZS ISO 3834 (Quality Requirements for Fusion Welding of Metallic Materials)**: Relevant for the construction of reactor components.

Planning for the construction of an SMR power plant in Australia requires careful adherence to a wide range of building codes, construction regulations, safety standards, and environmental guidelines. The country's regulatory framework, led by ARPANSA and other federal and state agencies, is designed to ensure that nuclear projects meet stringent safety and environmental standards.

Given the unique challenges associated with SMRs, including modular construction and passive safety systems, Australia would need to leverage international best practices while adapting its existing regulatory framework to accommodate the specific characteristics of SMR technology. By doing so, Australia can position itself to safely and effectively integrate SMRs into its energy mix, contributing to its decarbonization goals and energy security.

Construction Timing

The construction timeline for a Small Modular Reactor (SMR) power plant is a critical aspect of its deployment, influenced by various factors including design complexity, regulatory requirements, site conditions, and resource availability. The modular design of SMRs allows for a potentially expedited construction process compared to traditional large nuclear reactors, which typically have longer timelines due to their scale and complexity. This response synthesizes relevant literature to outline the expected phases and durations involved in constructing an SMR power plant.

1. Pre-Construction Phase (2 to 5 years)

The pre-construction phase is essential for laying the groundwork for the SMR project. This phase typically includes feasibility studies, site selection, regulatory approvals, environmental assessments, and detailed design work. The duration of this phase can vary significantly depending on the regulatory environment and the specific requirements of the jurisdiction.

Feasibility studies, which assess the technical, economic, and environmental viability of the project, generally take about 6 months to 1 year [15]. Following this, site selection and licensing processes can extend from 1 to 3 years, as they involve securing regulatory permits and conducting environmental impact assessments [67]. The design and engineering phase, which includes finalizing reactor designs and completing safety analyses, can take an additional 1 to 2 years [210]. Collectively, these steps contribute to the estimated pre-construction duration of 2 to 5 years.

2. Manufacturing and Fabrication (1 to 3 years)

One of the significant advantages of SMRs is their reliance on factory-based modular construction. This method allows for key components to be prefabricated off-site, enhancing quality and reducing risks associated with on-site construction. The manufacturing of SMR modules, such as reactor pressure vessels and steam generators, typically takes about 1 to 2 years, depending on the complexity of the components and supply chain logistics [221]. Additionally, rigorous quality control and testing of these components before shipment can add another 6 months to 1 year to the timeline [66]. Thus, the manufacturing and fabrication phase is estimated to last between 1 to 3 years.

3. On-Site Construction and Assembly (2 to 3 years)

The on-site construction phase involves the assembly of prefabricated modules and the integration of various systems. Initial site preparation and foundation work can take approximately 6 months to 1 year [13]. The modular assembly process, which is significantly faster than traditional methods due to the pre-manufactured components, typically requires around 1 to 2 years [170]. Following assembly, system integration and testing, including safety checks and commissioning activities, can take an additional 6 months to 1 year [222]. Therefore, the total duration for on-site construction and assembly is estimated to be between 2 to 3 years.

4. Commissioning and Startup (6 months to 1 year)

The final phase involves rigorous commissioning tests to ensure all systems operate as intended. This includes loading nuclear fuel into the reactor and conducting initial tests to verify that the reactor can achieve criticality safely. This stage may take 3 to 6 months [123]. Operational readiness, which involves final checks and regulatory inspections, can take another 3 to 6 months [70]. Consequently, the commissioning and startup phase is expected to last between 6 months to 1 year.

Total Estimated Timeline

In summary, the overall timeline for constructing an SMR power plant, from initial planning to full operation, generally ranges between 5 to 10 years. The breakdown is as follows:

- Pre-Construction: 2 to 5 years

- Manufacturing and Fabrication: 1 to 3 years

- On-Site Construction and Assembly: 2 to 3 years

- Commissioning and Startup: 6 months to 1 year

Factors Influencing the Timeline

Several factors can influence the timeline for SMR construction. Regulatory approvals are often the most significant bottleneck, particularly for first-of-a-kind reactors in jurisdictions with stringent safety requirements [223]. The availability of specialized manufacturing capabilities and site conditions, such as environmental factors and local infrastructure, also play crucial roles [224]. Efficient project management and a reliable supply chain are essential for minimizing delays, particularly in the delivery of prefabricated modules [225]. Countries with established nuclear industries may experience faster timelines due to streamlined regulatory processes and a skilled labour force [226].

Comparison with Traditional Large Nuclear Reactors

Traditional large nuclear reactors typically require 10 to 15 years from planning to commissioning due to their extensive on-site construction and complex regulatory requirements. In contrast, the modular construction and factory fabrication of SMRs can reduce construction times by up to 50% [227]. This efficiency positions SMRs as a promising alternative for meeting the growing global demand for clean energy solutions.

Reducing the SMR Construction Timeline

The construction timeline for Small Modular Reactors (SMRs) can potentially be reduced from current estimates of 5 to 10 years to a more feasible 3 to 6 years. This reduction can be achieved through a combination of strategies that focus on streamlining regulatory approvals, advancing modular and factory-based construction, improving on-site assembly and project management, and accelerating testing and commissioning.

1. Streamlining Regulatory Approvals (Potential Reduction: 1 to 2 years)

The regulatory approval process is a significant bottleneck in SMR construction. Streamlining this process can lead to substantial time savings. One effective approach is the adoption of a harmonized regulatory framework. By collaborating internationally to establish standardized safety regulations, countries can facilitate the acceptance of designs certified in one jurisdiction by others, thereby minimizing additional reviews [19]. Furthermore, pre-licensing engagement with regulators allows for early identification of potential issues, which has been successfully implemented in countries like Canada and the UK [14]. Regulatory sandboxes, where developers can test SMR designs under regulatory supervision, can also expedite approvals, potentially reducing the licensing phase by 6 to 12 months [14]. Consequently, the pre-construction and licensing phases could be compressed from 2 to 5 years to approximately 1.5 to 3 years.

2. Advancing Modular and Factory-Based Construction (Potential Reduction: 1 to 1.5 years)

The modular nature of SMRs inherently allows for some construction time savings, but further advancements can enhance this benefit. Increased factory fabrication of components can significantly reduce on-site construction time, as factory settings are less susceptible to weather-related delays [228]. The use of advanced robotics and automation in manufacturing can streamline production, potentially shortening the manufacturing phase by up to 6 months [14]. Additionally, standardizing reactor components across various SMR projects enables bulk production, which reduces lead times and costs [19]. The incorporation of 3D printing technologies for complex parts can further accelerate manufacturing processes by enhancing prototyping efficiency and minimizing material waste [19]. Overall, the manufacturing and construction phases can be reduced from 3 to 5 years to around 1.5 to 2.5 years.

3. Improving On-Site Assembly and Project Management (Potential Reduction: 6 months to 1 year)

Optimizing on-site assembly processes is crucial for reducing construction timelines. Implementing parallel construction techniques allows for simultaneous assembly of different modules, which can significantly expedite the overall timeline [228]. The integration of digital twins and AI-based project management tools can optimize scheduling and monitor real-time progress, thus predicting and mitigating potential delays [14]. Moreover, delivering larger, pre-assembled modules to the site can minimize on-site labour requirements and reduce assembly time, allowing for a potential assembly period of just 12 to 18 months [228]. This optimization could cut the on-site construction and assembly phases from 2 to 3 years down to approximately 1 to 1.5 years.

4. Accelerating Testing and Commissioning (Potential Reduction: 6 months)

The commissioning phase, which involves extensive safety checks and testing, is another area where time savings can be realized. Utilizing advanced simulation software for virtual testing can help identify and resolve issues before physical testing begins, thereby reducing commissioning time [14]. Additionally, implementing automated quality assurance systems can expedite the verification of safety protocols, further shortening the commissioning phase [228]. This could allow for a reduction in the commissioning timeline from 6 to 12 months to about 3 to 6 months.

Summary of Timeline Reductions

By implementing these strategies, the overall timeline for SMR construction could realistically be reduced as follows:

- Pre-Construction and Licensing: From 2-5 years → 1.5-3 years

- Manufacturing and Fabrication: From 1-3 years → 1-2 years

- On-Site Construction and Assembly: From 2-3 years → 1-1.5 years

- Commissioning and Startup: From 6 months-1 year → 3-6 months

This results in a total compressed timeline of approximately 3 to 6 years.

Case Study Examples of Development Times

The following are three case studies of Small Modular Reactors (SMRs) that have either been completed or are in advanced stages of development, illustrating their timelines from planning to commissioning.

1. Akademik Lomonosov (KLT-40S, Russia)

- **Overview:** The Akademik Lomonosov is a floating nuclear power plant equipped with two KLT-40S reactors. It was developed by Rosatom to provide electricity to remote Arctic regions in Russia.

- **Reactor Type:** Block-type SMR (KLT-40S, 35 MWe per reactor).

- **Location:** Pevek, Chukotka Region, Russia.

Timeline

- **Planning & Design:** The project began in 2006, with detailed design work taking approximately 4 years.

- **Construction Start:** Physical construction of the floating platform commenced in 2009.

- **Assembly & On-Site Installation:** After delays due to financial and logistical challenges, the reactors were installed on the floating platform in 2013.

- **Testing & Commissioning:** The plant was launched for sea trials in 2018 and connected to the grid in December 2019.

- **Total Time:** Approximately 13 years from initial planning to commissioning.

Key Takeaways

The timeline was prolonged due to the project's complexity and logistical challenges of building a floating platform. However, the unique approach of deploying the SMR on a barge reduced on-site construction time significantly, making it ideal for remote locations.

2. NuScale Power Module (United States)

- **Overview:** NuScale is a leading SMR developer in the U.S., aiming to deploy its first reactor at the Idaho National Laboratory site. The project uses an innovative design featuring 77 MWe reactor modules.

- **Reactor Type:** Integral Pressurized Water Reactor (iPWR).

- **Location:** Idaho National Laboratory (INL), Idaho, USA.

Timeline

- **Planning & Design**: The concept for the NuScale reactor was developed in the early **2000s**, with intensive design work starting in **2007**.

- **Licensing & Regulatory Approval**: NuScale submitted its design certification application to the U.S. Nuclear Regulatory Commission (NRC) in 2016. The NRC approved the design certification in 2020 after a thorough review process.

- **Construction Planning**: The first site permit was approved in 2021, with construction projected to start in 2025.

- **Expected Commissioning**: The first module is scheduled to be operational by 2029.

- **Total Estimated Time**: Approximately 22 years from initial concept to commissioning.

Key Takeaways

NuScale's project highlights the significant time required for regulatory approval in countries with stringent safety standards. The use of factory-built modules is expected to reduce on-site construction time, with the goal of achieving a build time of approximately 3 years for subsequent reactors once the first unit is operational.

3. HTR-PM (High-Temperature Gas-Cooled Reactor, China)

- **Overview**: The HTR-PM project is a high-temperature gas-cooled SMR developed by China. It consists of two reactors driving a single 210 MWe steam turbine.

- **Reactor Type**: Pebble-bed High-Temperature Gas-Cooled Reactor (HTGR).

- **Location**: Shidaowan, Shandong Province, China.

Timeline

- **Planning & Design**: The project was conceptualized in the early 2000s with research and design work completed by 2008.

- **Construction Start**: Construction began in 2012 with the aim of showcasing China's capabilities in nuclear technology.

- **Testing & Commissioning**: After several delays related to testing and safety validations, the reactors achieved first criticality in September 2021 and were fully operational by the end of 2022.

- **Total Time**: Approximately 14 years from planning to commissioning.

Key Takeaways

China's experience with the HTR-PM project shows that despite having a robust domestic supply chain and regulatory framework, developing new nuclear technologies still faces significant challenges, particularly in testing and validation phases. The modular design of HTR-PM is expected to significantly reduce construction times for future reactors.

Lessons Learned from Case Studies

- **Regulatory Approval**: A major bottleneck in all projects is obtaining regulatory approvals, especially in countries with stringent safety standards like the U.S. This step alone can take **5 to 7 years** or more.

- **Modular Construction**: The shift to factory-built modular components, as seen in the NuScale and HTR-PM projects, has the potential to drastically reduce on-site construction time, aiming for timelines as short as 2-3 years for future units.

- **Complex Logistics**: Projects like Akademik Lomonosov demonstrate that innovative solutions like floating power plants can address infrastructure challenges in remote regions but may face unique logistical hurdles, extending the overall timeline.

Reducing the time from planning to commissioning for SMRs will require streamlining regulatory processes, leveraging advanced construction techniques, and adopting digital technologies for project management and design validation.

There are a few examples of Small Modular Reactors (SMRs) that have been constructed and commissioned in relatively short timeframes, with some achieving completion in less than 10 years and even under 5 years. These examples often involve streamlined regulatory processes, optimized construction techniques, and modular design principles, particularly in countries with strong government support for nuclear energy. Two examples include:

1. HTR-10 (China)

The HTR-10 is a high-temperature gas-cooled reactor (HTGR) developed by Tsinghua University's Institute of Nuclear and New Energy Technology in China. This reactor was conceived as a demonstration project aimed at testing the feasibility of pebble-bed technology and validating the safety benefits of modular HTGR systems. It was primarily designed to explore advanced nuclear technologies that could enhance both safety and efficiency. The HTR-10 is classified as a pebble-bed HTGR, with a modest electrical generation capacity of 10 MWe. Strategically located on the Tsinghua University campus in Beijing, China, it serves as both a research tool and a technological showcase.

The planning and initial research for the HTR-10 began in the early 1990s, laying the groundwork for the subsequent design phase. Construction officially started in 1995, following several years of design refinements and feasibility studies. The project progressed efficiently, and by 2000, the reactor had achieved its first criticality, marking the successful completion of its commissioning phase. Thus, from the start of construction to commissioning, the entire process took roughly 5 years.

This project highlights the advantages of focusing on smaller, research-oriented SMRs. By leveraging modular designs and utilizing streamlined regulatory processes, China was able to achieve significant reductions in construction timelines. The relatively simple and compact design of the HTR-10 allowed for faster construction and regulatory approval, demonstrating that SMRs, especially those designed for research and demonstration purposes, can be developed and deployed efficiently.

2. Akademik Lomonosov (KLT-40S, Russia)

The *Akademik Lomonosov* is a floating nuclear power plant developed by Russia, featuring two KLT-40S reactors. This innovative project was designed to provide a reliable power source for remote Arctic regions, demonstrating the practicality and versatility of transportable small modular reactor (SMR) technology. The floating plant employs block-type SMR reactors, with each unit capable of generating 35 MWe, resulting in a total power output of 70 MWe. Strategically positioned in the port city of Pevek in the Chukotka Region, it addresses the challenge of providing electricity to isolated areas where traditional infrastructure would be difficult and costly to establish.

The planning and design phase for *Akademik Lomonosov* began in 2006, setting the foundation for a highly specialized nuclear power solution. By 2009, construction commenced on the floating platform, which housed the reactors and auxiliary systems. After overcoming various technical and logistical challenges, the project reached a critical milestone in 2019 when the reactors were connected to the grid, marking the beginning of full-scale operation.

Although the entire project spanned approximately 13 years from initial planning to full commissioning, the physical construction phase was completed in about 10 years. The use of modular and transportable technology was a significant advantage, allowing the reactors to be assembled off-site and then transported to their final location. This approach minimized the need for extensive on-site construction and expedited the project timeline, making it possible to deliver power to remote regions in a more efficient manner. The success of the *Akademik Lomonosov* project underscores the potential of modular SMRs for use in areas with challenging environments and limited infrastructure.

Potential SMRs Targeting Less than 5-Year Construction Times

Notably, NuScale Power, based in the U.S., aims for a construction timeline of 3-4 years for its first full-scale SMR deployment, contingent upon the completion of regulatory approvals and initial groundwork. This ambitious timeline is supported by the modular design of the reactor, which allows for factory fabrication and streamlined assembly on-site, thereby reducing overall construction time and costs [134].

Similarly, the BWRX-300, developed by GE Hitachi and currently planned for deployment in Canada and the U.S., targets a construction timeline of 3-5 years once all necessary approvals are secured. The BWRX-300 benefits from a simplified and modular design that leverages existing light-water reactor technology, which enhances safety and efficiency while facilitating quicker construction processes [229, 230]. The modular approach allows for parallel construction of multiple units, further expediting the deployment timeline [134].

The economic implications of these reduced construction times are significant. Research indicates that the modular nature of SMRs can lead to lower capital costs and improved economic viability, as the sequential deployment of units can be optimized based on market conditions and demand [231]. This flexibility in construction and operation is crucial for addressing the growing energy needs while minimizing the risks associated with large-scale nuclear projects, which have historically faced budget overruns and extended timelines [134].

Sample Construction Schedule for a 4000 MW SMR Power Station Comprising 40 X 100 MW SMRs.

This schedule is designed to optimize modular construction, reduce time, and ensure compliance with regulatory, safety, and environmental standards. Given the modular nature of SMRs, many activities can be performed in parallel, allowing for faster deployment.

Below is a high-level sample construction schedule covering a span of approximately **6 to 8 years**, which is considered realistic for a large-scale SMR project. This schedule includes key phases such as planning, licensing, procurement, construction, testing, and commissioning.

Year 1 - 2: Pre-Construction Phase

1. Project Planning & Feasibility Study (6 months)

- Conduct feasibility studies, site assessments, and environmental impact analyses.

- Develop project budget, financing plan, and risk assessments.

- Select project management and engineering design teams.

2. Regulatory Approvals & Licensing (12 - 18 months)

- Apply for and secure initial permits and licenses from regulatory authorities.

- Submit Preliminary Safety Analysis Report (PSAR) to nuclear regulatory body.

- Conduct public consultations and address stakeholder concerns.

- Obtain environmental clearances and land acquisition.

3. Design Engineering (18 months)

- Complete conceptual, plant-level, and system-level engineering designs.

- Develop detailed component designs for modular fabrication.

- Finalize procurement strategy and supplier agreements.

- Submit Final Safety Analysis Report (FSAR) for approval.

Year 2 - 3: Procurement & Manufacturing Phase

4. Procurement & Supplier Contracting (12 months)

- Finalize contracts with suppliers for reactor modules, turbines, cooling systems, and other major components.

- Procure long-lead items such as reactor vessels, steam generators, and control systems.

- Establish quality control and inspection protocols with suppliers.

5. Factory-Based Manufacturing of Modules (18 - 24 months)

- Begin manufacturing key components in controlled factory environments.

- Fabricate prefabricated modules, including reactor pressure vessels, heat exchangers, and steam turbines.

- Pre-assemble systems and conduct quality assurance checks before transportation.

Year 3 - 5: Site Preparation & Civil Works

6. Site Preparation (6 - 9 months)

- Clear land, prepare foundation, and establish access roads.

- Set up temporary facilities for workers, equipment storage, and logistics.

- Install utility connections (water, electricity, telecommunications).

7. Civil Construction (24 months)

- Construct reactor buildings, turbine halls, control rooms, and cooling towers.

- Install underground piping, electrical cabling, and auxiliary systems.

- Build containment structures for each reactor module.

- Perform structural integrity tests on buildings and support structures.

Year 4 - 6: Installation & Assembly of SMRs

8. Modular Transport & Assembly (18 - 24 months)

- Transport prefabricated modules to the site using road, rail, or barge.

- Use heavy-lift cranes to position reactor modules, steam generators, and cooling systems.

- Connect modules using pre-configured interfaces for electrical, control, and cooling systems.

9. Mechanical, Electrical, and Plumbing (MEP) Installation (12 - 18 months)

- Install electrical systems, instrumentation, and control (I&C) systems.

- Connect reactor units to the power grid and cooling water intake/outlet systems.

- Perform leak tests, pressure tests, and system integrity checks.

Year 5 - 7: Testing & Commissioning

10. Pre-Commissioning Tests (6 months)

- Conduct individual tests on reactor modules, turbines, and auxiliary systems.

- Perform cold hydrostatic tests, electrical system checks, and control system calibration.

- Simulate emergency shutdown scenarios to validate safety systems.

11. Fuel Loading & Initial Startup (6 - 9 months)

- Obtain regulatory approval for fuel loading.

- Load nuclear fuel into reactor cores and seal pressure vessels.

- Conduct initial startup tests, including low-power testing and hot functional testing.

12. Final Commissioning & Grid Connection (6 months)

- Perform full-power testing to validate performance under operating conditions.

- Gradually connect reactors to the power grid, starting with the first set of modules.

- Conduct final regulatory inspections and obtain operating license.

Year 6 - 8: Full Operation & Handover

13. Phased Operation Start-Up (12 months)

- Bring each group of 10 reactors online in phases to optimize grid integration.

- Monitor initial operations to ensure stable power generation and system reliability.

- Provide training for operating staff on-site.

14. Project Close-Out (6 months)

- Finalize project documentation, regulatory compliance reports, and safety certifications.

- Conduct a post-construction review to identify lessons learned.

- Handover plant operation to the utility company.

Estimated Timeline Summary

- **Total Project Duration**: 6 - 8 years

- **Construction & Commissioning**: 4 - 5 years

- **Modular Assembly & Testing**: 2 - 3 years

This schedule optimizes the use of modular construction techniques, parallel activities, and streamlined regulatory processes to reduce overall project duration. By leveraging factory-based manufacturing, off-site prefabrication, and phased commissioning, the timeline for an SMR power plant can be realistically reduced compared to traditional large nuclear reactors.

The construction schedule for a 4,000 MW Small Modular Reactor (SMR) power station, which involves 40 individual 100 MW SMR units, is based on several key assumptions to create an efficient and realistic timeline. The project aims to leverage the modular nature of SMRs to streamline construction, reduce costs, and meet regulatory requirements. Below is a detailed explanation of these assumptions and how they impact the planning and execution of the project.

Location and Regulatory Context: The project is assumed to take place in a country with an established yet relatively efficient nuclear regulatory framework, such as the United States, Canada, or the United Kingdom. These regions already have comprehensive but streamlined nuclear licensing processes that can accommodate innovative SMR technology, thereby potentially shortening the overall approval timeline. The site for the power plant is envisioned to be on flat, stable terrain that has good access to essential transport infrastructure like roads, railways, or waterways. This logistical advantage is crucial for efficiently transporting large prefabricated modules to the site. Additionally, the plant is planned to be situated in a semi-remote location that, while somewhat isolated, still benefits from access to existing utility infrastructure. This setup is particularly suitable for coastal or sparsely populated areas where grid stability is a concern and reliable, independent power sources like SMRs are needed.

Reactor Type and Design: The reactors selected for this project are assumed to be pressurized water reactor (PWR)-based SMRs, drawing on proven designs such as the NuScale Power Module, SMART (System-integrated Modular Advanced Reactor), or BWRX-300. These designs incorporate advanced passive safety systems, making them well-suited for modular construction. The reactors are fully modular, meaning that most of their components are prefabricated in controlled factory environments. This modular approach significantly reduces the time needed for on-site construction, allowing for efficient parallel assembly and installation of multiple units simultaneously.

Human Resources and Workforce Requirements: The project anticipates the availability of a highly skilled labour force, including nuclear engineers, construction specialists, crane operators, and safety experts. Given the specialized nature of the project, the workforce is expected to include a mix of local and international talent. During peak construction phases—particularly during civil engineering works and the assembly of modules—the project will require an on-site workforce of approximately 1,500 to 2,000 individuals. In addition, a separate team of 500 to 700 skilled workers will be dedicated to off-site module fabrication, including the manufacturing of reactor vessels, steam generators, and other critical components. The project's reliance on off-site prefabrication is aimed at reducing the intensity of on-site labour, thereby lowering costs and construction times.

Construction and Engineering Approach: A significant portion of the construction process will involve off-site prefabrication, leveraging modern manufacturing techniques to produce nuclear-grade components in factory settings. By minimizing on-site construction activities, the project reduces potential delays associated with weather or site-specific challenges. The modular nature of SMRs allows for parallel construction activities; for example, while some reactors are being assembled on-site, others are undergoing pre-commissioning tests. This parallel approach is designed to optimize the project timeline, enabling phased commissioning of reactor units. The schedule also assumes the availability of specialized construction equipment, such as heavy-lift cranes and transport vehicles, which are essential for handling large prefabricated modules.

Licensing and Regulatory Compliance: Regulatory engagement is assumed to be proactive, with opportunities for pre-licensing discussions that address potential concerns early in the process. The regulatory framework is expected to accommodate concurrent reviews of safety assessments, thus accelerating the overall licensing timeline. The project assumes that the chosen SMR design already holds preliminary certification in at least one country, allowing for streamlined approvals in other jurisdictions. Additionally, pre-licensing activities, such as safety design reviews, are conducted prior to formal construction approvals, reducing delays during the critical early phases of the project.

Timeline Considerations: The estimated total project duration is between 6 to 8 years. This timeline leverages modular construction techniques and the ability to carry out parallel construction activities, which are both essential for a project of this scale. The phased commissioning approach involves bringing groups of 10 reactors online sequentially, enabling the generation of revenue earlier while construction on additional units continues. This approach not only reduces financial risk but also accelerates the overall deployment of the power plant.

Financial and Logistical Assumptions: The project assumes that financing is secured early, ensuring minimal delays due to funding gaps. Favourable financial arrangements are anticipated, with the goal of reducing interest rates and other borrowing costs. Efficient logistics are crucial for transporting prefabricated modules from factories to the construction site, and the schedule assumes minimal delays in shipping and customs clearance processes. This efficiency is particularly critical given the scale and complexity of the project.

Environmental and Social Considerations: The planning phase includes comprehensive environmental impact assessments (EIA) and public consultations, which are expected to be completed within the first 18 months of the project. The project also incorporates robust mitigation strategies to address environmental concerns, particularly those related to waste management and emissions. Early engagement with local communities is prioritized to gain public support, especially in regions where nuclear projects may face resistance. Transparent communication and community involvement are critical to ensuring smooth project execution.

Summary of Key Assumptions: To summarize, the project is based on several critical assumptions:

- The location is a semi-remote area with existing infrastructure support.

- The reactors are modular PWR-based SMRs with advanced passive safety features.

- The workforce includes a peak on-site presence of up to 2,000 workers, supplemented by off-site manufacturing staff.

- Extensive use of prefabrication and modular construction techniques aims to minimize on-site activities.

- Regulatory processes are streamlined, with pre-licensing engagements to reduce approval times.

- The total project duration is expected to be between 6 to 8 years, with a phased commissioning plan.

These assumptions are essential for optimizing the construction schedule, ensuring regulatory compliance, and minimizing risks. The integration of advanced modular construction methods, a skilled labour force, and strategic regulatory engagement is key to delivering the project on time and within budget.

As an introduction to financial aspects of SMR power plant implementation, the cost of constructing a 4,000 MW Small Modular Reactor (SMR) power plant, consisting of 40 individual 100 MW SMRs, can vary significantly based on factors such as location, reactor technology, regulatory environment, and the efficiency of the construction process. However, we can estimate the cost using available data from existing SMR projects and factoring in economies of scale.

Estimated Total Project Cost:

The overall cost of a 4,000 MW SMR power station can range between $16 billion to $24 billion USD. This translates to approximately $4,000 to $6,000 per kilowatt (kW) of installed capacity. The wide range in cost estimates is due to various uncertainties, including site-specific factors, regulatory hurdles, labour costs, and the chosen reactor technology.

Cost Breakdown:

1. **Reactor Module Costs:**

   o **Reactor Unit Cost:** The cost for each 100 MW SMR unit is estimated to be between $300 million to $500 million USD.

   o **Total Reactor Module Cost (40 units):** This would result in a total reactor module cost ranging from $12 billion to $20 billion USD.

2. **Site Preparation and Civil Works:**

   o **Cost Estimate:** Approximately $1 billion to $1.5 billion USD.

   o This includes land acquisition, site clearing, excavation, foundations, and construction of support buildings.

3. **Balance of Plant (BOP) and Infrastructure:**

   o **Cost Estimate:** Around $1.5 billion to $2.5 billion USD.

   o BOP includes the power conversion systems, cooling systems, electrical infrastructure, control rooms, and auxiliary facilities.

4. **Off-site Prefabrication and Module Assembly:**

   o **Cost Estimate:** Between $500 million to $1 billion USD.

   o This covers the cost of manufacturing and transporting prefabricated modules from factories to the site.

5. **Regulatory, Licensing, and Permitting:**

   o **Cost Estimate:** Approximately $300 million to $500 million USD.

   o This includes the costs for obtaining licenses, regulatory approvals, environmental impact assessments, and legal fees.

6. **Engineering, Procurement, and Construction (EPC) Management:**

   o **Cost Estimate:** Around $500 million to $800 million USD.

o   This covers project management, detailed engineering design, procurement, and construction management.

7. **Contingencies and Risk Management:**

o   **Cost Estimate:** Roughly 10% of total project cost, equating to around $1.5 billion to $2 billion USD.

o   This accounts for unforeseen expenses, potential delays, and cost overruns.

Factors Affecting Cost:

1. **Location and Regulatory Environment:**

o   In countries with streamlined regulatory processes (e.g., Canada, the UK, or the U.S.), costs may be on the lower end due to efficient licensing. In contrast, stricter regulatory environments or those with limited experience in nuclear projects can increase costs.

2. **Economies of Scale:**

o   The use of modular construction techniques and the ability to standardize components can reduce costs. However, being a first-of-a-kind (FOAK) project may require additional resources, driving up costs compared to nth-of-a-kind (NOAK) units.

3. **Labor Costs:**

o   Labor costs can vary significantly based on the location of the project. For example, regions with higher wages or limited availability of skilled labour will incur higher costs.

4. **Supply Chain and Logistics:**

o   The availability of manufacturing facilities for prefabricated modules, transportation logistics, and infrastructure for heavy-lift equipment can impact the cost.

5. **Financing Costs:**

o   The timeline of 6 to 8 years means that the project will likely incur significant interest costs. Securing low-interest financing or government-backed loans can mitigate this expense.

Cost Comparison with Traditional Nuclear Plants:

For comparison, traditional large nuclear power plants (e.g., 1,000 MW units) typically cost between $6,000 to $10,000 per kW. The cost advantage of SMRs primarily stems from the modular construction approach, shorter construction timelines, and reduced regulatory risks, particularly for projects involving multiple smaller units.

Conclusion:

- **Estimated Total Project Cost**: $16 billion to $24 billion USD

- **Cost per kW**: $4,000 to $6,000 USD

- **Estimated Project Timeline**: 6 to 8 years (with potential reductions to 5 years through optimization)

The final cost can vary based on site-specific factors, technological advancements, and the efficiency of project management. By optimizing modular construction techniques and streamlining regulatory approvals, the cost and timeline can potentially be further reduced.

The cost per kilowatt (kW) for Small Modular Reactors (SMRs) varies significantly based on design, technology, and market conditions. Recent studies indicate a range of costs, reflecting both the potential for cost reduction through modularization and the challenges posed by initial capital expenditures.

According to Asuega et al. [175], the mean cost for advanced SMR designs is approximately $3,782 per kW, with some designs reported to be as low as $2,500 per kW. This aligns with findings from Abdulla et al. [232], which suggest that median estimates for SMRs can range widely, from $4,000 to $16,300 per kW for smaller reactors, specifically a 45 MWe SMR, and from $3,200 to $7,100 per kW for a larger 225 MWe SMR. These variations highlight the uncertainty in estimating costs due to factors such as design complexity, location, and the scale of production.

Further analysis by Hong and Brook [19] indicates that the 'nth-of-a-kind' cost for SMRs is expected to be below $100 per MWh of electricity output, suggesting that SMRs could become economically viable for microgrid applications. This is particularly relevant as the levelized cost of electricity (LCOE) for SMRs is projected to be competitive with other energy sources, especially when considering the potential for reduced construction times and costs through factory production and standardized components [233].

Moreover, the economic landscape for SMRs is influenced by the need for capital cost minimization, as highlighted by Webbe-Wood and Nuttall [234], who emphasize that high capital costs can jeopardize the feasibility of SMR projects. This sentiment is echoed in the work of Kessides and Kuznetsov [3], who note that while SMRs may have higher specific capital costs compared to larger reactors, their shorter build times and flexibility can lead to significant economic advantages.

To effectively compare the costs of constructing a 4,000 MW Small Modular Reactor (SMR) power plant to that of a 4,000 MW renewable energy power plant (comprised of solar and wind), we'll take into account the typical capital expenditure (CapEx) for each technology. This comparison will include all necessary assumptions, as the cost structures and factors influencing each type of power generation can differ significantly.

Estimated Cost of Constructing a 4,000 MW Renewable Power Plant (Solar and Wind)

1. Capital Cost Estimate:

- **Solar Power Cost:** The average cost for utility-scale solar power plants in 2024 is approximately $1,000 to $1,500 per kW of installed capacity.

- **Wind Power Cost:** Onshore wind power plants typically cost around $1,200 to $1,800 per kW of installed capacity.

- For this comparison, we'll assume a mix of 50% solar and 50% wind to achieve the 4,000 MW capacity:

    o   2,000 MW of solar power

    o   2,000 MW of wind power

- **Cost Breakdown:**

    o   Solar (2,000 MW): $2 billion to $3 billion USD

    o   Wind (2,000 MW): $2.4 billion to $3.6 billion USD

- **Total Cost Estimate (Renewables):** $4.4 billion to $6.6 billion USD

2. Land and Infrastructure Costs:

- Solar and wind farms require substantial land, especially for wind due to spacing requirements for turbines. Estimated land costs and infrastructure expenses (e.g., transmission lines, substations, access roads) can add an additional $500 million to $1 billion USD.

3. Energy Storage Costs:

- To ensure grid stability, particularly with intermittent sources like solar and wind, battery storage is essential. The cost of utility-scale battery storage is approximately $400 to $500 per kWh of storage capacity.

- To provide backup for intermittent generation, we'll assume storage sufficient for 4 hours of peak capacity (16 GWh).

    o   Estimated storage cost: $6.4 billion to $8 billion USD

4. Total Estimated Cost for Renewables (Including Storage):

- **Without storage:** $4.4 billion to $6.6 billion USD

- **With storage:** $10.8 billion to $14.6 billion USD

Comparative Analysis: SMR vs. Renewable Energy Costs

Category	SMR (4,000 MW)	Renewables (4,000 MW)
Capital Cost	$16 billion to $24 billion USD	$4.4 billion to $6.6 billion USD
With Storage	N/A	$10.8 billion to $14.6 billion USD
Construction Time	6 to 8 years	2 to 4 years
Land Requirement	Moderate	High (solar/wind farms require extensive land)
Capacity Factor	90%+	25-35% (solar), 35-45% (wind)
Grid Stability	High (continuous power output)	Low without storage; requires batteries
Lifetime	60+ years	20-30 years (panels/turbines)

Assumptions for the Comparison:

- **Location & Site Preparation**:

  o SMR power plant assumes access to a semi-remote, stable site with existing infrastructure.

  o Solar and wind projects assume a combination of onshore wind farms and large solar farms, typically located in areas with high wind and solar irradiance.

- **Regulatory and Licensing**:

  o SMR projects require extensive regulatory approvals, contributing to higher upfront costs and longer construction times.

  o Solar and wind projects have fewer regulatory hurdles, resulting in shorter lead times.

- **Capacity Factor**:

  o SMRs have a capacity factor of over 90%, meaning they can produce power continuously.

  o Solar has a capacity factor of about 25-35%, and wind ranges from 35-45%, depending on location.

  o To match the 4,000 MW continuous output of an SMR, a renewable plant would require overbuilding capacity or relying on energy storage.

- **Energy Storage**:

- Renewable sources require energy storage to handle intermittency and ensure a stable power supply. The costs for battery storage significantly increase the overall project cost.

- **Lifespan & Maintenance**:

  - SMRs are designed for a lifespan of 60 years or more with periodic refuelling.

  - Solar panels and wind turbines typically have lifespans of 20-30 years, requiring periodic replacement and higher maintenance over time.

Conclusion:

- **Cost Efficiency**: Without factoring in storage, renewables are significantly cheaper to build than SMRs. However, when storage is included to provide grid stability and mimic the reliability of nuclear power, the cost advantage diminishes.

- **Construction Time**: Renewables have a shorter construction timeline compared to SMRs, which require more time for regulatory approvals and construction. However, SMRs provide continuous, reliable power once operational.

- **Suitability for Baseline Power**: SMRs are more suitable for providing stable, 24/7 power, making them ideal for baseload electricity supply. In contrast, renewables are more suited for variable power needs unless paired with significant storage solutions.

- **Long-Term Investment**: While SMRs require higher initial investments, their long operational life and high capacity factor can provide stable power and financial returns over decades. Renewables, while cheaper initially, require ongoing investments in storage, replacements, and maintenance.

Ultimately, the choice between SMRs and renewables depends on the specific needs of the region, grid stability requirements, financial constraints, and environmental priorities.

Modular Assembly and On-Site Integration

Modular Assembly: Off-Site Fabrication in Factories

The construction of Small Modular Reactors (SMRs) leverages a modular assembly approach, primarily focused on off-site fabrication in specialized factories. This strategy is a significant departure from traditional nuclear reactor construction, which often involves extensive on-site fabrication and assembly. By shifting most of the construction process to controlled factory environments, SMR projects can achieve greater efficiency, cost savings, and consistency in quality.

Factory-Based Manufacturing

One of the most innovative aspects of SMR construction is the extensive use of factory-based manufacturing techniques. Unlike conventional large nuclear reactors, which require many components to be built and assembled on-site, SMRs are designed to be modular. Key components, such as the reactor pressure vessel, steam generators, heat exchangers, turbines, containment structures, and piping systems, are prefabricated off-site. This approach is advantageous because it allows these critical parts to be manufactured in a controlled environment, free from the variability of outdoor construction sites.

Factory settings provide a stable environment where factors like temperature, humidity, and air quality can be precisely controlled. This ensures that the components are manufactured to the highest standards without the risk of environmental disruptions. Additionally, off-site fabrication optimizes labour efficiency as specialized workers can focus on specific tasks in a centralized location. This reduces the reliance on highly skilled labour at the often remote and challenging construction sites of nuclear power plants. By reducing the on-site workload, factory-based manufacturing also helps mitigate weather-related delays, which are common in traditional construction projects.

Assembly Line Production

The modular components of SMRs are produced using assembly line techniques, a process inspired by industries like automotive and shipbuilding. In these industries, standardized production lines allow for mass production with high efficiency. Similarly, SMR components are produced in a repetitive, standardized manner, allowing manufacturers to achieve economies of scale. This means that the more units produced, the lower the cost per unit becomes, which can drive down the overall cost of the project.

The use of assembly line production is not just about efficiency but also about ensuring repeatability and consistency. By standardizing the production process, each module is built to identical specifications, ensuring uniform quality and performance. This consistency is critical in nuclear reactor construction, where even minor deviations can lead to significant safety concerns. Moreover, by producing components in large quantities, the modular approach reduces the likelihood of production delays, which can often occur when components are custom-built on-site. This also simplifies the logistics of sourcing materials, as factories can procure supplies in bulk, further reducing costs.

Quality Control and Testing

A major advantage of off-site modular fabrication is the ability to conduct rigorous quality control and testing in a factory setting before the components are shipped to the construction site. In the traditional on-site construction of large reactors, quality control can be challenging due to the complexity of the environment and the difficulty of performing comprehensive tests in real-time.

However, in a factory, components can undergo extensive testing under controlled conditions to ensure they meet nuclear safety standards.

Quality control processes in the factory include pressure tests, integrity checks, and simulations to confirm that each module will perform as required once installed on-site. For instance, reactor pressure vessels and containment systems are subjected to high-pressure tests to verify their structural integrity, ensuring they can withstand the operational stresses they will encounter. Heat exchangers and steam generators are tested for efficiency and leak-proof performance, while control systems are validated for responsiveness and reliability.

These quality assurance measures are critical because they significantly reduce the need for extensive testing during on-site installation. By resolving any issues in a controlled environment, the risk of delays and costly rework at the construction site is minimized. This not only shortens the overall construction timeline but also enhances safety by ensuring that every module performs to specification before it is integrated into the reactor system.

Break Down of the Factory Manufacturing Process

The manufacturing of Small Modular Reactors (SMRs) is a complex process that involves several meticulously planned stages to ensure compliance with nuclear safety standards and quality control measures. Each stage, from design finalization to preparation for transport, is critical to the overall safety and efficacy of the reactor.

Design and Engineering Finalization: The initial phase of SMR manufacturing is the design and engineering finalization, which necessitates the creation of detailed design plans, including precise 3D models and technical blueprints for all components. This design process must consider various factors such as material properties, thermal dynamics, safety requirements, and modular compatibility. The engineering team collaborates closely with nuclear safety experts to ensure that all designs comply with both national and international regulations [235, 236]. Advanced digital simulations and computer-aided design (CAD) software play a crucial role in optimizing component layouts and identifying potential issues before physical production begins [235, 236].

Procurement of Raw Materials: The procurement of raw materials is another critical stage in the manufacturing of SMRs. High-quality, nuclear-grade materials are essential to withstand the extreme conditions present in a nuclear reactor, including high temperatures, pressure, and radiation exposure. Key materials include high-strength steel for reactor pressure vessels, nickel-based alloys for heat exchangers, and specialized ceramics for components in high-temperature gas reactors [237, 238]. For instance, graphite is extensively used as a moderator and structural component in reactors due to its excellent irradiation performance and mechanical properties at high temperatures [237, 238]. Additionally, control rod materials, often made from boron carbide or hafnium, are vital for regulating the nuclear reaction, ensuring that the reactor operates safely and efficiently [239, 240].

The procurement process is designed to ensure that all materials meet stringent specifications, which is vital for the durability and safety of the reactor components. The integrity of materials such as graphite is particularly important, as it can undergo significant physical modifications under neutron flux and high temperatures, which could affect reactor performance [237, 241]. Thus, careful selection and testing of materials are paramount to the successful manufacturing of SMRs.

Fabrication Processes: The fabrication of Small Modular Reactor (SMR) components involves a series of sophisticated processes that ensure precision, durability, and safety. This synthesis will detail the key fabrication processes, including precision machining and welding, modular assembly and pre-fitting, and coating and surface treatment.

The production of critical components such as reactor vessels, steam generators, and containment shells is primarily executed using Computer Numerical Control (CNC) machines. These machines allow for high precision in the machining of pressure vessels, which are typically made from solid steel billets. The billets undergo multiple heat treatments to enhance their strength and durability, ensuring they can withstand the operational pressures and temperatures associated with nuclear reactors. Advanced welding techniques, such as electron beam welding and automated arc welding, are employed to create strong, defect-free joints. These methods are crucial for maintaining the integrity of pressure vessels and other components under extreme conditions.

The construction of Small Modular Reactors (SMRs) has reached a breakthrough with recent advancements in welding technology. The UK-based company, Sheffield Forgemasters, has demonstrated a significant leap forward in reactor vessel fabrication, completing the welding of a full-size nuclear reactor vessel in under 24 hours — a process that traditionally takes up to 12 months [242]. This innovative approach marks a potential turning point in the nuclear industry, enabling the faster rollout of SMRs and reducing costs substantially.

Traditionally, nuclear reactors have been built as one-off, large-scale civil engineering projects that require years of construction, complex infrastructure, and substantial investments. The development of SMRs, however, offers a paradigm shift by transforming nuclear plants into factory-produced units that can be standardized and mass-produced. Unlike conventional reactors, SMRs do not require the same massive containment buildings, making them quicker to install and more adaptable to varying power needs. By shifting from bespoke constructions to modular, standardized designs, SMRs can be scaled to fit local energy demands, allowing for more flexible deployment [242].

One of the critical bottlenecks in nuclear reactor construction has historically been the time-consuming process of welding the reactor vessels. These vessels are essential as they house the reactor core and ensure the isolation of radioactive materials from the environment. Using traditional welding techniques, it can take over a year to complete the welding of these thick-walled, high-grade steel vessels. Sheffield Forgemasters has overcome this challenge by employing Local Electron-Beam Welding (LEBW), a revolutionary welding technology that has drastically cut the welding time to less than a day [242].

The LEBW technique uses a high-powered electron gun to generate a concentrated beam of electrons in a localized vacuum environment. This process enables a high-energy density fusion, which melts and fuses the metal components together with exceptional efficiency. The technology achieves a welding efficiency of up to 95%, with deep penetration and an impressive depth-to-width ratio, allowing it to handle thick materials with ease [242]. For this project, Sheffield Forgemasters welded a reactor vessel with a diameter of three meters (approximately 10 feet) and walls 200 millimetres (8 inches) thick. The results were impressive: the welds were completed with zero defects, a significant achievement given the stringent safety requirements of nuclear components [242].

The benefits of using LEBW extend beyond speed. This method not only reduces the time required for welding but also lowers the cost of production. The precision and efficiency of the technique mean fewer weld inspections are necessary, as the weld joints are nearly indistinguishable from the original material. Additionally, the welding machine incorporates advanced techniques like sloping-in and sloping-out, which refine the start and finish of each weld, ensuring even greater integrity and reliability [242].

This milestone is particularly significant for the UK's nuclear sector, which has been largely dormant for decades apart from advancements in submarine reactors and a few showcase power plants. The UK government is now looking to reinvigorate its nuclear industry, with plans to construct new plants, including 15 modular reactors in collaboration with Rolls-Royce. The success of the LEBW technology could play a crucial role in accelerating these projects, reducing the cost and time required to deploy SMRs [242].

The development of Local Electron-Beam Welding represents a disruptive technological advance that aligns perfectly with the modular, scalable nature of SMR construction. By drastically reducing welding times and ensuring high-quality, defect-free welds, this innovation promises to make nuclear energy more accessible and cost-effective. As countries look toward low-carbon energy solutions, the ability to rapidly deploy SMRs using such groundbreaking technologies could play a vital role in meeting global energy demands and achieving decarbonization goals [242].

Local Electron-Beam Welding (LEBW) is a cutting-edge welding technology that utilizes a high-energy electron beam to fuse metal components together with exceptional precision and efficiency. Unlike traditional welding methods, which typically require extensive heat input and large-scale equipment, LEBW focuses a highly concentrated stream of electrons onto a localized area of the metal, causing it to heat up and melt instantly. This process allows for deeper penetration with minimal heat dispersion, resulting in stronger and more precise welds. The localized nature of the process significantly reduces the risk of thermal distortion or damage to surrounding areas, which is particularly crucial in industries where the structural integrity of components is paramount, such as in nuclear power, aerospace, and automotive sectors.

One of the key advantages of LEBW is its ability to operate in a controlled vacuum environment. The electron beam is generated by a high-powered electron gun within a vacuum chamber, which not only prevents oxidation but also enables the beam to remain highly focused. The vacuum setting ensures that the electrons do not scatter, thus maintaining the integrity of the weld and allowing for exceptional precision. This capability is especially beneficial when working with high-grade materials that are sensitive to contamination, such as those used in the construction of reactor pressure vessels in Small Modular Reactors (SMRs). By maintaining a pristine environment, LEBW produces welds that are nearly free of defects, resulting in stronger joints and longer-lasting components.

The efficiency of LEBW is another significant advantage over conventional welding methods. Traditional welding, especially for thick and high-strength steel used in nuclear applications, can take months to complete due to the need for multiple passes, pre-heating, and post-weld heat treatments to relieve stress. In contrast, LEBW can complete complex welds in a matter of hours. For example, a project that would typically require over a year to weld using conventional methods was completed in under 24 hours using LEBW. This is achieved due to the high energy density of the electron beam, which can penetrate thick materials (up to several inches) in a single pass. The result is a dramatic reduction in construction time, which can be a game-changer for industries that rely on timely project delivery, such as the nuclear power sector.

The high precision of LEBW also leads to cost savings. Since the process achieves near-perfect welds that replicate the parent material, it reduces the need for extensive inspections and rework that are common with traditional welding methods. The ability to produce defect-free welds eliminates the need for costly non-destructive testing (NDT) procedures, which are usually required to verify the integrity of conventional welds. Additionally, the efficiency of the electron-beam welding process means that less material is wasted, further driving down costs. These savings are particularly impactful in large-scale projects like SMR construction, where even small reductions in time and cost can lead to significant financial benefits.

Another noteworthy feature of LEBW is its versatility in handling different geometries. The technology is capable of welding complex shapes and configurations that would be challenging or impossible with traditional methods. The electron beam can be precisely controlled to follow intricate paths, making it suitable for applications requiring high precision, such as joining reactor vessels with sloped or curved surfaces. The use of advanced techniques like sloping-in and sloping-out ensures a smooth start and finish to the weld, reducing stress concentrations that could lead to potential weak points. This flexibility allows LEBW to adapt to various design requirements, making it ideal for modular reactor components that may have unique geometrical constraints.

In terms of scalability, LEBW holds significant potential for the future of modular reactor construction. As the nuclear industry shifts toward smaller, more flexible designs like SMRs, the ability to mass-produce standardized components with consistent quality becomes crucial. By leveraging the speed and precision of LEBW, manufacturers can achieve higher throughput in factory settings, thus accelerating the overall construction timeline for nuclear power plants. This technology not only supports the modular assembly approach but also enhances the reliability and safety of nuclear reactors by ensuring that critical components are welded to the highest standards.

Moreover, the fabrication of piping systems involves precision cutting and pre-fabrication with connectors that facilitate quick assembly on-site. This approach not only streamlines the construction process but also minimizes the potential for errors during assembly, which is critical in maintaining safety standards in nuclear facilities.

Once the individual components are fabricated, they are pre-assembled into larger modules within the factory. This modular assembly process includes pre-fitting critical components such as the reactor core, heat exchangers, and control systems to ensure compatibility and functionality before shipment. Factory Acceptance Tests (FATs) are conducted to verify that each module meets the required specifications and performance standards. This modular approach enhances efficiency by allowing for parallel construction of multiple units, significantly reducing the overall construction time for SMRs.

Components that are exposed to high radiation or corrosive environments undergo specialized coating processes to enhance their durability and performance. Anti-corrosive coatings are applied to heat exchangers and other components to protect them from degradation over time. Additionally, radiation shielding materials are applied to critical surfaces to protect personnel during maintenance operations. The effectiveness of these coatings is paramount, as they not only extend the lifespan of the components but also ensure the safety of the operational environment.

Recent advancements in coating technologies, such as the development of high-density polymer-resin composites, have shown promise in improving the shielding performance of these materials. Techniques like spray-coating have been explored to ensure uniform distribution of shielding materials, which is essential for maintaining consistent performance across the coated surfaces.

Quality Assurance and Testing: The production of Small Modular Reactors (SMRs) necessitates stringent quality assurance and testing protocols to ensure the highest levels of safety and reliability. Given the critical nature of nuclear energy production, where even minor defects can lead to significant safety concerns, manufacturers implement rigorous quality control measures at every stage of the manufacturing process. This comprehensive approach not only enhances the safety profile of SMRs but also fosters public confidence in nuclear technology.

One of the cornerstones of quality assurance in SMR manufacturing is Non-Destructive Testing (NDT). NDT encompasses a variety of techniques designed to identify internal flaws in components without causing any damage. Among the most widely utilized methods are ultrasonic testing, radiographic testing, and magnetic particle inspections. Ultrasonic testing employs high-frequency sound waves to detect imperfections within materials, while radiographic testing utilizes X-rays or gamma rays to visualize internal structures. Magnetic particle inspections are particularly effective for detecting surface and near-surface discontinuities in ferromagnetic materials.

In the context of SMR production, critical components such as pressure vessels and reactor cores undergo rigorous testing under simulated operational conditions. This process is essential for verifying the integrity of these components, ensuring that they can withstand the extreme conditions encountered during operation. Furthermore, leak tests and hydrostatic pressure tests are conducted on piping systems to confirm their ability to endure high pressures without failure. These tests are vital for preventing potential leaks that could compromise the safety of the reactor and its surrounding environment.

In addition to NDT, functional testing and simulation play a pivotal role in the quality assurance process for SMRs. Before the modules are shipped, fully assembled units are subjected to a series of functional tests conducted in controlled environments. These tests are designed to confirm that the modules perform as intended under various operational scenarios. For instance, thermal cycling tests are employed to simulate temperature fluctuations that the reactor may experience during its operational life. This testing ensures that materials and components can withstand thermal stresses without degrading or failing.

Operational simulations are another critical aspect of functional testing. These simulations utilize advanced control systems to replicate the operational environment of the reactor, allowing engineers to verify that all safety protocols function correctly. By simulating real-world conditions, manufacturers can identify potential issues and rectify them before the modules are deployed in the field. This proactive approach to quality assurance not only enhances the reliability of SMRs but also contributes to the overall safety of nuclear energy production.

Human Resource Requirements: The manufacturing of Small Modular Reactors (SMRs) necessitates a diverse and skilled workforce, comprising various specialized roles essential for the design, fabrication, testing, and safety oversight of these advanced nuclear systems. This synthesis will explore the human resource requirements across three primary domains: engineering and design experts, skilled labour for fabrication, and technical specialists for testing and safety.

The design of SMRs is critically dependent on a range of engineering disciplines. Nuclear engineers and mechanical engineers play pivotal roles in the development of reactor components, ensuring that designs comply with stringent safety standards. These engineers are responsible for integrating advanced safety features into reactor designs, which is particularly crucial given the heightened safety concerns following incidents such as the Fukushima disaster [13]. Electrical engineers are also integral, focusing on the design of control systems and instrumentation that are vital for the safe operation of SMRs [243]. The complexity of these systems necessitates a workforce that is not only technically proficient but also well-versed in regulatory compliance and safety protocols [244].

In addition to engineering expertise, the fabrication of reactor components requires skilled labour, including welders, machinists, and pipefitters. These tradespeople are essential for the precision fabrication of reactor components, where adherence to nuclear safety standards is non-negotiable [244]. Quality control inspectors play a crucial role in this process, ensuring that all fabricated parts meet the rigorous safety and quality standards required in the nuclear industry [244]. The integration of advanced manufacturing techniques, such as modular construction, further emphasizes the need for skilled labour capable of adapting to innovative fabrication processes [245].

The safety of SMRs during the manufacturing phase is overseen by technical specialists, including non-destructive testing (NDT) technicians who conduct critical inspections on reactor components. Their work is essential to identify potential flaws that could compromise safety [244]. Safety officers are also vital, as they ensure compliance with regulatory requirements throughout the manufacturing process. This oversight is crucial for maintaining a culture of safety within the organization, which has been shown to significantly impact operational reliability and public trust in nuclear technology [246]. The estimated workforce for an SMR factory ranges from 500 to 700 skilled personnel, highlighting the extensive human resource requirements necessary to support the safe and efficient production of these reactors [229].

Special Requirements for SMR Manufacturing: The manufacturing of Small Modular Reactors (SMRs) in a factory setting necessitates adherence to several specialized requirements that ensure safety, compliance, and logistical efficiency. These requirements can be categorized into three main areas: nuclear-grade facilities, compliance with international safety standards, and logistics planning for transport.

Factories producing SMRs must meet stringent nuclear-grade certification standards. This includes the establishment of clean rooms, specialized ventilation systems, and adequate radiation shielding to protect workers and the environment. The design and construction of these facilities are crucial as they must accommodate the unique safety and operational requirements of nuclear components. The modular design of SMRs allows for factory-based production, which can enhance safety and quality control compared to traditional onsite construction methods [15, 228]. The integration of advanced safety systems in SMR designs further emphasizes the need for specialized manufacturing environments that can support these technologies [18, 247].

Compliance with international safety standards is another critical requirement for SMR manufacturing. Factories must adhere to guidelines set by organizations such as the International Atomic Energy Agency (IAEA), the American Society of Mechanical Engineers (ASME), and various ISO standards pertinent to nuclear components. These standards ensure that the manufacturing processes are aligned with global safety protocols, which is essential for gaining regulatory approval and public trust [14, 70]. The integration of safety, operations, security, and safeguards into the design of SMRs is a foundational aspect of their development, as highlighted in various studies that outline the regulatory frameworks necessary for safe nuclear operations [70, 248].

Given the size and weight of modular components, meticulous logistics planning is required for their transport from the manufacturing facility to the construction site. This involves coordinating heavy-lift equipment, specialized transport vehicles, and carefully planned routes that can accommodate oversized loads. The modular nature of SMRs facilitates their transportation, as they can be manufactured in sections and assembled onsite, which can significantly reduce the logistical challenges associated with traditional large-scale nuclear plants [134]. Effective logistics planning not only ensures the safe delivery of components but also contributes to the overall cost-effectiveness and efficiency of the SMR deployment process [3, 15].

Transporting Modules to the Construction Site

The process of transporting prefabricated modules to the construction site is a critical phase in the construction of Small Modular Reactors (SMRs). By leveraging modular construction techniques, a significant portion of the reactor assembly is completed off-site in controlled factory environments, allowing for more efficient and streamlined on-site assembly. However, transporting these large and complex modules presents its own set of challenges and requires meticulous planning to ensure timely and secure delivery. This process is divided into two main areas: modular transport logistics and minimizing site disruption.

Modular Transport Logistics

Once the SMR modules are prefabricated, they need to be transported to the construction site using a well-coordinated logistics plan. The modules are often large, heavy, and highly sensitive due to the nuclear-grade components they contain. To address these challenges, modules are typically designed to fit within standard shipping containers or specialized transport vehicles. This allows for flexibility in utilizing various modes of transport, such as road trucks, railcars, or barges, depending on the proximity of the site to existing transportation infrastructure.

Figure 34: SMRs are easily transportable from factory to site [249].

Before transport, a comprehensive logistics plan is developed to account for factors such as module size, weight, and fragility, as well as the route's accessibility. For example, transporting a large reactor vessel might require a custom flatbed truck with reinforced support structures, while smaller, lighter components could be shipped using standard containers. The logistical planning also includes assessing the capacity of roads, bridges, and tunnels to handle the heavy loads, which may involve temporary road modifications or permits for oversized loads.

In coastal or river-adjacent locations, barge transport can be utilized to deliver large modules directly to the site, reducing road traffic and potential wear on infrastructure. Rail transport, on the other hand, can be ideal for long-distance deliveries, particularly when road access is limited or costly. To avoid damage during transport, each module undergoes secure packaging, vibration testing, and impact protection, ensuring that delicate components arrive at the site in optimal condition.

Minimizing Site Disruption

One of the key advantages of modular construction is the ability to significantly reduce on-site activities, thereby minimizing environmental and social disruptions. By transporting nearly complete modules to the construction site, the need for extensive on-site fabrication is minimized. This approach is particularly beneficial in remote or environmentally sensitive areas where traditional construction could have adverse impacts.

The modular transport method reduces the amount of heavy equipment, materials, and personnel needed on-site, which directly lowers noise levels, dust generation, and traffic congestion. This is especially important when constructing in areas near residential communities or protected ecosystems. By reducing the volume of on-site activities, the project can operate with a smaller footprint, thereby mitigating the impact on the local environment and minimizing the need for site preparation, such as clearing land or altering natural landscapes.

Additionally, this approach helps reduce the disruption to local communities, as the number of workers and construction vehicles traveling to and from the site is significantly decreased. Fewer on-site activities also mean shorter construction timelines, which benefits nearby residents by reducing the duration of noise and dust emissions. In some cases, modular transport may be coordinated to occur during off-peak hours to further minimize the impact on local traffic and daily routines.

On-Site Integration and Assembly for SMR Construction

The on-site integration and assembly of Small Modular Reactors (SMRs) are critical phases in the construction process that leverage the modular design to streamline installation, reduce construction time, and improve project efficiency. The key stages include site preparation, modular assembly using heavy-lift cranes, parallel construction activities, and pre-commissioning tests and system integration. Each of these phases is designed to maximize efficiency while maintaining the highest standards of safety and quality.

Site Preparation

Before any modular components arrive at the construction site, extensive site preparation work is necessary to ensure a smooth and efficient assembly process. This involves ground leveling, soil stabilization, and constructing the necessary foundations to support the heavy reactor modules. Additionally, infrastructure development such as access roads, water supply systems, drainage, and power connections must be completed. These preparations are critical because once the modules start arriving, the assembly process must proceed without delays.

Proper site preparation also includes setting up temporary facilities like storage areas, worker accommodations, and on-site offices for project management. The construction team ensures that foundations and basic civil structures, such as reinforced concrete bases for reactor units and auxiliary buildings, are ready by the time the modules are delivered. This preparation phase is crucial in modular construction since it enables a seamless transition to the assembly stage, minimizing potential delays that could arise from incomplete groundwork.

Modular Assembly Using Heavy-Lift Cranes

Once the site is fully prepared, the modular components are transported to the site and assembled using heavy-lift cranes. This phase involves precisely positioning large prefabricated modules like reactor vessels, steam generators, and containment structures onto the prepared foundations. Due to the substantial size and weight of these components, the assembly process is meticulously planned and executed to ensure safety and efficiency.

The use of pre-configured connections, such as bolt-and-weld joints, allows for the rapid and secure integration of modules. This method eliminates the need for extensive on-site fabrication, reducing construction time and minimizing human error. The heavy-lift cranes are operated by skilled personnel to position each module accurately, ensuring that the connections align perfectly. This level of precision is essential, as even slight misalignments could lead to costly delays and potential safety issues.

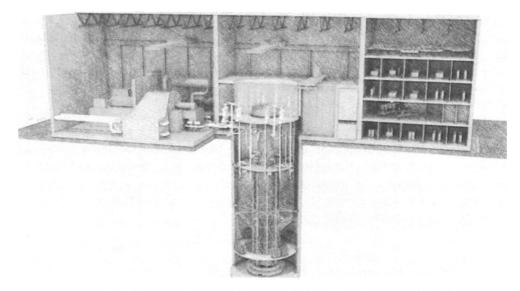

Figure 35: SMRs can be partially or totally buried underground [249].

Parallel Construction Activities

One of the major advantages of using a modular construction approach is the ability to conduct parallel construction activities. While one team is focused on assembling reactor modules, other teams can simultaneously work on separate sections of the plant, such as cooling towers, turbine halls, electrical substations, and control rooms. This parallel workflow significantly reduces the overall project timeline, allowing multiple phases of construction to progress simultaneously.

For example, while the primary reactor modules are being integrated, teams can lay piping, install electrical wiring, and complete auxiliary structures. This concurrent construction approach not only shortens the project schedule but also optimizes resource utilization, ensuring that labour and equipment are continuously employed without downtime. By breaking down the construction into parallel streams, the project can achieve substantial time savings, which is critical for reducing costs and accelerating the plant's operational readiness.

Pre-Commissioning Tests and System Integration

After the physical assembly of all modules is complete, the next step is to conduct pre-commissioning tests and system integration. This phase is crucial to ensure that all systems are functioning correctly before the plant is fully operational. The testing process involves extensive checks on electrical systems, control systems, cooling circuits, and safety mechanisms to verify that they meet design specifications and safety standards.

Pre-commissioning tests are typically conducted in stages, starting with individual modules and progressing to integrated systems. For example, the reactor core cooling system might be tested independently before being connected to the steam turbine system. Control systems and automation are also rigorously tested to ensure seamless operation. Any issues identified during this stage are addressed before the plant moves to full commissioning, thus preventing potential operational risks.

System integration is an intricate process that ensures all subsystems, such as instrumentation and control (I&C), emergency cooling systems, and power conversion units, work together harmoniously. These tests confirm that the reactor modules, once connected, will function as a cohesive unit. The success of these pre-commissioning activities is vital to securing the necessary regulatory approvals to proceed to the final operational phase, thus enabling the SMR plant to generate power safely and efficiently.

On-Site Integration and Assembly for SMR Construction

The on-site integration and assembly of Small Modular Reactors (SMRs) are critical phases in the construction process that leverage the modular design to streamline installation, reduce construction time, and improve project efficiency. The key stages include site preparation, modular assembly using heavy-lift cranes, parallel construction activities, and pre-commissioning tests and system integration. Each of these phases is designed to maximize efficiency while maintaining the highest standards of safety and quality.

Site Preparation

Before any modular components arrive at the construction site, extensive site preparation work is necessary to ensure a smooth and efficient assembly process. This involves ground leveling, soil stabilization, and constructing the necessary foundations to support the heavy reactor modules. Additionally, infrastructure development such as access roads, water supply systems, drainage, and power connections must be completed. These preparations are critical because once the modules start arriving, the assembly process must proceed without delays.

Proper site preparation also includes setting up temporary facilities like storage areas, worker accommodations, and on-site offices for project management. The construction team ensures that foundations and basic civil structures, such as reinforced concrete bases for reactor units and auxiliary buildings, are ready by the time the modules are delivered. This preparation phase is crucial in modular construction since it enables a seamless transition to the assembly stage, minimizing potential delays that could arise from incomplete groundwork.

Modular Assembly Using Heavy-Lift Cranes

Once the site is fully prepared, the modular components are transported to the site and assembled using heavy-lift cranes. This phase involves precisely positioning large prefabricated modules like reactor vessels, steam generators, and containment structures onto the prepared foundations. Due to the substantial size and weight of these components, the assembly process is meticulously planned and executed to ensure safety and efficiency.

The use of pre-configured connections, such as bolt-and-weld joints, allows for the rapid and secure integration of modules. This method eliminates the need for extensive on-site fabrication, reducing construction time and minimizing human error. The heavy-lift cranes are operated by skilled personnel to position each module accurately, ensuring that the connections align perfectly. This level of precision is essential, as even slight misalignments could lead to costly delays and potential safety issues.

Parallel Construction Activities

One of the major advantages of using a modular construction approach is the ability to conduct parallel construction activities. While one team is focused on assembling reactor modules, other teams can simultaneously work on separate sections of the plant, such as cooling towers, turbine halls, electrical substations, and control rooms. This parallel workflow significantly reduces the overall project timeline, allowing multiple phases of construction to progress simultaneously.

For example, while the primary reactor modules are being integrated, teams can lay piping, install electrical wiring, and complete auxiliary structures. This concurrent construction approach not only shortens the project schedule but also optimizes resource utilization, ensuring that labour and equipment are continuously employed without downtime. By breaking down the construction into parallel streams, the project can achieve substantial time savings, which is critical for reducing costs and accelerating the plant's operational readiness.

Pre-Commissioning Tests and System Integration

After the physical assembly of all modules is complete, the next step is to conduct pre-commissioning tests and system integration. This phase is crucial to ensure that all systems are functioning correctly before the plant is fully operational. The testing process involves extensive checks on electrical systems, control systems, cooling circuits, and safety mechanisms to verify that they meet design specifications and safety standards.

Pre-commissioning tests are typically conducted in stages, starting with individual modules and progressing to integrated systems. For example, the reactor core cooling system might be tested independently before being connected to the steam turbine system. Control systems and automation are also rigorously tested to ensure seamless operation. Any issues identified during this stage are addressed before the plant moves to full commissioning, thus preventing potential operational risks.

System integration is an intricate process that ensures all subsystems, such as instrumentation and control (I&C), emergency cooling systems, and power conversion units, work together harmoniously. These tests confirm that the reactor modules, once connected, will function as a cohesive unit. The success of these pre-commissioning activities is vital to securing the necessary regulatory approvals to proceed to the final operational phase, thus enabling the SMR plant to generate power safely and efficiently.

Phased Commissioning and Start-Up for SMR Power Plants

The construction and operational launch of a large Small Modular Reactor (SMR) power station, particularly one comprising multiple units, involves a strategic process known as phased commissioning and start-up. This approach is designed to optimize the timeline, minimize financial risk, and ensure a steady progression towards full operational capacity. Below, we delve into the key steps of this process, including sequential commissioning of reactor units, integration with the grid, and final safety validation.

Sequential Commissioning of Reactor Units

In the case of a large-scale SMR power station, such as a 4,000 MW plant consisting of 40 reactors each with a capacity of 100 MW, commissioning is typically executed in a phased manner. This means that reactors are brought online in groups rather than all at once. For example, the plant may commission the reactors in batches of 10 units. This sequential approach allows for a more controlled and manageable ramp-up of power generation, enabling early revenue generation even as construction and assembly of the remaining modules continue.

By commissioning the reactors in phases, the project can begin producing electricity and generating income before the entire plant is fully completed. This incremental revenue stream helps offset initial construction costs and supports the financial stability of the project. Additionally, this phased method provides an opportunity to identify and resolve any unforeseen technical issues in the early units, which can then be applied to the subsequent modules, thereby reducing delays and improving efficiency in the later stages of commissioning.

Integration with the Grid

Once a reactor module is commissioned, it must be seamlessly integrated with the national power grid. This integration process involves synchronizing the reactor's output with the grid's frequency and voltage requirements. Ensuring that each reactor can adjust to grid fluctuations and contribute stable power is crucial, especially in regions where demand can vary significantly.

The synchronization process includes a series of tests to verify the reactor's load-following capabilities, which are essential for SMRs designed to support grids with high levels of renewable energy sources. These tests ensure that the SMR can adjust its output in response to changes in electricity demand, thus enhancing grid stability. Additionally, emergency protocols are thoroughly tested to confirm that each reactor can safely disconnect from the grid in the event of faults or other emergency scenarios. This step is vital to prevent cascading failures that could affect the wider electrical network.

Final Safety Validation

Before the entire SMR power station is fully operational, regulatory authorities conduct a final safety validation to ensure that the plant meets all required safety, environmental, and operational standards. This comprehensive review is conducted after all the modules have been integrated and tested. It includes evaluating the performance of the plant's integrated safety systems, such as cooling circuits, emergency shutdown procedures, and radiation containment measures.

The final safety validation process involves both on-site inspections and the review of extensive documentation to confirm that the plant adheres to national and international nuclear safety standards. Regulatory bodies assess whether the plant's design, construction, and operational procedures align with the approved safety plans and licensing conditions. Only after the regulators are fully satisfied that all safety protocols have been met can the plant be granted permission for long-term operation.

This final validation is critical not only for ensuring public and environmental safety but also for gaining the trust of stakeholders, including investors, the government, and the local community. The rigorous nature of this process ensures that the SMR power station can operate reliably over its projected lifespan, providing a safe and consistent source of low-carbon energy.

Quality Control in SMR Construction

Quality Control in Small Modular Reactor (SMR) Construction is a comprehensive, multi-stage process designed to ensure that all components, systems, and construction practices meet stringent safety, performance, and regulatory standards. Given the complexity and critical nature of nuclear technology, the emphasis on quality control (QC) in SMR projects is even greater than in conventional infrastructure projects. A robust quality control process is essential for preventing defects, ensuring operational safety, extending the lifecycle of reactor components, and ultimately gaining regulatory approval.

The success of a Small Modular Reactor (SMR) construction project fundamentally hinges on meticulous quality control throughout the design and engineering phases. This phase is crucial for establishing a foundation that ensures the reactor's safety, efficiency, and compliance with nuclear regulations. Quality control during this stage encompasses rigorous design validation, robust document control systems, and comprehensive peer reviews and audits. Below, I provide a detailed breakdown of these critical quality control aspects, complete with specific examples.

Design validation is an intensive process that involves a series of computational simulations, stress tests, and thermal hydraulics modelling to confirm that the proposed reactor design can withstand operational demands. For instance, Finite Element Analysis (FEA) is commonly used to assess the structural integrity of key components like reactor pressure vessels and steam generators under various stress conditions. This method helps identify areas prone to stress concentrations that could lead to potential failures.

In addition, Computational Fluid Dynamics (CFD) is utilized to model the flow of coolant within the reactor core, ensuring that heat is effectively dissipated to prevent overheating. For example, in the design of the NuScale Power Module, CFD simulations were extensively used to optimize the reactor's passive cooling system. Thermal hydraulics analysis, such as using the RELAP5 or TRACE software, is also applied to simulate transient conditions and evaluate the reactor's behaviour during potential accident scenarios.

These design validation tools are complemented by probabilistic risk assessments (PRA) to evaluate the likelihood of various failure modes. This multi-layered validation approach ensures that the reactor design is robust, safe, and compliant with international safety standards before any physical construction begins.

The management of technical documents is critical for maintaining consistency and traceability throughout the SMR project lifecycle. A centralized document control system, such as SharePoint, ProjectWise, or EDMS (Electronic Document Management System), is used to manage all project documentation. These platforms allow for version control, access management, and audit trails, ensuring that only the most up-to-date information is available to project teams.

Key documents that must be controlled include:

- Design Basis Documents: Outlines the technical requirements and regulatory standards that guide the design of the SMR.

- Engineering Drawings: Detailed blueprints of components like the reactor vessel, piping layouts, and electrical systems.

- Safety Analysis Reports (SARs): Includes Preliminary Safety Analysis Reports (PSAR) and Final Safety Analysis Reports (FSAR), which are crucial for regulatory submissions.

- Materials Specifications: Defines the types of materials used in construction, including certifications for nuclear-grade components.

- Change Request Forms and Revision Logs: Track any modifications to the design, ensuring changes are documented, reviewed, and approved by relevant stakeholders.

By employing a centralized system, all changes, approvals, and updates are systematically documented, which reduces the risk of inconsistencies and ensures compliance with regulatory requirements. For instance, during the construction of the Hinkley Point C nuclear plant, a comprehensive document control system was employed to manage over 100,000 technical documents, ensuring seamless coordination between designers, contractors, and regulatory bodies.

Peer reviews and independent audits are vital components of the quality control process, providing an additional layer of scrutiny to identify potential design weaknesses. In the case of SMRs, peer reviews typically involve experts from different fields, such as thermal hydraulics, structural engineering, and safety systems, to assess the design's robustness.

For example, before the NuScale Power Module design was approved by the U.S. Nuclear Regulatory Commission (NRC), it underwent extensive peer review by independent experts to validate its passive safety features. These peer reviews often focus on key safety systems, such as emergency cooling systems and containment structures, to ensure they meet international standards like those set by the International Atomic Energy Agency (IAEA).

Audits are also conducted by third-party organizations, such as Lloyd's Register, DNV GL, or regulatory agencies, to confirm compliance with both national and international standards. The audit process includes a thorough review of design documentation, quality control procedures, and manufacturing practices. For instance, during the construction of Canada's Darlington Refurbishment project, independent audits were conducted to verify compliance with the Canadian Nuclear Safety Commission (CNSC) standards.

Moreover, Stage-Gate Reviews are commonly employed in SMR projects. At key milestones, such as the transition from the conceptual design phase to the detailed engineering phase, an independent review board assesses the project's readiness to proceed. This ensures that any identified risks are mitigated before moving on to the next phase of construction.

The selection of materials for SMR components is a meticulous process that focuses on ensuring durability, radiation resistance, and overall structural integrity. The reactor pressure vessels, steam generators, heat exchangers, and piping systems must be constructed from high-strength alloys, such as SA-508/SA-533 steel for pressure vessels, which are known for their excellent toughness and resistance to radiation-induced embrittlement. For components exposed to extreme temperatures and corrosive environments, materials like Inconel 690 or Alloy 800H are commonly used due to their superior high-temperature strength and corrosion resistance.

To verify that these materials meet stringent specifications, rigorous testing is conducted. This includes:

- **Non-Destructive Testing (NDT)**: Techniques such as ultrasonic testing (UT) and radiographic testing (RT) are employed to detect internal flaws like cracks or voids without damaging the material. For example, ultrasonic testing is used to inspect the integrity of welds in reactor vessels to ensure they are free from internal defects that could compromise safety.

- **Mechanical Testing**: Tests like tensile strength, impact toughness, and hardness testing are performed to assess the material's ability to withstand operational stresses. In particular, impact tests are critical for pressure vessels to confirm that they can endure sudden stress changes without fracturing.

- **Corrosion Resistance Assessment**: Materials that will be exposed to coolant systems undergo corrosion testing in simulated reactor environments to confirm their longevity. For example, Inconel alloys used in steam generators are tested for resistance to stress corrosion cracking, which can be accelerated in high-temperature water environments.

These rigorous testing protocols ensure that only materials that meet the highest nuclear industry standards are used in SMR construction, thereby reducing the risk of failures during operation.

A critical aspect of SMR construction is the welding and fabrication of key components like reactor pressure vessels, heat exchangers, and piping systems. Welding in the nuclear industry requires the highest level of precision and quality assurance, given the critical role that welded joints play in maintaining the structural integrity of pressure-containing components.

Once welded, the joints undergo stringent quality checks, including:

- **X-ray and Ultrasonic Inspections**: These techniques are used to verify the internal integrity of welds, ensuring that there are no hidden defects that could compromise the reactor's safety.

- **Pressure Testing**: After assembly, components like steam generators are subjected to hydrostatic pressure tests to confirm their ability to withstand operational pressures. For instance, a reactor pressure vessel may be tested at pressures 1.5 times higher than its normal operating conditions to ensure its resilience.

- **Dye Penetrant Testing**: For surface crack detection, dye penetrant inspection is often used on critical welds to identify any surface-breaking flaws that could lead to leaks or structural failure over time.

These quality assurance processes are essential for ensuring that the components meet the stringent safety requirements of nuclear reactors, thereby minimizing the risk of failures during both construction and operation.

Before the prefabricated modules leave the factory, they undergo a comprehensive series of Factory Acceptance Tests (FAT). These tests are designed to validate that each module functions as intended and adheres to the design specifications under simulated operational conditions. The FAT process is critical in identifying and rectifying any defects before the modules are transported to the construction site, thereby reducing costly delays during on-site assembly.

Examples of Factory Acceptance Testing include:

- **Hydraulic Testing for Pressure Vessels**: To verify the strength and leak-tightness of reactor pressure vessels, they are filled with water and pressurized to levels well above their normal operating pressure. This ensures that the vessel can handle pressure fluctuations during operation without leaking or bursting.

- **Electrical and Control Systems Testing**: Modules that include instrumentation and control (I&C) systems undergo tests to ensure that sensors, control loops, and automated safety shutdown mechanisms are fully functional. For example, digital control modules for NuScale's SMR are tested for fail-safe operations and load-following capabilities to confirm their reliability in regulating reactor power.

- **Thermal Performance Tests**: Heat exchangers and steam generators are tested to confirm their efficiency in transferring heat under varying loads. Simulations are run to mimic the thermal conditions that the modules will experience in the reactor environment.

Once the prefabricated Small Modular Reactor (SMR) modules reach the construction site, a rigorous quality control process is set in motion to ensure that the site preparation and civil engineering works adhere to stringent standards. This phase is crucial for the safe and efficient assembly of the modular components and involves meticulous inspections and testing to guarantee the structural integrity and environmental compliance of the project.

One of the first steps in site preparation is to ensure that the land can adequately support the reactor's weight and remain stable throughout its operational life. Comprehensive soil testing is conducted to assess load-bearing capacity, porosity, and soil composition. For example, before the construction of the HTR-PM reactors in China, extensive geotechnical surveys were conducted to assess the site's resistance to earthquakes. Seismic assessments are crucial, especially for reactors built in regions prone to seismic activity. This involves performing vibration analysis and soil liquefaction tests to confirm that the ground is stable enough to prevent shifting or settling under the heavy load of reactor components. Any necessary ground stabilization measures, such as soil compaction or the use of deep foundation piles, are implemented based on these findings.

As construction progresses, quality control focuses on the foundational and structural elements critical to supporting the reactor modules. This involves regular inspections during concrete pouring and curing processes to ensure that the foundations meet the required strength and durability standards. For instance, during the construction of the NuScale Power Plant, inspectors used ultrasonic sensors to monitor the curing of concrete in real-time, ensuring that it reached the necessary compressive strength before any heavy modules were installed.

Additionally, reinforced steel structures are carefully inspected for correct placement and alignment according to design specifications. The reactor building and containment structures undergo detailed checks to ensure that they meet both engineering and safety standards. This includes verifying the alignment of load-bearing walls, inspecting the quality of welds for structural steel frameworks, and ensuring compliance with blueprints.

Quality control also extends to mitigating the environmental impact of construction activities, particularly if the site is near populated areas. Continuous environmental monitoring systems are set up to measure parameters like dust, noise, and emissions. For example, during the construction of the Akademik Lomonosov floating SMR in Russia, noise levels were closely monitored to ensure they remained within permissible limits, minimizing disturbances to nearby communities. Additionally, water runoff from the construction site is tested for contaminants to prevent pollution of local waterways.

Real-time monitoring tools, such as air quality sensors and noise meters, are deployed to ensure that construction activities do not exceed environmental impact thresholds. This is especially important in locations with sensitive ecosystems or stringent regulatory requirements. By maintaining strict oversight over environmental factors, the construction process ensures compliance with local and international environmental standards.

Once the prefabricated Small Modular Reactor (SMR) modules are transported to the construction site, the focus shifts to the careful integration and assembly of these components. This phase involves the use of heavy-lift cranes to position large modules, such as reactor vessels and containment structures, followed by securing them with pre-configured bolt-and-weld connections. Quality control during this on-site assembly is crucial to ensure that each module is correctly integrated into the overall plant, maintaining both safety and performance standards.

As soon as the modules are placed, quality control teams begin verifying their alignment and fit-up before finalizing connections. Precision is essential because even minor deviations from design specifications can cause operational inefficiencies or, worse, create safety risks. To achieve this level of accuracy, technologies like laser scanning and 3D modelling are employed to check that each component aligns perfectly with the design blueprints. For instance, during the construction of the NuScale SMR, laser alignment systems were used to ensure that the reactor pressure vessels and steam generators were positioned with millimetre-level precision, minimizing any potential misalignments that could affect reactor operations.

Once the structural components are secured, attention turns to the integration of electrical systems and instrumentation. This phase is critical for ensuring that all control systems, sensors, and circuits function correctly as designed. Comprehensive tests are conducted to confirm that sensors detect the correct parameters, alarms trigger under specific conditions, and control systems respond accurately to operator inputs. During the construction of the HTR-PM reactors in China, extensive testing was performed on control circuits to ensure that all reactor safety interlocks were functional. Calibration of instrumentation is another critical step, involving precise adjustments to sensors that monitor temperature, pressure, and radiation levels.

For systems designed to handle high-pressure fluids, such as the reactor coolant loops and steam circuits, pressure testing is an essential quality control measure. This involves subjecting pipes, valves, and welded joints to pressures higher than normal operating levels to identify any potential leaks or weaknesses. For example, in the construction of the Akademik Lomonosov floating SMR, pressure tests were conducted on the reactor coolant systems to confirm the integrity of the welded joints. Any detected leaks are addressed immediately through repairs or adjustments to ensure that the system can withstand operational stresses without failure.

Special attention is given to areas where modules connect, as these joints are critical points that could develop leaks under high pressure. Non-destructive testing methods, such as ultrasonic and radiographic inspections, are frequently used to verify the integrity of these connections, ensuring that no internal defects are present.

By implementing stringent quality control protocols at each stage of the on-site integration, SMR projects can avoid costly delays and ensure the safety and efficiency of the plant. These measures are particularly important given the modular nature of SMR construction, where the reliability of each prefabricated module and its integration into the larger system is essential for the plant's success. Through a combination of advanced alignment technologies, rigorous electrical and instrumentation checks, and robust pressure testing, the project team ensures that every module operates seamlessly as part of a fully integrated power station.

Once the physical construction and assembly of the Small Modular Reactor (SMR) units are completed, the focus shifts to an intensive phase of pre-commissioning tests. This stage is crucial for verifying that all integrated systems are fully functional, safe, and ready for continuous operation. The testing process is comprehensive, covering everything from static assessments of individual components to dynamic trials of fully integrated systems.

The first step in pre-commissioning involves system integration tests, which are designed to ensure that all interconnected modules—such as cooling systems, turbines, steam generators, and control mechanisms—work together seamlessly. These tests are not just limited to confirming the mechanical fit but also involve checking the interoperability of the software systems that control reactor functions. For instance, during the pre-commissioning of the NuScale SMR, extensive tests were conducted to verify that the reactor's passive cooling system could automatically kick in during emergency conditions without any external power. Similarly, emergency shutdown mechanisms are rigorously tested to confirm that they respond correctly to simulated faults, ensuring that the reactor can be safely shut down if needed.

Once the system integration tests are successfully completed, the focus shifts to load testing and synchronization with the national power grid. In this phase, each reactor unit is gradually brought up to its full power output to confirm that it can generate electricity at its designated capacity without experiencing any performance issues. For example, the Akademik Lomonosov floating SMR underwent staged load testing to ensure its twin reactors could reliably produce a combined output of 70 MWe, supplying power to remote Arctic regions. Grid synchronization tests are equally critical; they confirm that the plant can connect smoothly to the existing grid infrastructure, handle load fluctuations, and respond to changes in electricity demand. This process involves fine-tuning the plant's load-following capabilities to prevent disruptions to the grid.

Before the SMR plant can transition to full commercial operations, it must pass a series of final safety checks and regulatory inspections. This stage is conducted by national nuclear regulatory authorities to ensure that the plant meets all required safety, environmental, and operational standards. For instance, in Canada, SMRs like the ARC-100 must comply with the stringent guidelines set by the Canadian Nuclear Safety Commission (CNSC). Inspectors carry out comprehensive assessments to validate that all systems conform to the specifications outlined in the Final Safety Analysis Report (FSAR). This includes verifying that emergency response protocols, radiation protection measures, and environmental safeguards are fully operational.

These final inspections are critical for securing an operating license. Any deficiencies identified during this stage must be rectified before the plant can be declared operational. The rigorous nature of this process ensures that the SMR operates safely from day one, protecting both the environment and the surrounding communities while contributing to a stable power supply.

Quality control for a Small Modular Reactor (SMR) does not cease once it becomes operational. Ensuring the long-term safety, efficiency, and reliability of the reactor requires ongoing monitoring, periodic inspections, and adherence to a robust maintenance schedule. This phase is crucial in safeguarding against potential issues that could affect performance or compromise safety, especially given the extended operational lifespan of nuclear power plants.

SMRs are often outfitted with sophisticated real-time monitoring systems that utilize a network of sensors and artificial intelligence (AI) algorithms to detect anomalies. For example, the NuScale Power Module is equipped with a suite of sensors that continuously monitor reactor core temperatures, pressure levels, coolant flow rates, and other critical parameters. These AI-driven systems can identify deviations from normal operating conditions, enabling operators to quickly address minor issues before they escalate into major concerns. This proactive approach significantly reduces the risk of unplanned shutdowns and enhances the overall safety profile of the reactor.

To ensure optimal performance and compliance with safety standards, SMRs follow a strict schedule of maintenance and inspections. Regular preventive maintenance activities might include replacing aging components like pumps, valves, and seals, as well as recalibrating sensors to maintain accurate readings. These inspections are vital for extending the reactor's operational life, reducing the likelihood of mechanical failures, and maintaining adherence to regulatory requirements. Additionally, shutdown periods for scheduled maintenance allow engineers to conduct comprehensive safety tests, ensuring that all systems remain in peak condition.

Periodic audits conducted by third-party experts and regulatory authorities play a crucial role in maintaining high safety and quality standards throughout the reactor's operational life. For example, the Canadian Nuclear Safety Commission (CNSC) mandates regular independent reviews for reactors operating in Canada, such as the ARC-100 and Terrestrial Energy's IMSR. These audits involve detailed assessments of the reactor's safety systems, operational protocols, and environmental impact measures. Independent reviews help to identify potential areas for improvement, ensuring that the plant remains in compliance with evolving safety regulations and international best practices.

Workforce Training and Skill Requirements

Constructing a Small Modular Reactor (SMR) power station involves multiple stages, each requiring specialized human resources to ensure efficient and safe project execution. Below is a comprehensive overview of the key roles required throughout the lifecycle of an SMR construction project, from initial planning and design through to construction, commissioning, and eventual operation.

1. Project Management and Planning

- **Project Director**: Oversees the entire project to ensure it stays on schedule, within budget, and meets quality standards.

- **Project Managers**: Coordinate the work of different teams, manage schedules, and resolve on-site issues.

- **Construction Managers**: Supervise on-site construction activities and ensure compliance with safety regulations.

- **Planning and Scheduling Engineers**: Develop detailed project timelines and monitor progress.

- **Cost Estimators**: Assess the costs of materials, labour, and equipment needed for the project.

- **Contract Administrators**: Manage contracts with suppliers, subcontractors, and other third parties.

- **Logistics Coordinators**: Plan the transportation of prefabricated modules and equipment to the site.

2. Engineering and Design

- **Nuclear Engineers**: Design the reactor core, cooling systems, and safety features to ensure safe operation.

- **Mechanical Engineers**: Develop systems like heat exchangers, steam turbines, and piping networks.

- **Electrical Engineers**: Design power distribution systems, control circuits, and instrumentation.

- **Civil Engineers**: Plan and oversee the construction of foundations, buildings, and structural supports.

- **Structural Engineers**: Ensure the structural integrity of containment buildings, cooling towers, and support structures.

- **Control and Instrumentation Engineers**: Design and implement control systems for reactor monitoring and automation.

- **Thermal-Hydraulics Engineers**: Analyse heat transfer and fluid flow in the reactor systems.

- **Materials Engineers**: Select materials that can withstand high temperatures, radiation, and corrosion.

- **Environmental Engineers**: Conduct assessments to minimize the environmental impact of construction and operation.

3. Procurement, Quality Control, and Supply Chain

- **Procurement Specialists**: Source and purchase materials, components, and services required for construction.

- **Supply Chain Managers**: Coordinate the delivery of components, especially prefabricated modules, from factories to the site.

- **Quality Assurance (QA) Managers**: Ensure that construction processes meet industry standards and regulations.

- **Quality Control Inspectors**: Conduct inspections to verify compliance with engineering specifications.

- **Non-Destructive Testing (NDT) Technicians**: Perform ultrasonic, radiographic, and other testing methods to detect defects in materials and welds.

- **Welding Inspectors**: Ensure high-quality welds for critical components like reactor vessels.

4. Construction and On-Site Assembly

- **Site Supervisors**: Oversee day-to-day construction activities on-site.

- **Heavy Equipment Operators**: Operate cranes, excavators, and other machinery to lift and position modules.

- **Welders and Fabricators**: Assemble components and modules, especially for the reactor pressure vessels.

- **Electricians**: Install electrical wiring, panels, and control systems.

- **Plumbers and Pipefitters**: Assemble and install cooling and steam piping systems.

- **Concrete Technicians**: Ensure the proper mixing, pouring, and curing of concrete for foundations.

- **Laborers and General Construction Workers**: Assist with various construction tasks.

- **Health and Safety Officers**: Monitor on-site safety and compliance with regulations.

- **Security Personnel**: Maintain site security and control access.

5. Commissioning and Testing

- **Commissioning Engineers**: Conduct tests to validate that all systems are operational before the plant is brought online.

- **Control Room Operators**: Test control systems, automation processes, and safety protocols.

- **System Integration Engineers**: Ensure that all subsystems work together seamlessly.

- **Grid Synchronization Engineers**: Connect the SMR plant to the national grid and test load-following capabilities.

- **Radiation Protection Officers**: Monitor radiation levels during commissioning and ensure adherence to safety protocols.

6. Environmental and Regulatory Compliance

- **Regulatory Affairs Specialists**: Work with regulatory bodies to ensure compliance with nuclear safety laws.

- **Environmental Impact Assessors**: Conduct studies to assess the project's impact on the local environment.

- **Permit Coordinators**: Obtain necessary permits and approvals from local, regional, and national authorities.

7. Information Technology and Cybersecurity

- **IT Infrastructure Engineers**: Set up and manage IT systems and networks for plant operations.

- **Cybersecurity Specialists**: Protect the SMR plant's digital infrastructure from cyber threats.

- **Data Analysts**: Monitor and analyse operational data to optimize performance.

8. Support Services and Administration

- **Human Resources (HR) Managers**: Handle recruitment, training, and management of the workforce.

- **Administrative Assistants**: Provide support with scheduling, documentation, and coordination.

- **Finance and Accounting Specialists**: Manage project budgets, payroll, and financial reporting.

- **Legal Advisors**: Provide legal support for contracts, regulatory compliance, and dispute resolution.

- **Public Relations Officers**: Engage with local communities, stakeholders, and media to ensure transparency and build trust.

9. Operation and Maintenance (Post-Construction)

- **Operations Manager**: Oversee the day-to-day operation of the SMR plant once it becomes operational.

- **Maintenance Engineers**: Perform routine maintenance to ensure the plant runs efficiently.

- **Nuclear Safety Officers**: Continuously monitor safety systems and ensure adherence to nuclear safety standards.

- **Training Coordinators**: Provide training for staff on new equipment, safety protocols, and operational procedures.

As an example, constructing a 4,000 MW Small Modular Reactor (SMR) power station (using a previous example), consisting of multiple 100 MW SMR units, is an extensive project that requires a wide range of specialized human resources.

Below is a detailed table outlining the various job roles required for constructing a 4,000 MW Small Modular Reactor (SMR) power station. The table includes estimated numbers of personnel needed for each role and the duration for which they would be employed throughout the project.

Category	Job Role	Number of Personnel Required	Duration of Employment
Project Management and Planning	Project Director	1	Full Project Duration (8-10 years)
	Project Managers	10	Full Project Duration (8-10 years)
	Construction Managers	15	6-8 years
	Planning & Scheduling Engineers	8	4-6 years
	Cost Estimators	5	2-3 years
	Contract Administrators	6	3-5 years

Category	Job Role	Number of Personnel Required	Duration of Employment
	Logistics Coordinators	10	5-6 years
Engineering and Design	Nuclear Engineers	20	4-6 years
	Mechanical Engineers	30	4-6 years
	Electrical Engineers	25	4-6 years
	Civil Engineers	30	4-6 years
	Structural Engineers	20	3-5 years
	Control & Instrumentation Engineers	15	4-6 years
	Thermal-Hydraulics Engineers	12	4-6 years
	Materials Engineers	10	3-5 years
	Environmental Engineers	8	2-4 years
Procurement, Supply Chain & Quality Control	Procurement Specialists	10	3-5 years
	Supply Chain Managers	6	3-5 years
	Quality Assurance Managers	5	Full Project Duration (8-10 years)
	Quality Control Inspectors	25	5-7 years
	NDT Technicians	20	4-6 years
	Welding Inspectors	15	3-5 years
	Site Supervisors	30	6-8 years

Category	Job Role	Number of Personnel Required	Duration of Employment
Construction and On-Site Assembly	Heavy Equipment Operators	40	4-6 years
	Rigging Specialists	20	4-6 years
	Welders & Fabricators	50	4-6 years
	Concrete Technicians	25	3-5 years
	Pipefitters & Plumbers	30	4-6 years
	Electricians	40	4-6 years
	Scaffolders & Laborers	100	3-5 years
	Health & Safety Officers	15	Full Project Duration (8-10 years)
	Security Personnel	25	Full Project Duration (8-10 years)
Commissioning & Testing	Commissioning Engineers	20	2-4 years
	Control Room Operators	10	2-3 years
	System Integration Engineers	10	3-4 years
	Grid Synchronization Engineers	8	1-2 years
	Radiation Protection Specialists	12	2-4 years
Environmental & Regulatory Compliance	Regulatory Affairs Specialists	8	4-6 years
	Environmental Impact Assessors	10	2-4 years

Category	Job Role	Number of Personnel Required	Duration of Employment
	Permit Coordinators	6	3-5 years
	Waste Management Specialists	10	2-4 years
Information Technology & Cybersecurity	IT Infrastructure Engineers	12	3-5 years
	Cybersecurity Specialists	10	3-5 years
	Data Analysts	8	2-3 years
Support Services	Human Resources Managers	5	Full Project Duration (8-10 years)
	Administrative Assistants	15	5-7 years
	Finance & Accounting Specialists	8	Full Project Duration (8-10 years)
	Legal Advisors	5	4-6 years
	Public Relations Officers	6	4-6 years
Operation & Maintenance (Post-Construction)	Plant Operations Manager	2	Ongoing
	Maintenance Engineers	20	Ongoing
	Nuclear Safety Officers	10	Ongoing
	Training Coordinators	5	Ongoing

Key Insights:

1. **Total Workforce**: Approximately 750-1,000 people at peak construction phase, with specialized roles added as needed throughout different phases.

2. **Duration**: The overall project duration is estimated at 8-10 years, including design, construction, commissioning, and initial operations.

3. **Phased Approach**: The project utilizes a phased construction strategy, where certain roles are scaled up or down depending on the stage of the project. For example, more welders and fabricators are needed during the peak construction phase, while the need for commissioning engineers increases towards the end of the project.

This detailed breakdown provides insight into the extensive human resource planning required to successfully build an SMR power station, ensuring that all phases are adequately staffed to meet timelines and safety standards.

Several countries possess the capability, expertise, and workforce necessary to support the construction of a large-scale project like a 4,000 MW Small Modular Reactor (SMR) power station. Here's a detailed overview of the countries best positioned to undertake such projects:

The United States has a well-established nuclear industry supported by a robust regulatory framework managed by the Nuclear Regulatory Commission (NRC). The country has a vast pool of experienced nuclear engineers, commissioning experts, and control room operators. The U.S. also benefits from skilled labour, including welders, heavy equipment operators, and non-destructive testing (NDT) technicians, owing to its extensive history in power plant construction. Additionally, the U.S. actively promotes nuclear innovation, including SMR development, with leading companies such as NuScale Power, Westinghouse Electric, and GE Hitachi Nuclear Energy [16]. The U.S. also benefits from a strong educational infrastructure that produces a continuous supply of skilled labour, including welders and technicians, essential for nuclear construction projects [137].

Canada is a strong player in nuclear energy, especially with its experience in CANDU reactor designs. It has a skilled workforce, particularly in nuclear engineering, regulatory compliance, welding, and quality control. The country's well-established supply chains for nuclear components and government support for SMRs in provinces like Ontario and Saskatchewan are crucial enablers [138]. Major companies involved in nuclear projects include SNC-Lavalin, Bruce Power, and Ontario Power Generation [3].

The United Kingdom boasts a robust nuclear workforce, bolstered by recent investments in projects like Hinkley Point C and plans for SMRs led by Rolls-Royce [134]. The UK's regulatory framework, managed by the Office for Nuclear Regulation (ONR), supports the deployment of new nuclear technologies. Skilled engineers, project managers, and specialized tradespeople are readily available, particularly in civil and mechanical engineering sectors. Key industry players include Rolls-Royce, EDF Energy, and Balfour Beatty.

France remains a global leader in nuclear power with its extensive expertise in constructing and operating nuclear plants. EDF leads the industry, supported by companies like Framatome and

Areva. The country has a deep pool of nuclear engineers, control room operators, and commissioning experts. Its nuclear regulatory framework is well-defined, overseen by the French Nuclear Safety Authority (ASN), which ensures that projects meet stringent safety standards [11, 250].

China is rapidly expanding its nuclear fleet and has developed substantial expertise in both large reactors and SMRs. The country has a highly skilled workforce in civil construction, mechanical engineering, and nuclear commissioning, with a track record of completing large-scale projects efficiently. Government-backed support and streamlined regulatory processes have further boosted nuclear development. Leading companies include the China National Nuclear Corporation (CNNC) and the China General Nuclear Power Group (CGN) [56, 63].

South Korea is recognized for its efficiency in nuclear construction, especially with its APR1400 and SMART SMR technologies. The country benefits from a highly skilled labour force in nuclear engineering, construction management, and electrical engineering. Strong government backing and a well-regulated industry support its nuclear ambitions. Key players include Korea Hydro & Nuclear Power (KHNP) and Doosan Heavy Industries [232].

Japan has deep expertise in nuclear technology, particularly in advanced reactor designs and safety systems. The country has a skilled workforce of engineers and technicians experienced in nuclear plant construction and maintenance. Despite regulatory challenges post-Fukushima, Japan remains capable of scaling up nuclear projects, particularly in modular construction. Key industry players are Mitsubishi Heavy Industries, Hitachi, and Toshiba [11, 102].

Russia is a leader in modular reactor technology, as demonstrated by its floating SMR, Akademik Lomonosov. The country has a large and experienced workforce skilled in nuclear construction, welding, and reactor commissioning [3, 13]. The centralized support from state enterprises like Rosatom and streamlined regulatory approvals facilitate rapid project execution [56].

India's nuclear industry is expanding, with a focus on utilizing advanced nuclear technologies, including SMRs, to enhance its energy portfolio. The country has a substantial pool of skilled engineers, project managers, and construction personnel experienced in large infrastructure projects. Strong government backing for nuclear energy aligns with India's strategy for energy security. Key industry players include the Nuclear Power Corporation of India Limited (NPCIL) and Larsen & Toubro [3, 16, 243].

Countries such as the U.S., Canada, the UK, France, China, South Korea, and Japan are well-equipped to execute a large-scale SMR project due to their established nuclear industries, skilled workforce, and supportive regulatory environments. Emerging nuclear nations like India and Russia also have the necessary expertise, particularly with their recent focus on modular reactor technologies.

Countries without a well-established nuclear infrastructure may face challenges in sourcing the specialized workforce needed for SMR projects. These nations may need to rely on international partnerships or bring in specialized labour from countries with established nuclear expertise to meet project demands.

As an example, If Australia were to embark on constructing a 4,000 MW Small Modular Reactor (SMR) power station, it would face substantial challenges given the country's current limited experience in nuclear power infrastructure. However, with a strategic approach involving both short-term and long-term investments, Australia can develop the necessary capabilities to undertake such a large-scale nuclear project. Below is a comprehensive breakdown of strategies that could help bridge the skills gap and successfully achieve this goal:

To build up expertise quickly, Australia could form strategic partnerships with countries that have well-established nuclear industries, such as the United States, Canada, the United Kingdom, and South Korea. These partnerships could involve technology transfer agreements, joint ventures, and knowledge-sharing initiatives to accelerate the development of local capabilities. Additionally, creating exchange programs where Australian engineers, technicians, and managers are seconded to nuclear power projects overseas would allow them to gain hands-on experience. This approach would be instrumental in rapidly building up Australia's nuclear workforce, leveraging global best practices.

A crucial step in ensuring Australia can meet the workforce demands of an SMR project is to expand local educational and training programs. This would involve establishing specialized nuclear engineering courses in Australian universities, focusing on areas such as reactor design, nuclear safety, thermal hydraulics, and materials science. Leading institutions like the University of New South Wales (UNSW) and the Australian National University (ANU) could play key roles in developing curricula tailored to SMR technology.

Additionally, vocational training and apprenticeships would be necessary to develop technical skills like welding, non-destructive testing (NDT), and electrical engineering. Collaboration with the TAFE (Technical and Further Education) system could ensure these programs align with industry needs. To further enhance expertise, Australia could partner with leading international universities such as MIT in the U.S., Imperial College London, and KAIST in South Korea, providing Australian students and professionals access to cutting-edge research and advanced nuclear courses.

Australia has a strong talent pool in industries such as mining, oil and gas, and power generation. Many of these skills, particularly in project management, heavy equipment operation, and safety compliance, are transferable to nuclear projects. Tailored training programs could be developed to upskill professionals from these sectors. For instance, workers with experience in high-pressure systems, welding, and process engineering could be re-trained for SMR construction and operations. Additionally, online and blended learning platforms like Coursera, edX, and LinkedIn Learning could deliver specialized nuclear courses to expedite workforce development.

Given the immediate need for specialized skills, Australia could address short-term gaps by recruiting experienced nuclear engineers, project managers, and technicians from countries with established nuclear sectors. This could be supported through special visa programs and incentives to attract skilled professionals willing to relocate. By leveraging Australia's reputation for high living standards and stable work environments, the country could establish a talent pipeline that draws experts from around the world. Simplifying visa processes and offering attractive relocation packages would further enhance this effort.

To support modular construction, Australia would need to invest in local manufacturing capabilities for key components like reactor vessels, heat exchangers, and control systems. This would require collaboration between the government and private sector to establish modern manufacturing facilities. Encouraging small and medium enterprises (SMEs) to enter the nuclear supply chain through grants, tax incentives, and training programs would also help develop a robust domestic industry capable of supporting the construction and maintenance of SMRs.

Establishing a dedicated nuclear regulatory body is critical, as Australia currently lacks a comprehensive regulatory framework for nuclear power. A body similar to the U.S. Nuclear Regulatory Commission (NRC) or the UK's Office for Nuclear Regulation (ONR) would be essential. This body would need to develop guidelines and standards specific to SMRs, balancing safety with innovation. Specialized training programs for regulators and safety inspectors would be required to build expertise in nuclear safety, environmental assessments, and licensing processes. Participation in organizations like the International Atomic Energy Agency (IAEA) could provide valuable guidance in adopting global best practices.

To stimulate interest in nuclear careers, the government could offer scholarships, grants, and internships for students pursuing nuclear engineering and related fields. Public-private partnerships between the government, industry, and educational institutions would be crucial in ensuring a steady pipeline of skilled labour. Additionally, investing in research and development (R&D) centres focused on SMR technology could attract talent and drive innovation in reactor design, materials science, and safety systems.

The construction of a 4,000 MW SMR power station in Australia would require a multi-faceted approach to workforce development, blending international collaboration, local education initiatives, and targeted recruitment. By leveraging existing capabilities in related industries and investing in specialized training programs, Australia can build the necessary skills base to support its nuclear ambitions. With strategic planning, robust government support, and collaboration with global experts, Australia could position itself as a leader in the deployment of SMR technology, contributing to the region's energy security and carbon reduction goals.

Developing a skilled local workforce is critical, especially given Australia's limited nuclear experience. These costs would need to be factored into any nuclear power station project. Estimating the cost of constructing a 4,000 MW Small Modular Reactor (SMR) power station in Australia requires a detailed analysis, factoring in various elements such as the cost of modular reactor units, workforce development, regulatory setup, supply chain establishment, and long-term operational planning. Below, I break down the major cost components involved in such a large-scale project, along with realistic estimates.

1. Cost of SMR Units and Modular Construction

The bulk of the expenses will stem from the purchase, manufacturing, and installation of SMR modules. Assuming the use of 40 SMRs each with 100 MW capacity, the cost per SMR module varies based on technology, suppliers, and location. As a general estimate:

- **Cost per 100 MW SMR module**: Approximately USD 300-500 million, depending on factors such as technology type (e.g., pressurized water reactors like NuScale, BWRX-300) and market conditions.

- **Total cost for 40 modules**: USD 12-20 billion.

Modular construction offers cost efficiencies compared to traditional nuclear reactors. However, given Australia's limited nuclear infrastructure, initial projects may see higher costs due to the need for local manufacturing setup and technology adaptation.

2. Site Preparation and Infrastructure Development

Australia's lack of existing nuclear power infrastructure means significant investment in **site preparation**, including:

- **Land acquisition and site development**: USD 500 million - 1 billion, depending on location and existing infrastructure.

- **Grid integration and power lines**: USD 300-500 million to connect the plant to the national grid.

- **Cooling systems, water supply, and other utilities**: USD 200-400 million.

3. Workforce Development and Training Programs

Developing a skilled local workforce is critical, especially given Australia's limited nuclear experience. Costs will include:

- **International partnerships and training**: Establishing exchange programs with countries like the U.S., Canada, and the UK could cost around USD 100-200 million.

- **Educational programs and local training**: Establishing or expanding nuclear engineering courses and vocational training programs through universities and TAFE could require an investment of around USD 50-100 million.

- **Recruitment of international experts**: Special visa programs, relocation incentives, and competitive salaries could cost an additional USD 50 million.

4. Establishing a Regulatory Framework

Setting up a **dedicated nuclear regulatory body** in Australia to oversee SMR projects would be essential:

- **Regulatory body establishment and training**: USD 100-200 million over 5-7 years, including recruitment, training, and infrastructure setup.

- **Compliance with international standards (IAEA guidelines)**: USD 50 million for consultations, audits, and certifications.

5. Developing Domestic Nuclear Supply Chains

Building a domestic supply chain for nuclear components is a long-term investment:

- **Setting up manufacturing facilities** for reactor vessels, control systems, and heat exchangers: USD 1-2 billion.

- **Supporting SMEs through grants and tax incentives**: USD 200-300 million to encourage local participation in the nuclear supply chain.

- **Logistics and transport infrastructure**: USD 100-200 million to facilitate the movement of large modular components.

6. Long-Term Operational Costs and Maintenance

After construction, the focus shifts to operation and maintenance:

- **Commissioning and initial operational setup**: USD 200-300 million.

- **Maintenance contracts and ongoing training**: USD 100-200 million per year.

- **Waste management and decommissioning fund**: USD 500 million set aside for safe waste handling and future decommissioning.

7. Estimated Total Project Cost

Taking into account all of the above components:

Category	Estimated Cost (USD)
SMR Modules (40 x 100 MW)	12-20 billion
Site Preparation and Infrastructure	1-2 billion
Workforce Development	200-350 million
Regulatory Framework	100-200 million
Domestic Supply Chain	1.5-2.5 billion
Operational Setup and Maintenance	200-300 million (initial setup)
Waste Management and Decommissioning Fund	500 million
Total Estimated Cost	**15.5-25.5 billion**

8. Comparative Analysis with Renewable Energy Projects

For comparison, constructing a 4,000 MW solar or wind power plant would typically cost between USD 4-6 billion. However, renewable sources require large land areas, depend on weather conditions, and have higher operational variability. The levelized cost of electricity (LCOE) for SMRs, although higher upfront, offers benefits like consistent baseload power, smaller land footprint, and reduced intermittency compared to renewables.

Constructing a 4,000 MW SMR power station in Australia would involve a substantial initial investment of USD 15.5-25.5 billion (approximately AUD 24.03 to 39.53 billion, using an exchange rate of 1.55 AUD per USD), driven by the need to establish a regulatory framework, develop local capabilities, and build supply chains. While this is significantly higher than the cost of a comparable renewable energy project, the strategic benefits of reliable, low-carbon baseload power could justify the investment, especially as Australia seeks to diversify its energy portfolio and reduce its carbon footprint.

Overcoming Construction Challenges

Constructing a Small Modular Reactor (SMR) power station presents a myriad of unique challenges and risks, primarily due to its innovative technology, complex regulatory landscape, and intricate project dynamics. This response outlines the key construction challenges, associated risks, mitigation strategies, and methods to anticipate and overcome these challenges, supported by relevant literature.

The construction of SMRs is heavily influenced by stringent regulatory scrutiny. The licensing processes for nuclear projects are often complex and can lead to significant delays and increased costs. Many countries lack a standardized regulatory framework for SMRs, which exacerbates these challenges [231, 251]. Delays in securing necessary permits and regulatory non-compliance are notable risks that can arise from this environment [1].

To mitigate these risks, early engagement with regulatory bodies is essential. Establishing communication during the planning phase can clarify requirements and streamline the approval process [251]. Pre-licensing consultations can further address concerns and facilitate smoother transitions through regulatory hurdles [231]. Employing regulatory experts familiar with the nuclear landscape of the host country can also enhance compliance efforts [1]. Regular updates to risk assessments based on evolving regulatory landscapes will help maintain project flexibility [1, 251].

The modular construction approach of SMRs relies heavily on prefabricated components, making the supply chain critical to project success. Disruptions in the supply chain can lead to significant delays due to manufacturing delays, transportation issues, and quality control failures [252, 253].

To mitigate these risks, establishing a diversified supplier base is important. This reduces dependency on single suppliers and enhances resilience against supply shortages [252]. Robust logistics planning, including detailed assessments of transport routes and access roads, is necessary to ensure the viability of the supply chain [253]. On-site inventory management of critical components can further alleviate delays caused by supply chain disruptions [252]. Regular audits of the supply chain and strong supplier relationships are essential for anticipating potential issues [253].

The assembly of prefabricated modules on-site presents its own set of challenges, particularly concerning alignment and integration. Misalignment or quality issues during assembly can lead to project delays and increased costs [17, 254].

Utilizing detailed 3D modelling through Building Information Modelling (BIM) can ensure precise alignment of modules before they arrive on-site, thus reducing integration failures [17]. Employing a skilled workforce, including experienced welders and assembly teams, is vital for maintaining high construction quality [17]. Rigorous quality control inspections, including ultrasonic testing, can identify issues early in the assembly process [251]. Conducting mock assemblies and factory acceptance tests can further prepare teams for on-site deployment [17].

The construction of SMRs necessitates a highly specialized workforce, leading to potential shortages that can impact project timelines and quality [17]. Risks associated with inadequate staffing and skills gaps can result in higher labour costs and project delays [17].

To address these challenges, investing in training programs in collaboration with local educational institutions can help build a skilled labour pool [17]. Additionally, international recruitment strategies can attract talent from countries with established nuclear expertise [17]. Upskilling existing workers from related industries, such as oil and gas, can also mitigate workforce shortages [17]. Regular assessments of workforce capabilities will allow for timely adjustments to recruitment and training plans [17].

The innovative designs of SMRs introduce technical challenges, as many components have not been extensively tested in real-world conditions [251]. This can lead to engineering design flaws and unexpected technical issues during commissioning [251].

Prototyping and testing pilot SMR units can validate designs before full-scale construction begins [251]. The use of digital twins allows for simulation of reactor operations, helping to identify potential issues and optimize performance [251]. Engaging independent experts for third-party peer reviews of engineering designs can further enhance the reliability of technical assumptions [251]. Continuous updates to risk assessments based on new technical insights are essential for managing these challenges [251].

Large-scale nuclear projects often face public opposition regarding safety and environmental impacts [251]. Risks include delays due to community protests and increased scrutiny from environmental groups [251].

To mitigate these risks, early community engagement is crucial. Informing and involving local communities in the planning process can foster support and reduce opposition [251]. Transparent communication about safety measures and environmental protection practices is essential [251]. Conducting comprehensive Environmental Impact Assessments (EIAs) demonstrates compliance with regulations and can alleviate public concerns [251]. Monitoring public sentiment will allow for timely adjustments to communication strategies [251].

Nuclear projects are capital-intensive and subject to cost overruns due to inflation, regulatory delays, and supply chain issues [251]. Risks include budget overruns and funding gaps that can jeopardize project viability [251].

Phased funding based on project milestones can reduce financial risk [251]. Implementing strict budget tracking and cost control measures is essential for maintaining financial discipline [251]. Seeking government support through incentives or grants can help offset high upfront costs [251]. Regular financial risk assessments will ensure that budget allocations remain flexible to adapt to changing economic conditions [251].

Ensuring the safety of SMRs and preventing cybersecurity threats is paramount, especially given the interconnected nature of modular construction [251]. Risks include safety incidents and cyber-attacks on control systems [251].

Developing robust safety protocols and conducting regular safety drills can enhance preparedness [251]. Implementing cybersecurity frameworks to protect control systems is critical [251]. Engaging independent safety auditors for regular reviews can further strengthen safety practices [251]. Conducting regular security risk assessments will help adapt protocols to emerging threats [251].

Chapter 6
Infrastructure and Systems for SMR Power Stations

Cooling Systems and Heat Management

The construction of a Small Modular Reactor (SMR) power station, especially one with advanced cooling systems and heat management capabilities, involves building specialized infrastructure to support its unique operational characteristics. Below is a detailed breakdown of the critical infrastructure, systems, and support structures needed to implement efficient cooling and heat management for SMRs, covering everything from construction to operational readiness.

Example of SMR Power Plant Infrastructure: GEN4 Energy Module

The GEN4 Energy Module, formerly known as the Hyperion Power Module, exemplifies the next generation of SMR infrastructure by integrating advanced technologies to enhance safety, efficiency, and adaptability. This small modular reactor design, see **Error! Reference source not found.**, focuses on innovative approaches to cooling, refuelling, and modular deployment, making it a versatile option for various energy applications, especially in regions with limited infrastructure [3].

1. Cooling Systems and Heat Management

A key feature of the GEN4 Energy Module is its use of natural circulation for the primary coolant. Unlike traditional reactors that rely on pumps to circulate coolant, the GEN4 module utilizes passive cooling systems that naturally circulate coolant through the reactor core. This eliminates the need for mechanical pumps, reducing the risk of mechanical failure and maintenance demands. The natural circulation method enhances the plant's safety by ensuring continuous heat removal, even during power outages, thus simplifying the overall operational requirements [3].

The reactor's containment structure includes integrated passive heat dissipation systems, such as air-cooled heat exchangers, which help to manage residual heat efficiently. These passive systems ensure that in the unlikely event of an emergency shutdown, decay heat can still be dissipated without the need for external power sources, contributing to a high level of safety [3].

2. Factory-Based Refuelling and Modularity

One of the distinctive features of the GEN4 Energy Module infrastructure is its approach to fuelling and refuelling. Unlike traditional reactors, which require on-site refuelling operations, the GEN4 design is fully factory-fuelled and de-fuelled. This off-site refuelling strategy significantly reduces downtime during maintenance cycles and minimizes the complexities associated with handling nuclear fuel on-site. By transporting pre-fuelled modules to the plant and returning them to the factory for de-fuelling, this method lowers the risks related to fuel handling and enhances overall safety [3].

This factory-based refuelling capability also simplifies the plant's design and infrastructure needs, as on-site facilities for core refuelling are not required. This reduces the footprint of the nuclear installation and simplifies compliance with regulatory safety standards, especially in remote or developing areas.

3. Modular Construction and Flexible Deployment

The GEN4 Energy Module is designed with a focus on compact, modular construction, allowing for quick and efficient deployment. The modular nature of the design means that multiple GEN4 units can be installed in a single location to scale power output as needed. This flexibility supports both single-module installations for isolated grids and multi-module configurations for larger power demands.

The reactor modules are prefabricated in controlled factory environments, ensuring consistent quality control and reducing on-site construction time. This approach allows for rapid deployment, particularly in areas where building large traditional nuclear power plants would be challenging. The modular design also enables easy transportation of units to various locations, making it ideal for remote or off-grid applications where grid stability is a concern [3].

4. Enhancing Energy Security and Sustainability

The GEN4 Energy Module's infrastructure is designed to enhance energy security by providing reliable, low-carbon power in diverse environments. The compact and modular design allows for deployment in regions with limited infrastructure, supporting both industrial and community power needs. Additionally, the ability to use passive cooling systems and factory-based refuelling aligns with sustainability goals by reducing environmental impact and minimizing the risk of operational accidents [3].

This infrastructure model, driven by research from Los Alamos National Laboratory, leverages cutting-edge nuclear technology to provide a scalable, safe, and efficient energy solution. Its design is particularly beneficial for countries aiming to enhance energy independence while transitioning to greener energy sources, making it a strategic asset for global energy security [3].

The integration of passive safety features, factory-fuelled modules, and flexible deployment positions the GEN4 Energy Module as a pioneering solution in the field of SMR technology, promising to reshape the future of nuclear energy by making it safer, more efficient, and accessible to a wider range of applications around the world [3].

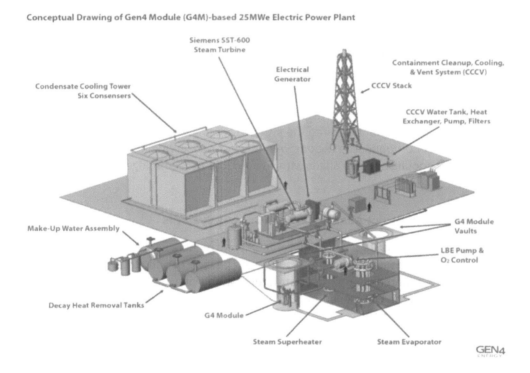

Figure 36: Layout of the GEN4 Energy plant [3].

Cooling Water Intake and Distribution Systems

In Small Modular Reactors (SMRs) that utilize water as either a primary or secondary coolant—particularly in designs like Light Water Reactors (LWRs)—establishing robust cooling water infrastructure is crucial for efficient and safe operation. These systems must be meticulously designed to ensure reliable cooling while adhering to environmental standards. The facility requires specialized intake structures to draw water from natural sources such as rivers, lakes, or oceans, along with efficient systems for distributing and returning the water once it has been used in the cooling process.

The intake structures are designed to filter out debris and prevent it from entering the cooling circuits. This filtration is essential to maintain the efficiency and safety of the reactor's cooling systems. Additionally, these structures must be carefully engineered to minimize any negative environmental impact, particularly thermal pollution, which can affect local ecosystems. Adhering to strict environmental regulations ensures that the intake process does not disrupt the surrounding natural habitat.

Pumping stations play a key role in circulating water through the reactor's heat exchangers. These stations are equipped with high-capacity pumps capable of moving significant volumes of water. To ensure uninterrupted cooling, even in cases of power outages, the system is designed with redundant pumps and backup generators. This redundancy is critical for maintaining continuous cooling and preventing overheating, especially during emergency situations.

For SMR installations in inland areas where water resources are limited, cooling towers are implemented to manage heat dissipation. These towers use evaporative cooling techniques to release heat into the atmosphere, thereby reducing the amount of water consumed. By recirculating the cooled water back into the system, cooling towers help optimize water usage, making them a vital component in regions where water conservation is a priority. This not only supports efficient cooling but also aligns with sustainable water management practices.

Reactor Containment Structures and Passive Cooling Systems

In Small Modular Reactors (SMRs), ensuring efficient and reliable cooling is a top priority, and this is achieved through a combination of both active and passive cooling technologies. At the heart of this setup is the reactor containment structure, which plays a critical role in housing and supporting these systems while ensuring the safety of the reactor during normal operations as well as in emergency situations.

The Reactor Containment Building (RCB) is a highly fortified structure, typically constructed from reinforced concrete and steel. It is specifically designed to withstand extreme external impacts, such as earthquakes, explosions, or other catastrophic events. The RCB not only serves as a physical barrier to contain radioactive materials but also integrates various cooling systems to manage residual heat. One of the key features of modern SMR designs is the inclusion of passive cooling mechanisms, such as air-cooled heat exchangers, which are capable of dissipating heat even if active systems are compromised. These systems are particularly valuable in maintaining safety during unexpected shutdowns or power failures.

To enhance safety further, SMRs utilize passive heat dissipation systems that do not rely on external power sources or mechanical pumps. For instance, some designs incorporate large water pools surrounding the reactor vessel, which act as heat sinks to absorb and dissipate excess heat through natural convection. Additionally, cooling fins or heat exchangers attached to the containment structure facilitate heat transfer to the surrounding air. An exemplary model of this approach is the NuScale reactor design, which employs a containment pool that passively cools the reactor core in case of an emergency shutdown. This design ensures that the reactor can safely cool itself down without requiring operator intervention, thus reducing the risk of overheating and potential core damage.

Secondary Heat Transfer and Steam Generation Systems

For Small Modular Reactors (SMRs) that utilize steam-driven turbines, such as the NuScale and CAREM reactors, secondary heat transfer systems are vital to optimizing thermal efficiency. These systems play a crucial role in transferring heat from the reactor core to generate steam, which is then used to drive the turbines and produce electricity. The design of these systems must be highly efficient to ensure maximum energy output while minimizing thermal losses.

A key component in these reactors is the Helical Coil Steam Generator. These compact units are specifically engineered to provide superior heat transfer efficiency within a limited physical space. The helical coil design allows for a high surface area contact between the coolant and the secondary steam circuit, thereby enhancing the rate of heat exchange. This configuration not only maximizes the amount of heat transferred but also reduces the overall size of the system, making it well-suited for modular reactor designs that prioritize compactness and efficiency.

Helical coil steam generators are increasingly becoming a preferred choice for Small Modular Reactor (SMR) projects due to their compact design, high heat transfer efficiency, and ability to support passive safety systems. These steam generators are specifically designed to meet the unique needs of SMRs, which prioritize modularity, safety, and cost-effectiveness. Below, we describe some of the commercially available helical coil steam generators that are currently utilized or being developed for SMR applications:

1. NuScale Power Module - Helical Coil Steam Generator

- **Overview**: The NuScale Power Module is one of the most advanced and widely recognized SMR designs incorporating a helical coil steam generator. This system is integrated directly within the reactor pressure vessel, making it highly efficient and compact.

- **Key Features**:

    o The helical coil steam generator is submerged in the reactor's primary coolant, allowing for efficient heat transfer from the reactor core to the secondary steam circuit.

    o Utilizes natural circulation for coolant flow, eliminating the need for primary coolant pumps and reducing the risk of mechanical failure.

    o Provides passive safety by ensuring continued heat removal even during a loss of external power.

- o   Designed to operate at pressures of around 2,000 psi (138 bar) and temperatures of approximately 300°C (572°F).
- **Application**: Ideal for remote or off-grid applications where reliability and low maintenance are crucial. The compact design allows for deployment in regions with limited infrastructure.

2. Rolls-Royce SMR - Integrated Helical Coil Steam Generator

- **Overview**: The Rolls-Royce SMR project integrates helical coil steam generators into its modular design to optimize heat transfer and efficiency. This steam generator is central to its strategy of reducing costs and simplifying the construction process.

- **Key Features**:

    - o   Designed to be fully integrated within the reactor module, reducing the overall footprint and allowing for modular assembly.

    - o   The helical coils are made from corrosion-resistant materials like Inconel 690, which can withstand high temperatures and pressures while preventing material degradation over time.

    - o   Uses a secondary heat transfer loop to generate superheated steam for power turbines, optimizing the thermal efficiency of the reactor.

- **Application**: The Rolls-Royce SMR aims to provide a cost-competitive solution for countries looking to decarbonize their energy grids. Its modular construction approach also allows for scalability and quicker deployment compared to conventional reactors.

3. Korea's SMART Reactor - Helical Coil Steam Generator

- **Overview**: The System-integrated Modular Advanced Reactor (SMART) developed by Korea Hydro & Nuclear Power (KHNP) includes a helical coil steam generator as part of its integrated reactor design.

- **Key Features**:

    - o   The SMART steam generator is compact and optimized for high thermal efficiency with natural convection flow.

    - o   The design supports a modular approach, enabling easier transport and installation on-site.

    - o   The helical coil configuration ensures a large surface area for heat transfer, allowing efficient steam generation without the need for extensive mechanical pumping.

- o Built to handle high pressure (over 1,500 psi) and high temperature conditions, making it suitable for a variety of applications, including desalination and district heating.

- **Application**: The SMART reactor is specifically designed for combined heat and power (CHP) applications, making it ideal for both electricity generation and industrial heat supply in regions with growing energy demands.

4. Czech Republic's Energy Well Project - Helical Coil Steam Generator

- **Overview**: The Energy Well project focuses on developing smaller SMRs for decentralized power production. Their helical coil steam generators are designed to optimize heat transfer in compact reactor designs.

- **Key Features**:

  - o Modular steam generator design, which can be scaled up or down depending on the power output requirements.

  - o Utilizes high-efficiency helical coils to maximize the heat transfer coefficient while maintaining a compact footprint.

  - o Focuses on flexibility and adaptability, allowing the steam generator to be integrated into various SMR designs, such as light water or gas-cooled reactors.

- **Application**: Targeted at industries and remote areas requiring reliable, low-carbon energy solutions. The design is adaptable for use in both electricity generation and industrial applications.

In addition to steam generators, heat exchangers are employed in reactors that operate with non-water coolants, such as gas-cooled reactors (GCRs) and molten salt reactors (MSRs). These reactors require specialized heat exchangers capable of transferring heat to secondary circuits without mixing the coolants. Given the extremely high operating temperatures—sometimes reaching up to 950°C—these heat exchangers must be constructed from materials that can withstand both intense heat and corrosive environments. Alloys like Hastelloy and Inconel are commonly used due to their exceptional corrosion resistance and structural integrity at elevated temperatures. These materials ensure that the heat exchangers maintain their performance over prolonged operational periods, thereby supporting the reactor's overall efficiency and reliability.

Thermal Energy Storage and Waste Heat Utilization Systems

Small Modular Reactors (SMRs) are designed not only to generate electricity but also to maximize efficiency through the utilization of thermal energy storage (TES) and waste heat recovery systems. These technologies are especially beneficial in industrial applications where there is a consistent demand for both power and heat.

One of the key strategies to enhance efficiency in SMRs is the integration of thermal batteries. These systems typically involve molten salt tanks or similar heat storage mediums that can store excess thermal energy produced during periods of low electricity demand. The stored heat can be retained for extended periods due to the high thermal capacity of molten salts, and it can be released when needed to generate additional power or to support district heating networks. For example, during peak energy demand, the stored thermal energy can be converted into steam to drive turbines, thereby increasing the reactor's output without ramping up nuclear fission processes. This not only optimizes the energy utilization but also enhances the overall economic efficiency of the plant.

In addition to thermal storage, SMRs can capitalize on waste heat recovery to further improve efficiency. This involves integrating heat exchangers and pipelines within the reactor infrastructure to capture residual heat that would otherwise be lost. The recovered waste heat can be diverted to nearby industrial facilities where it can be used for processes like chemical manufacturing, or to desalination plants for producing freshwater. Such systems are particularly beneficial for SMRs located in remote or arid regions where water resources are scarce. By channelling waste heat into secondary applications, SMRs add significant economic value to their installation, supporting a variety of energy-intensive industries and contributing to sustainability goals.

The combination of thermal energy storage and waste heat utilization not only enhances the thermal efficiency of SMRs but also supports their role in providing reliable, low-carbon energy for both electricity generation and industrial processes. This integrated approach allows SMRs to operate flexibly, adapting to fluctuations in grid demand while maintaining high efficiency and reducing environmental impact.

Decay Heat Removal Systems

Even after an SMR (Small Modular Reactor) is shut down, decay heat continues to be generated within the reactor core due to the radioactive decay of fission products. This residual heat, although significantly lower than during active operation, must still be effectively managed to prevent overheating and ensure safety. SMRs incorporate various advanced cooling systems to handle decay heat removal, leveraging both passive and active mechanisms to enhance reliability.

One of the key strategies used in SMRs is the deployment of gravity-fed cooling loops. These systems are designed to function without the need for mechanical pumps, making them highly effective during power outages or emergency situations. Elevated tanks filled with coolant are positioned above the reactor, allowing gravity to drive the flow of water into the reactor vessel when needed. This gravity-fed approach ensures a continuous and reliable supply of coolant to the core, even if external power sources are lost. This passive cooling method is critical for maintaining safe reactor temperatures, especially during unplanned shutdowns.

For gas-cooled SMRs, passive air cooling towers are commonly used to dissipate decay heat. These systems leverage the principles of natural convection to cool the reactor core. Air cooling towers are designed with extensive surface areas to maximize heat transfer and are often equipped with cooling fins or extended surfaces to enhance efficiency. The use of passive air cooling not only reduces dependence on electrical systems but also provides a robust and maintenance-free solution for managing decay heat. This design is particularly advantageous for SMRs deployed in remote or off-grid locations where access to water or power may be limited.

Overall, these decay heat removal systems are crucial for ensuring the long-term safety of SMRs, allowing them to achieve high levels of reliability and resilience under various operating conditions.

Electrical and Control Systems Integration

The successful operation of a Small Modular Reactor (SMR) depends on the seamless integration of its cooling and control systems, ensuring safe and efficient reactor performance. This integration involves the use of advanced monitoring, automation, and backup systems that work together to maintain optimal reactor conditions.

1. SCADA Systems (Supervisory Control and Data Acquisition)

SMRs rely on SCADA systems to monitor critical reactor parameters in real time, such as temperatures, pressures, flow rates, and coolant levels. These automated systems collect data from a network of sensors distributed throughout the reactor [255]. The SCADA system continuously analyses this data, alerting operators to any deviations from normal operating conditions. By providing a comprehensive, real-time overview of the reactor's status, SCADA systems allow for rapid response to any changes, ensuring the reactor remains within safe operational limits. For example, if a rise in core temperature is detected, the system can automatically adjust coolant flow rates to stabilize temperatures.

2. Redundant Power Supplies

Given the critical importance of cooling systems in preventing reactor overheating, SMR facilities must be equipped with redundant power supplies to maintain continuous operation. This includes backup diesel generators and Uninterruptible Power Supplies (UPS) to ensure that the cooling systems remain functional even during grid outages. The redundancy strategy is particularly vital in remote locations where power reliability may be less stable. For instance, during a loss of external power, the UPS can instantly supply electricity to essential cooling pumps until the diesel generators take over, preventing any interruptions in cooling.

3. Cybersecurity Measures

With the increasing use of digital control systems, SMRs are designed with robust cybersecurity protections to safeguard against potential cyber threats. Control systems are fortified with firewalls, encryption, and intrusion detection systems to prevent unauthorized access. These measures ensure that only authorized personnel can control reactor operations and make adjustments to critical systems. By securing the control network, SMRs can protect against cyberattacks that might attempt to disrupt cooling systems, potentially leading to unsafe conditions. Advanced cybersecurity protocols are particularly crucial in light of the growing risks of cyber threats to critical infrastructure.

By combining advanced SCADA systems, redundant power solutions, and cybersecurity protections, the electrical and control systems of an SMR are optimized to ensure both safety and operational efficiency. These integrated systems enable SMRs to operate reliably, even in challenging environments, thereby supporting their role in providing stable, low-carbon energy.

Site Preparation and Civil Engineering Requirements

Proper site preparation is crucial for the stability, safety, and long-term operation of Small Modular Reactor (SMR) power stations. Given the varied environments in which these reactors may be deployed, comprehensive civil engineering measures must be in place to address both natural and operational challenges.

In regions with high seismic activity, SMR reactors are constructed on base isolators designed to absorb and mitigate the effects of earthquakes. These isolators work by decoupling the reactor building from ground movements, thereby reducing the transmission of seismic shocks to critical components like the cooling systems and reactor containment structures. For instance, isolators and seismic dampers are commonly used in Japan, where reactors must be engineered to withstand significant tremors. This protective measure is essential to ensure the integrity of the reactor, especially in areas prone to seismic disturbances.

SMR power stations often require deep foundations to support the substantial weight of reactor vessels, steam generators, and cooling systems. These foundations, typically made of reinforced concrete, provide a stable base that can handle the heavy loads associated with nuclear power infrastructure. Additionally, vibration dampening systems are integrated into the structural design to minimize the impact of external vibrations, such as those caused by nearby industrial activities or heavy equipment operation. This is especially important for protecting sensitive reactor control and monitoring equipment, ensuring the plant operates smoothly without disruptions.

For coastal SMR installations, cooling water intake systems are a vital component of the site infrastructure. Specially constructed channels or pipes are used to draw seawater into the plant for cooling purposes. These channels are lined with anti-corrosion coatings to withstand the harsh marine environment and prevent degradation over time. Additionally, filtration systems are installed to prevent biofouling, where marine organisms like algae and shellfish could clog the intake pipes. By ensuring a continuous and reliable flow of cooling water, these channels help maintain optimal reactor temperatures and prevent overheating.

These civil engineering and site preparation measures are integral to the safe and efficient operation of SMR power stations. By addressing site-specific challenges such as seismic risks, structural stability, and efficient cooling, these systems ensure that SMRs can be deployed in a wide range of environments while maintaining the highest safety standards.

Environmental Protection and Waste Management Systems

Small Modular Reactors (SMRs) are designed with advanced environmental protection and waste management systems to ensure they meet stringent safety and regulatory standards. These systems are essential not only for maintaining compliance but also for minimizing the environmental footprint of nuclear power generation.

One of the key aspects of environmental protection in SMR operations is the management of water that comes into contact with radioactive materials. Effluent treatment systems are employed to treat, purify, and monitor this water before it is released back into the environment. These systems typically include decontamination units that filter radioactive isotopes from the coolant loops, ensuring that no harmful substances are discharged. Continuous monitoring systems are integrated to detect any potential leaks or contamination, allowing for immediate corrective actions. This approach ensures that SMRs can operate safely while protecting surrounding ecosystems.

Proper management of nuclear waste is critical to the long-term sustainability of SMR operations. SMRs are equipped with spent fuel pools, which provide a secure and controlled environment for cooling newly removed nuclear fuel rods. These pools are lined with radiation-shielding materials and are continuously monitored for temperature and radiation levels to prevent overheating. Once the fuel has sufficiently cooled, it is transferred to dry storage casks for longer-term containment. These casks are constructed from materials like reinforced concrete and steel, designed to withstand extreme environmental conditions and prevent radiation leaks. The combination of wet and dry storage options ensures a flexible and scalable waste management strategy, particularly for modular plants where fuel management needs may evolve over time.

Integration of SMR Systems with Municipal Services and Civil Infrastructure

The integration of Small Modular Reactor (SMR) systems with municipal services and civil infrastructure is a multifaceted process that necessitates careful planning and execution. This integration is crucial for ensuring the safe, efficient, and sustainable operation of SMRs. The various systems involved—such as cooling, waste management, effluent treatment, and control systems—must be interconnected with municipal utilities to support their continuous operation.

For SMRs utilizing water for cooling, establishing a reliable connection to local water sources is essential. The cooling systems typically require a consistent supply of fresh or seawater, necessitating the construction of large intake pipelines that connect to municipal water systems, rivers, or coastal sources. These pipelines must be designed to filter out debris before the water enters the reactor cooling loops, ensuring operational integrity and safety. Furthermore, the treated water from the cooling system must be discharged back into the environment through effluent discharge channels that connect to municipal stormwater systems or directly to natural water bodies. This connection is vital for minimizing thermal pollution and ensuring compliance with environmental regulations.

SMRs generate wastewater that requires treatment before disposal. The integration with municipal sewage systems is critical, as wastewater, including effluents from decontamination processes, must be treated on-site before being released into these systems. SMRs are equipped with effluent treatment plants that utilize various technologies, including filtration and chemical neutralization, to remove radioactive contaminants. Additionally, emergency overflow systems must be established to link with municipal sewage networks, preventing untreated water from being released during unexpected surges in wastewater production.

The management of nuclear waste is a significant aspect of SMR operations, requiring secure handling and integration with municipal waste management services. Non-radioactive solid waste generated during operations can be disposed of through municipal solid waste services, provided that strict waste segregation protocols are followed. Moreover, spent nuclear fuel, after being stored on-site, may need to be transported to centralized long-term storage facilities. This process necessitates collaboration with local transport authorities to ensure safe transportation routes and compliance with regulatory standards.

Efficient distribution of the power generated by SMRs involves connecting to the regional electrical grid through high-voltage transmission lines. Coordination with local utilities is necessary to ensure grid compatibility and synchronization, with redundant connections established to enhance reliability. Additionally, on-site substations are constructed to step up the voltage of electricity before it is fed into the national grid, while also being connected to municipal power lines for emergency power needs.

The integration of SMRs with local telecommunications and internet services is vital for real-time monitoring and control. Supervisory Control and Data Acquisition (SCADA) systems are employed to monitor reactor conditions and effluent discharge, linking these systems to municipal communication networks for emergency response coordination [255]. Given the critical nature of nuclear power plants, advanced cybersecurity measures must be implemented to protect these control systems from potential cyber threats, necessitating collaboration with municipal IT infrastructure.

Integration with local emergency services is essential for ensuring rapid response capabilities in the event of incidents. SMRs must have access to municipal firefighting resources and develop evacuation routes and shelter-in-place plans in collaboration with local authorities. This integration ensures that public safety is prioritized and that emergency protocols are effectively communicated and practiced.

The construction and operation of SMR facilities require robust transport infrastructure. Heavy equipment transport during construction necessitates reinforced roads and coordination with local transport authorities. Additionally, ongoing operations will require secure roadways for the delivery of consumables and access for personnel, emphasizing the need for comprehensive logistical planning.

Electrical Distribution and Grid Connectivity

Small Modular Reactors are designed to generate electricity in a flexible and efficient manner, making them ideal for diverse applications such as supplementing national grids, supporting remote communities, and stabilizing power networks with renewable energy integration.

Power Generation and Step-Up Transformation

Small Modular Reactors (SMRs) are designed to generate electricity at medium voltages, typically ranging from 13.8 kV to 24 kV. This voltage level is particularly suitable for on-site power distribution, allowing for localized energy management and reducing the need for extensive transmission infrastructure. However, this medium voltage is insufficient for long-distance transmission, which necessitates higher voltages to minimize energy losses during transport. To facilitate efficient electricity transport through the grid, the generated power must be transformed to higher transmission voltages, generally between 138 kV and 500 kV [256, 257].

The integration of SMRs into the electrical grid involves the establishment of on-site substations that play a critical role in the power distribution process. These substations are equipped with step-up transformers that convert the medium-voltage output from the SMR to higher voltages that are compatible with the national grid. This transformation is essential not only for voltage adjustment but also for managing the integration of the SMR's output with the broader transmission network. Such integration ensures that the electricity generated can be effectively distributed to meet regional or national energy demands, thereby enhancing grid stability and reliability [257].

A pertinent example of this integration can be seen in the NuScale Power Plant Project in Idaho, which is set to utilize SMRs. The design of this facility includes on-site substations equipped with step-up transformers, enabling the connection of the modular reactors to the Western Interconnection grid. This strategic design aligns with existing grid infrastructure, ensuring that the power generated is efficiently fed into the grid. The NuScale project exemplifies how SMRs can be effectively integrated into existing electrical networks, particularly in regions that are increasingly reliant on renewable energy sources, which can exhibit fluctuations in output [256].

Redundant Power Paths and Grid Synchronization

Small Modular Reactors (SMRs) present significant advantages in enhancing grid reliability, primarily through their ability to provide redundant power paths. These redundant paths function as backup transmission lines, ensuring a continuous electricity supply even during failures caused by maintenance, natural disasters, or unexpected technical issues. The redundancy offered by SMRs is crucial for maintaining grid stability and minimizing power outages, particularly in industries and regions that demand uninterrupted power supply. The integration of SMRs into the grid can significantly improve frequency stability, which is essential for accommodating the increasing share of intermittent renewable energy sources like solar and wind [59]. This redundancy is not only beneficial for grid reliability but also aligns with the growing need for sustainable and resilient energy systems.

To effectively integrate SMRs into the existing power grid, sophisticated grid synchronization systems are employed. These systems ensure that the output of the SMR aligns with the grid's frequency (either 50 Hz or 60 Hz, depending on the region) and voltage levels, thereby ensuring compatibility. Proper synchronization is critical to prevent power surges or frequency fluctuations that could destabilize the grid and potentially lead to blackouts. Continuous monitoring of grid conditions allows these synchronization systems to adjust the SMR's output in real-time, matching demand and supply fluctuations. The importance of such synchronization systems is underscored by the need for reliable power generation, especially in isolated areas where traditional grid infrastructure may be lacking.

A real-life example of SMR application is the Akademik Lomonosov floating SMR in Russia, which supplies power to the remote Arctic town of Pevek. This floating reactor utilizes redundant power paths to ensure a stable power supply in the harsh Arctic environment, demonstrating the practical benefits of SMRs in challenging conditions. The synchronization systems employed in the Akademik Lomonosov are specifically designed to integrate seamlessly with the local microgrid, ensuring that power generation aligns with the existing grid's frequency and voltage requirements. By leveraging advanced control systems, the floating reactor guarantees a reliable electricity supply, even in regions where traditional grid infrastructure is limited.

Supervisory Control and Data Acquisition (SCADA) Systems

To optimize the performance of Small Modular Reactors (SMRs) and their integration with the electrical grid, Supervisory Control and Data Acquisition (SCADA) systems play a crucial role. SCADA systems facilitate real-time monitoring and control of various operational parameters such as temperatures, pressures, flow rates, and electrical output. This capability allows operators to swiftly respond to changes in reactor conditions or grid demand, thereby enhancing operational efficiency and safety [258, 259]. The integration of SCADA systems with SMRs not only streamlines operational processes but also ensures that the reactors can adapt to fluctuating grid requirements, which is essential for maintaining grid stability [260].

One of the significant advantages of SCADA systems is their ability to identify and address potential issues before they escalate into more significant problems. For instance, in scenarios where grid demand suddenly spikes, SCADA systems can automatically adjust the output of the SMR to provide additional power, thus stabilizing the grid [261]. Conversely, during periods of reduced demand, SCADA systems can decrease the reactor's output to prevent overloading the grid, thereby optimizing resource utilization and enhancing safety [262]. This dynamic response capability is vital for integrating renewable energy sources and ensuring that SMRs can operate effectively within a modern energy landscape [263].

A pertinent example of SCADA application in SMR technology is the CAREM project in Argentina. The CAREM SMR employs a sophisticated SCADA system that aligns its power output with the national grid's requirements. This system enables the reactor to adjust its generation based on real-time grid conditions, making it a reliable source of backup power during peak demand periods [260, 264]. Furthermore, the SCADA system enhances the safety of the CAREM plant by continuously monitoring key operational parameters, thereby ensuring that any deviations from normal operating conditions are promptly addressed [265, 266]. The integration of SCADA technology in the CAREM project exemplifies how advanced monitoring and control systems can significantly improve the operational reliability and safety of SMRs.

Impacts of SMR Power Plants on Electrical Grids

The integration of Small Modular Reactors (SMRs) into electrical grids presents significant advantages, particularly in enhancing grid stability, facilitating decentralized power generation, reducing grid congestion, and providing ancillary services.

One of the primary benefits of SMRs is their ability to provide load-following capabilities, which is crucial for maintaining grid stability. Unlike traditional large nuclear plants that operate optimally at a constant output, SMRs can dynamically adjust their power output to match demand fluctuations. For instance, the NuScale Power Module is capable of ramping its output by 20% per minute, which is particularly beneficial in grids with high shares of variable renewable energy sources like wind and solar [267]. This flexibility allows SMRs to act as a stable backup source, mitigating the risk of blackouts during periods of low renewable generation due to weather variability [268]. The ability of SMRs to support load following aligns with the increasing need for reliable power sources as countries pursue ambitious renewable energy targets [269].

Load-following refers to the capability of a power generation system to adjust its output in response to fluctuations in electricity demand. Unlike traditional large nuclear reactors, which are typically designed for steady baseload power generation, SMRs are engineered to be more flexible, allowing them to ramp up or down their power output in response to changes in grid demand. This adaptability makes SMRs particularly valuable in modern power grids that are increasingly dependent on intermittent renewables like wind and solar.

As more countries integrate renewable energy sources into their grids, managing the balance between supply and demand has become increasingly challenging. Solar and wind power are inherently variable, with their output depending on weather conditions and time of day. For instance, solar panels generate less power on cloudy days or at night, while wind turbines are dependent on wind speeds. This variability can lead to sudden fluctuations in power supply, which can destabilize the grid if not managed properly. To address this, grid operators require power sources that can quickly adjust their output to fill in the gaps when renewable generation drops or to curtail output when there is excess supply.

SMRs are well-suited to fulfill this role due to their modular design, advanced control systems, and passive safety features that allow them to adjust their power output more dynamically than traditional nuclear reactors. Their smaller core size, along with sophisticated reactor control systems, enables rapid adjustments to thermal output, which in turn controls electricity generation. This flexibility ensures that SMRs can respond quickly to grid demands, either by increasing output during peak demand periods or reducing generation during periods of low demand or high renewable output.

SMRs are designed with several features that enhance their load-following capabilities:

1.   Advanced Reactor Control Systems: SMRs are equipped with state-of-the-art control systems, such as Supervisory Control and Data Acquisition (SCADA) systems, which allow for real-time monitoring and automated adjustments to reactor output. These systems enable the SMR to respond quickly to changes in grid frequency and voltage, ensuring stable integration with the electrical network.

2.   Passive Cooling and Safety Systems: The use of passive cooling systems in SMRs, such as natural circulation loops, allows for rapid adjustments in reactor power levels without the need for complex mechanical systems. This reduces the response time required to adjust power output, making SMRs more agile in load-following scenarios. For instance, designs like the NuScale SMR utilize gravity-driven cooling systems that enhance the reactor's ability to quickly ramp up or down based on demand.

3.   Flexible Fuel and Reactor Core Design: Many SMRs utilize fuel designs that allow for more flexible operation. For example, the BWRX-300 and NuScale reactors can operate at partial power levels without compromising fuel integrity or efficiency. This allows these reactors to reduce power output when demand is low, conserving fuel and extending the operational lifespan of the reactor.

A practical example of SMR load-following capabilities is demonstrated by NuScale's power modules, which are designed to provide rapid adjustments to power output. NuScale reactors can change their power levels by up to 20% per minute, allowing them to quickly adapt to changes in grid demand. This capability is particularly beneficial in regions like the Pacific Northwest, where the grid relies heavily on variable renewable energy sources like hydroelectric and wind power. By using SMRs to stabilize the grid, utilities can reduce the need for fossil fuel-based peaking plants, thus lowering carbon emissions.

In grids with a high penetration of renewables, the ability of SMRs to provide load-following enhances grid reliability. For example, during periods of excess solar or wind generation, an SMR can reduce its output, thereby making room for renewable energy on the grid. Conversely, when renewable output drops, such as during a calm, cloudy day, the SMR can quickly increase its power output to compensate. This capability helps to smooth out the fluctuations in renewable energy generation, providing a more stable and reliable power supply.

SMRs can also play a critical role in microgrids or isolated grids where renewable energy penetration is high but the stability of the grid is difficult to manage. By offering flexible, reliable, and carbon-free power generation, SMRs can serve as a backbone for microgrids, ensuring that these systems remain resilient, especially during periods of fluctuating renewable supply.

SMRs are also well-suited for decentralized power generation, especially in remote or isolated regions where traditional large-scale power plants may not be feasible. The Akademik Lomonosov, a floating nuclear power plant utilizing SMRs, exemplifies this by providing energy to the Arctic town of Pevek in Russia, thereby enhancing local energy security [270]. Furthermore, SMRs can be integrated into microgrids, which are localized power systems capable of operating independently from the main grid. This integration is crucial for disaster resilience, as microgrids equipped with SMRs can maintain power supply even when the central grid is compromised [15]. The versatility of SMRs in microgrid applications is supported by their low maintenance costs and extended fuel availability, making them a promising option for future energy systems [269].

Strategically deploying SMRs near industrial hubs or areas with high electricity demand can significantly reduce grid congestion and the need for extensive transmission infrastructure. By generating power closer to consumption points, SMRs help alleviate transmission losses and enhance overall grid efficiency [267]. In Canada, for example, provinces like Ontario are exploring the use of SMRs near industrial sites to provide reliable, carbon-free energy directly where it is needed [267]. This localized generation not only addresses energy demands more efficiently but also contributes to reducing the environmental impact associated with long-distance electricity transmission.

SMRs can also play a vital role in providing ancillary services such as voltage control, frequency regulation, and spinning reserves, which are essential during peak demand periods when the grid is under stress. The rapid ramping capabilities of NuScale's SMR technology enable it to maintain grid stability during unexpected demand spikes or when other power sources fail [267, 268]. Additionally, the ability of SMRs to participate in peak shaving strategies can help manage demand effectively, ensuring that the grid operates within safe limits [271]. This capability is increasingly important as the integration of intermittent renewable energy sources creates challenges for maintaining a balanced and stable electricity supply [272].

Safety and Monitoring Systems

Small Modular Reactors (SMRs) represent a significant advancement in nuclear technology, particularly in terms of safety and monitoring systems. These systems are designed to enhance the reliability and security of nuclear power generation, addressing both operational safety and potential risks associated with nuclear energy. This synthesis will explore the key components of safety and monitoring systems in SMRs, including passive safety systems, advanced real-time monitoring technologies, automated control systems, cybersecurity measures, redundancy, and regulatory oversight.

One of the most notable attributes of SMRs is their reliance on passive safety systems, which utilize natural physical principles to maintain reactor safety without the need for external power or human intervention. For instance, systems such as natural circulation cooling are integral to designs like the NuScale Power Module. This system allows for heat removal from the reactor core through natural convection, ensuring that even in a total power loss scenario, the reactor can cool itself effectively [123]. Similarly, gravity-driven emergency cooling systems, as seen in the CAREM-25 reactor, utilize gravitational forces to release coolant, thereby maintaining core temperatures without reliance on electrically powered pumps [123]. Furthermore, fail-safe shutdown mechanisms, such as those employed in the BWRX-300, automatically insert control rods into the reactor core under abnormal conditions, ensuring a safe shutdown even if electronic controls fail [123].

SMRs are equipped with sophisticated real-time monitoring systems that continuously gather data from numerous sensors throughout the reactor. These systems utilize Supervisory Control and Data Acquisition (SCADA) technologies to provide operators with immediate insights into reactor performance. Critical sensors monitor temperature and pressure within the reactor core, enabling automated responses to any anomalies, such as unexpected temperature increases [191]. Additionally, radiation detection systems are essential for identifying any abnormal radiation levels, triggering alarms and containment protocols if necessary [191]. Structural integrity sensors also play a crucial role, especially in seismically active regions, by detecting vibrations and potential microfractures, thus allowing for pre-emptive maintenance [191].

The integration of Artificial Intelligence (AI) into the control systems of SMRs represents a transformative approach to reactor management. AI-powered anomaly detection systems can identify deviations from normal operating patterns, allowing for early intervention before issues escalate [191]. Automated response systems can adjust reactor conditions in real-time, such as modifying control rod positions or coolant flow rates, thereby stabilizing the reactor quickly in response to detected abnormalities [191]. This capability significantly reduces the risk of human error, enhancing overall safety.

As SMRs increasingly rely on digital technologies for monitoring and control, robust cybersecurity measures become essential. Multi-layered defences, including firewalls and intrusion detection systems (IDS), protect against unauthorized access to control systems [15]. Regular cybersecurity audits, mandated by regulatory bodies such as the U.S. Nuclear Regulatory Commission (NRC), ensure that vulnerabilities are identified and addressed, thereby maintaining the integrity of safety systems [15].

Redundancy is a critical aspect of SMR safety, ensuring that if one system fails, others can maintain reactor stability. This includes backup power supplies, such as Uninterruptible Power Supplies (UPS) and diesel generators, which ensure continuous operation of safety systems during grid outages. Additionally, multiple independent cooling loops provide layers of protection, as demonstrated in the NuScale design, where a backup passive cooling system activates if the primary system fails [15].

SMRs undergo rigorous safety assessments by regulatory bodies like the NRC and the International Atomic Energy Agency (IAEA). These organizations enforce stringent safety standards and require comprehensive Safety Analysis Reports (SAR) from developers, detailing the reactor's safety features and emergency response plans [15]. Ongoing safety inspections and audits ensure compliance with regulatory standards throughout the operational life of the reactor [15].

NuScale Power's SMR design exemplifies the integration of these safety and monitoring features. Its passive cooling system allows for safe operation even during complete power loss, while its advanced control room utilizes AI-driven monitoring systems for predictive analytics, enabling proactive maintenance and rapid anomaly response [15, 123]. This combination of passive safety and cutting-edge technology establishes a new benchmark for nuclear safety.

Waste Management Infrastructure

The successful deployment of Small Modular Reactors (SMRs) relies not only on the safety and efficiency of their operational systems but also on robust and reliable waste management infrastructure. Managing radioactive waste is a critical aspect of nuclear power generation, including SMRs, as it ensures the protection of human health and the environment. The infrastructure required for waste management in SMR plants involves a series of processes and facilities designed to handle various types of waste generated during the reactor's lifecycle.

Like all energy industries, nuclear power generation produces some waste byproducts. However, nuclear waste is carefully managed and categorized into three main types based on radioactivity: low-level, intermediate-level, and high-level waste. The vast majority (90% by volume) consists of lightly contaminated materials, such as tools, protective clothing, and filters, which contribute to only 1% of the total radioactivity. On the other hand, high-level waste—mostly used, or "spent" nuclear fuel—makes up just 3% of the total waste volume but accounts for a staggering 95% of its radioactivity [273].

A unique aspect of the nuclear industry is its complete responsibility for managing its waste. Unlike any other energy sector, nuclear operators take full accountability, ensuring that all waste is securely managed and disposed of. There are well-established disposal facilities for low- and intermediate-level waste, and several projects are underway to develop long-term solutions for high-level waste and spent nuclear fuel [273].

Nuclear fuel is incredibly energy-dense, meaning it requires a small amount of material to generate substantial amounts of electricity. As a result, the volume of waste produced is relatively minimal. For instance, the waste generated by a reactor supplying a person's electricity needs for a year would only amount to the size of a brick. Of this, only about 5 grams—roughly the weight of a sheet of paper—is high-level waste [273].

A typical 1,000-megawatt nuclear power station, which can power over a million homes, generates only about three cubic meters of vitrified high-level waste per year if the spent fuel is recycled. In stark contrast, a coal-fired power plant of the same capacity produces around 300,000 tonnes of ash and over 6 million tonnes of carbon dioxide annually [273].

Despite public concerns, the handling of nuclear waste has not caused harm to people since the inception of the civil nuclear industry. The misconception that nuclear waste poses a long-term threat stems from the fact that some radioactive materials can remain hazardous for thousands of years. However, the most concerning radioactive components decay to safe levels within a few hundred years. Moreover, the quantities involved are small; even in the event of a containment breach, the release would be minimal, having negligible impact on the environment or human health [273].

It's also worth noting that our environment, including our own bodies, is naturally radioactive. Radiation is a part of life on Earth, and doses from a well-managed nuclear waste repository would be significantly lower—almost 50 times less—than the average background radiation we are exposed to daily [273].

Spent nuclear fuel, which is both hot and radioactive when removed from reactors, is initially stored in wet pools to allow it to cool down. After this initial cooling period, the fuel can either remain in wet storage or be transferred to dry storage casks. These temporary storage solutions are designed to reduce heat and radiation levels, making the fuel easier to handle for eventual recycling or disposal [273].

However, temporary storage is not the ultimate solution. Countries adopt one of two main strategies for managing used nuclear fuel: recycling or direct disposal. This decision is driven by national policies, economic factors, and technological capabilities [273].

While some countries, like the United States, treat used fuel as waste, many others recognize the potential for recycling. Approximately 97% of the material in spent fuel—mostly uranium—can be reused as fuel in specific reactor types. Recycling focuses on extracting plutonium and uranium, which can be combined with fresh uranium to create new fuel rods [273].

Countries such as France, Japan, Germany, Belgium, and Russia have successfully utilized plutonium recycling to generate electricity while minimizing the radioactive footprint of their waste. The remaining byproducts, primarily fission products (around 4% of the waste), are solidified through a process called vitrification, where they are immobilized in glass for long-term storage [273]. For example, La Hague, France, has been a leader in recycling used nuclear fuel for decades, significantly reducing its radiological impact.

Some nations opt for direct disposal, where spent nuclear fuel is considered waste and is permanently stored in underground repositories without recycling. This involves encasing the used fuel in robust canisters, which are then placed in deep geological tunnels, sealed with layers of rock and clay [273]. Countries like Finland are nearing completion of state-of-the-art repositories designed to safely contain high-level waste for thousands of years, ensuring minimal environmental impact and safeguarding future generations [273].

Spent nuclear fuel, or HLW, presents significant challenges in waste management due to its radiotoxicity and heat generation. Although SMRs produce less spent fuel compared to traditional reactors, effective management remains essential.

SMR plants typically incorporate spent fuel pools designed to store newly discharged spent fuel under water, which serves both as a radiation shield and a cooling mechanism for the fuel rods. These pools are constructed from thick reinforced concrete lined with stainless steel to prevent leaks and are engineered to withstand seismic events and other natural disasters [274]. For instance, the NuScale SMR design features a dedicated spent fuel pool adjacent to the reactor vessel, capable of storing used fuel for several years, allowing for gradual cooling before transfer to dry storage [274].

After a period in spent fuel pools, cooled spent fuel is transferred to dry cask storage systems. These systems utilize sealed metal containers that provide robust shielding and containment against radiation leakage. Dry casks are typically constructed from steel and concrete, and are stored on specially designed outdoor pads equipped with controlled access and radiation monitoring systems [274]. The SMART SMR plant in South Korea exemplifies this approach, utilizing both spent fuel pools and dry cask storage for long-term waste management [274].

In addition to spent fuel, SMRs generate intermediate-level waste (ILW) and low-level waste (LLW), which includes contaminated equipment and materials from maintenance activities. Dedicated facilities for sorting, processing, and packaging waste are crucial for managing ILW and LLW. These facilities typically include shielded hot cells, compactors, and incinerators designed for volume reduction, along with HEPA filtration systems to control airborne contaminants [274]. Effective waste segregation reduces the volume of high-level waste and optimizes disposal strategies [274].

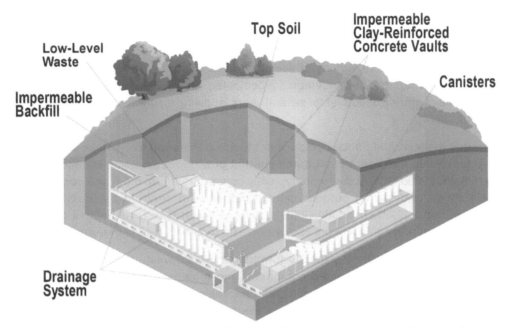

The LLW disposal site accepts waste from States participating in a regional disposal agreement.

Source: U.S. Nuclear Regulatory Commission - As of February 2023

Figure 37: Low-level Radioactive Waste Disposal. Nuclear Regulatory Commission, CC BY 2.0, via Flickr.

Intermediate-level waste is often solidified through cementation processes, creating a stable matrix that prevents the release of radionuclides. The encapsulated waste is stored in concrete-lined vaults or stainless steel drums, ensuring long-term containment and monitored for radiation and structural integrity [274].

SMRs also produce liquid and gaseous radioactive waste during operations. Radioactive liquids, such as coolant water, require treatment to remove contaminants before discharge or reuse. Common methods include evaporation, ion exchange, and filtration. Facilities for liquid waste treatment are equipped with evaporators, ion-exchange columns, and reverse osmosis systems, ensuring that treated water is stored in shielded tanks prior to release [274]. The BWRX-300, for example, employs advanced filtration and treatment systems to manage its liquid waste effectively [274].

Gaseous emissions, including tritium and noble gases, must be controlled to prevent environmental contamination. SMRs are equipped with gas treatment systems that utilize activated carbon filters and HEPA filters, with ventilation systems designed to monitor and control radiation levels before release [274].

Figure 38: Low-Level Waste Disposal site. Nuclear Regulatory Commission, CC BY 2.0, via Flickr.

Decommissioning at the end of an SMR's lifecycle generates substantial radioactive waste, including structural materials and contaminated soil. Facilities for managing decommissioning waste include temporary storage areas and processing units for sorting, decontaminating, and packaging materials for disposal [274].

High-level waste, including spent fuel that cannot be reprocessed, may require deep geological disposal in stable formations underground. Geological disposal facilities necessitate underground tunnels and engineered barriers to isolate radioactive waste from the biosphere for thousands of years. Finland's Onkalo Geological Repository serves as a pioneering example of such a facility [274].

Figure 39: Geological mapping in Onkalo spent nuclear fuel repository. kallerna, CC BY-SA 4.0, via Wikimedia Commons.

To mitigate environmental impacts, SMR plants must implement comprehensive monitoring systems. Continuous monitoring of radiation levels around waste storage facilities is essential for compliance with safety standards. This includes gamma radiation detectors, air sampling units, and groundwater monitoring wells [274]. In the event of a containment breach, SMR plants must have emergency response systems in place, including spill containment barriers and fire suppression systems. The Canadian Nuclear Safety Commission mandates stringent safety measures for waste management facilities to ensure rapid response capabilities [274].

Emergency Response and Contingency Planning

Emergency response and contingency planning are essential for the safety and preparedness of Small Modular Reactors (SMRs). Given their unique design features and operational contexts, these plans must be meticulously crafted to address the specific risks associated with nuclear power generation. The following synthesis discusses the integrated emergency response plans, passive safety systems, training protocols, communication strategies, infrastructure for contingencies, and post-incident recovery measures relevant to SMRs.

Emergency response plans for SMRs are developed in accordance with national and international standards, including guidelines from the International Atomic Energy Agency (IAEA). These plans begin with a comprehensive risk assessment to identify potential hazards unique to SMRs, such as coolant loss and core overheating. Research indicates that the smaller size and inherent passive safety features of SMRs significantly reduce the likelihood of severe accidents compared to traditional large reactors [5, 123]. The classification of emergencies into levels such as alert, site emergency, and general emergency is crucial for determining the appropriate response actions [244]. Furthermore, coordination with local authorities is vital, especially for SMRs located in proximity to small communities or remote areas, ensuring effective evacuation and public communication strategies [57, 244].

A defining characteristic of SMRs is their reliance on passive safety systems, which operate without external power or human intervention. For instance, the NuScale Power Module employs natural circulation for cooling, thereby minimizing the risk of overheating during emergencies [123]. Robust containment structures are designed to withstand natural disasters, and passive cooling systems ensure that even in the event of a power failure, the reactor can safely dissipate heat [5]. Automatic shutdown mechanisms further enhance safety by allowing the reactor to shut down autonomously if critical parameters exceed safe limits, thereby reducing the potential for human error during emergencies [123].

Regular emergency drills and training sessions are integral to ensuring preparedness among SMR personnel and local emergency responders. On-site drills simulate various emergency scenarios, allowing staff to practice immediate reactor shutdown and containment integrity checks [57, 244]. Community involvement in these drills is essential, as it familiarizes local populations with evacuation routes and safety protocols [244]. Additionally, tabletop exercises involving key decision-makers enhance strategic coordination and communication during high-pressure situations [244].

Effective communication during emergencies is paramount. SMR operators must have automated notification systems in place to alert both plant operators and local authorities of potential issues in real-time [57, 244]. Public communication plans should ensure that accurate information is disseminated to prevent panic and misinformation, utilizing local media and social platforms for updates on evacuation orders and safety measures [244]. Trained communication teams are essential for managing media relations and ensuring that the public receives timely and accurate information [244].

SMR facilities must be equipped with robust contingency systems to maintain critical functions during disruptions. This includes multiple layers of backup power, such as diesel generators and uninterruptible power supplies, to ensure the operation of essential cooling systems [57, 244]. Redundant communication networks, including satellite phones, are critical for maintaining contact with emergency responders even if local infrastructure is compromised [244]. Emergency shelters and medical facilities must also be established to protect personnel and provide immediate medical assistance if necessary [244].

Following an incident, a thorough review process is essential for assessing the effectiveness of the emergency response. This includes analysing feedback from personnel and local authorities to identify gaps in protocols and areas for improvement [57, 244]. Decontamination procedures must be implemented if radioactive materials are released, ensuring the safety of the site and surrounding environment [244]. Continuous improvement of emergency response plans based on insights gained from drills and real incidents is vital for enhancing future preparedness [244].

Chapter 7
Economic and Financial Aspects of SMR Projects

Cost-Benefit Analysis of SMR Implementation

The implementation of Small Modular Reactors (SMRs) presents a significant investment in the energy sector with potential long-term benefits. However, like any major infrastructure project, it comes with both substantial costs and benefits. Below is a detailed cost-benefit analysis, taking into account various factors, including capital expenditures, operational costs, benefits of scalability, safety, and economic impact.

1. Capital and Construction Costs

Costs:

- **High Initial Capital Investment:** The upfront capital cost for constructing a 4,000 MW SMR power station (consisting of multiple 100 MW units) is estimated to range between USD 15.5 billion to 25.5 billion (approximately AUD 23.5 billion to 38.7 billion).

- **Regulatory Compliance:** SMRs require extensive regulatory approvals, site assessments, and environmental impact studies, which can be time-consuming and costly.

- **Infrastructure Development:** Building SMRs often requires significant investments in site preparation, cooling systems, and grid connections, especially if deployed in remote areas.

- **Modular Construction Costs**: Although modular construction helps reduce on-site construction time, the prefabrication of components and transportation logistics can be expensive.

Benefits:

- **Lower Per-Unit Construction Costs Over Time**: As more SMR units are built (nth-of-a-kind units), the construction costs decrease due to economies of scale, standardized designs, and learning curves.

- **Faster Build Times**: Compared to traditional large reactors that can take 10-15 years to complete, SMRs can potentially be constructed in 5-8 years, reducing financing costs and enabling earlier returns on investment.

- **Modular Scalability**: The ability to add capacity incrementally reduces the financial risk compared to large, monolithic nuclear power plants.

Constructing a 4,000 MW SMR power station, made up of multiple 100 MW units, involves substantial capital investment, estimated between USD 15.5 to 25.5 billion (AUD 23.5 to 38.7 billion). The high costs are driven by the need for comprehensive regulatory compliance, site assessments, and environmental studies, all of which require significant time and resources. Additional expenses stem from infrastructure development, particularly for site preparation, cooling systems, and grid connections—critical for projects located in remote areas. While modular construction methods can reduce on-site assembly time, they involve substantial costs related to prefabrication, transportation, and logistics, adding to the project's overall budget.

Despite the hefty initial capital outlay, SMRs offer long-term economic advantages. As more units are built, the per-unit cost decreases due to economies of scale and the efficiency gains from standardized designs. Additionally, SMRs have significantly shorter construction timelines compared to traditional large reactors, which can take over a decade to complete. With SMRs potentially being built within 5-8 years, they allow for faster project turnaround and earlier revenue generation, reducing overall financing costs. The modular nature of SMRs also provides flexibility, enabling incremental capacity additions, which lowers the financial risk associated with large, single-unit nuclear plants.

2. Operational and Maintenance Costs

Costs:

- **Specialized Workforce**: Maintaining an SMR power station requires a highly skilled workforce, including nuclear engineers, safety inspectors, and control room operators.

- **Fuel and Waste Management**: Although SMRs use less fuel than traditional reactors, they still incur costs related to fuel procurement, handling, and waste management. High-level waste disposal remains an expensive and technically challenging process.

- **Periodic Refuelling and Maintenance**: The need for periodic inspections, refuelling, and replacement of critical components contributes to ongoing operational expenses.

Benefits:

- **Lower Operating Costs per MW**: Due to their smaller size, SMRs are more efficient and require less staff per unit of power generated. This reduces labour costs in the long term.

- **Higher Capacity Factors**: SMRs are designed for high reliability, with capacity factors above 90%, ensuring consistent power output and lower operational disruptions.

- **Passive Safety Features**: The inclusion of passive cooling systems and modular designs lowers the complexity of maintenance, reducing long-term operational costs.

Operating an SMR power station involves significant ongoing costs, primarily due to the need for a specialized workforce that includes nuclear engineers, safety inspectors, and skilled control room operators. Even though SMRs use less fuel compared to traditional large reactors, there are still considerable expenses associated with fuel procurement, handling, and the complex process of waste management, especially for high-level radioactive waste. Additionally, periodic refuelling, inspections, and maintenance activities—such as replacing key components—add to the operational expenditures, requiring careful budget planning to manage these recurring costs.

On the positive side, SMRs benefit from lower operating costs per megawatt (MW) due to their smaller size and higher efficiency, which translates to fewer personnel requirements for the same output compared to larger reactors. Their design enables higher capacity factors, often exceeding 90%, which ensures steady power generation with minimal interruptions. The inclusion of passive safety systems and simplified modular structures also reduces the complexity of maintenance, which in turn lowers long-term costs. These passive systems enhance safety while minimizing the need for extensive monitoring and maintenance, contributing to the overall cost-effectiveness of SMR operations.

3. Safety and Environmental Impact

Costs:

- **Safety Systems Implementation**: Advanced safety features such as passive cooling, seismic dampers, and backup power supplies add to the upfront and operational costs.

- **Waste Disposal**: Managing radioactive waste, particularly high-level waste, requires long-term storage solutions and regulatory compliance, which can be costly.

Benefits:

- **Enhanced Safety**: SMRs utilize passive safety features and modular designs that significantly reduce the risk of catastrophic failures, such as those seen in traditional large reactors.

- **Reduced Land Footprint**: SMRs require a smaller physical footprint compared to large nuclear reactors, reducing land use and environmental disruption.

- **Low Greenhouse Gas Emissions**: SMRs generate zero carbon emissions during operation, making them a key technology for achieving climate goals and reducing reliance on fossil fuels.

- **Waste Minimization**: Due to their high fuel efficiency, SMRs produce significantly less radioactive waste compared to traditional reactors.

Implementing SMR technology involves substantial costs, especially in the development of advanced safety systems. Features like passive cooling mechanisms, seismic dampers, and redundant power supplies are essential to meet stringent safety standards, but they also increase both capital and operational expenses. Additionally, the management of radioactive waste, particularly high-level waste, poses significant financial challenges due to the need for secure long-term storage and compliance with regulatory requirements. These factors contribute to the higher upfront investment required for SMR projects, even though they are designed to be more efficient and safer than traditional reactors.

On the benefits side, SMRs are engineered with cutting-edge passive safety features that reduce the likelihood of catastrophic failures, addressing major concerns associated with traditional large-scale reactors. Their modular design not only enhances safety but also minimizes the physical footprint, thereby reducing land use and environmental impacts. SMRs are particularly valuable in the transition to low-carbon energy systems since they generate zero carbon emissions during operation, contributing to global climate goals. Moreover, their high fuel efficiency results in less radioactive waste production, making SMRs a more sustainable option for nuclear energy by minimizing waste management challenges.

4. Economic and Social Benefits

Costs:

- **Community Engagement**: Gaining public acceptance and addressing concerns about nuclear safety can require significant investment in community outreach and education.

- **Decommissioning Costs**: At the end of their life cycle, SMRs require decommissioning, which involves dismantling the reactor and safely disposing of radioactive materials.

Benefits:

- **Job Creation**: Building and operating an SMR power station creates thousands of jobs in construction, engineering, manufacturing, and maintenance.

- **Energy Security**: SMRs can provide stable, reliable power in remote or isolated regions, reducing dependence on imported fuels.

- **Economic Development**: By creating a robust supply chain and investing in local infrastructure, SMR projects can stimulate regional economic growth.

- **Load-Following Capability**: SMRs offer load-following capabilities, which can help balance the grid as renewable energy sources like wind and solar fluctuate.

Implementing SMR projects comes with certain economic and social costs, particularly when it comes to engaging with communities and building public trust. Nuclear energy often faces public scepticism, necessitating substantial investment in community outreach, education campaigns, and transparent communication to address safety concerns. Additionally, once an SMR reaches the end of its operational life, decommissioning costs must be accounted for. This involves safely dismantling the facility, managing radioactive materials, and ensuring that decommissioning is carried out in compliance with regulatory standards—a process that can be both time-consuming and costly.

On the benefits side, SMR projects have the potential to generate significant economic and social value. Construction and operation can create thousands of jobs across various sectors, including construction, engineering, and skilled trades, providing a boost to local economies. Beyond employment, SMRs contribute to energy security by delivering reliable, stable power to remote or isolated areas, thus reducing reliance on imported fossil fuels. The development of a local supply chain for SMR components can further stimulate regional economic growth. Moreover, the load-following capabilities of SMRs make them well-suited to complement renewable energy sources like wind and solar, thereby enhancing grid stability and supporting the integration of variable renewables.

5. Comparative Analysis with Alternative Energy Sources

Costs:

- **Higher Initial Investment Compared to Renewables**: The upfront costs for SMRs are higher than those for wind and solar power installations. For example, a 4,000 MW solar power plant would cost significantly less in terms of capital expenditure, though it requires much more land and is subject to weather variability.

Benefits:

- **Baseload Power Supply**: Unlike intermittent renewables, SMRs can provide a constant baseload power supply, which is critical for grid stability.

- **Synergy with Renewables**: SMRs can complement renewables by providing stable power during periods when solar and wind resources are insufficient.

When comparing Small Modular Reactors (SMRs) to alternative energy sources, one of the most notable differences lies in the initial capital investment. SMRs generally require significantly higher upfront costs compared to renewable sources like solar or wind power. For instance, constructing a 4,000 MW solar power plant would involve lower capital expenditures, making it more economically attractive in the short term. However, solar and wind installations require extensive land use and are inherently variable, meaning their power output fluctuates based on weather conditions, which can pose challenges for consistent energy supply.

On the other hand, SMRs offer key advantages that renewables cannot match, particularly in providing reliable baseload power. This ability to supply a constant, uninterrupted flow of electricity is critical for grid stability, especially in regions with high energy demands. Additionally, SMRs can operate synergistically with renewable energy sources. They can serve as a backup to stabilize the grid when solar and wind generation is low, ensuring a continuous power supply. This complementary role helps reduce reliance on fossil fuels for backup power and enables a smoother transition to a cleaner, more sustainable energy mix.

6. Long-Term Strategic Benefits

- **Energy Independence**: By investing in SMR technology, countries can reduce their dependence on fossil fuel imports and improve their energy independence.

- **Decarbonization**: SMRs align with global climate goals by providing a low-carbon, scalable energy source that can replace coal and gas-fired power plants.

- **Future-Proof Technology**: As SMRs continue to advance, their flexibility, safety, and cost-effectiveness will likely improve, making them a viable solution for meeting future energy demands.

Investing in Small Modular Reactor (SMR) technology offers substantial long-term strategic benefits, particularly in terms of energy independence. For countries that rely heavily on fossil fuel imports, developing a domestic SMR power infrastructure can significantly reduce vulnerabilities related to energy supply disruptions and fluctuating global fuel prices. By providing a reliable and continuous energy source, SMRs enhance national energy security, allowing countries to control their power generation more effectively and reducing reliance on external energy markets.

Moreover, SMRs play a crucial role in global decarbonization efforts. As a low-carbon energy source, they align with international climate goals by helping to phase out coal and gas-fired power plants, which are major contributors to greenhouse gas emissions. The scalability and flexibility of SMR technology mean that it can be deployed incrementally to meet growing energy demands while maintaining a low environmental footprint. As advancements in SMR design and technology continue, their safety, efficiency, and cost-effectiveness are expected to improve, making them a future-proof solution for sustainable energy generation that can adapt to evolving global energy needs.

Overall, the implementation of SMRs involves significant upfront costs, but the long-term benefits in terms of safety, reliability, and environmental sustainability can outweigh these expenses. By leveraging economies of scale, optimizing modular construction, and integrating with renewable energy sources, SMRs can play a crucial role in achieving a stable, low-carbon energy future.

The decision to invest in SMRs should be based on a thorough analysis of national energy needs, grid stability, regulatory environment, and economic capacity. For countries committed to reducing their carbon footprint and enhancing energy security, SMRs offer a promising pathway, particularly when complemented by renewable energy sources.

Funding and Investment Models for SMR Projects

Funding and investment models for Small Modular Reactor (SMR) projects are critical for their successful deployment, given the substantial capital investments required and the long-term benefits they offer in terms of energy security and decarbonization. This overview synthesizes various funding strategies, highlighting the importance of government support, public-private partnerships, project financing, equity investment, green bonds, utility financing models, export credit agencies, and risk mitigation strategies.

Government funding plays a pivotal role in the early stages of SMR projects. Countries like the United States, Canada, and the United Kingdom have implemented financial assistance programs, including grants and tax incentives, to stimulate SMR development. For instance, the U.S. Department of Energy has invested significantly in companies such as NuScale Power, which is developing advanced SMR technologies [233]. Public-private partnerships (PPPs) are particularly effective in this context, as they allow for shared financial risks and leverage the expertise of private firms. This collaborative approach not only mitigates risks but also encourages private sector investment in a capital-intensive industry [56].

Given the high upfront costs associated with SMR projects, project financing through Special Purpose Vehicles (SPVs) has become a common strategy. An SPV is a legally distinct entity created to manage the financial aspects of an SMR project, thereby isolating its assets and liabilities. This structure is beneficial for attracting multiple stakeholders, including institutional investors and utilities, who can pool resources while limiting their financial exposure [275]. The SPV model is increasingly favoured for large infrastructure projects, as it facilitates risk-sharing and allows for a more manageable cash flow profile [30].

Equity investment is crucial for early-stage SMR companies seeking to develop innovative technologies. Venture capital firms and impact investors are increasingly interested in SMRs due to their potential to contribute to a sustainable energy future. For example, Breakthrough Energy Ventures has invested in TerraPower, which focuses on advanced nuclear reactor designs [233]. By providing equity funding, these investors align their financial returns with the success of the SMR projects, fostering a conducive environment for technological advancements and market competitiveness [56].

The rise of green bonds represents a promising financing mechanism for SMR projects, particularly as global interest in sustainable energy grows. These bonds are specifically earmarked for projects that deliver environmental benefits, such as reducing carbon emissions. Countries like Canada and the European Union are exploring the use of green bonds to fund nuclear projects, recognizing the role of nuclear energy in achieving net-zero emissions targets [56]. Additionally, climate finance initiatives, such as the Green Climate Fund, can provide concessional loans or grants to support SMR projects in developing countries, thereby alleviating financial burdens [56].

In regions dominated by regulated utilities, ratepayer financing models can be employed to fund SMR projects. This approach allows the costs of constructing SMRs to be included in the utility's rate base, enabling recovery of investments through electricity tariffs over time. The Tennessee Valley Authority (TVA) in the U.S. is an example of a utility exploring this financing model for SMRs [56]. However, this method requires regulatory approval and public support, as it ultimately shifts some financial burdens to consumers [56].

Export credit agencies (ECAs) play a vital role in financing SMR projects, especially for countries looking to deploy reactors in foreign markets. ECAs provide loans, guarantees, and insurance to domestic companies, facilitating their entry into high-risk international markets. For instance, Rosatom has successfully utilized ECA financing to export reactors to countries like Egypt and Turkey [56]. Furthermore, international collaboration through organizations like the International Atomic Energy Agency (IAEA) can standardize safety and regulatory frameworks, enhancing the flow of funding to SMR projects [56].

Given the inherent risks associated with nuclear energy, effective risk mitigation strategies are essential for securing funding for SMRs. Developers often utilize various insurance products to protect against construction and operational risks, while governments may offer loan guarantees to lower capital costs. The U.S. federal loan guarantee program for advanced nuclear projects exemplifies this approach, providing reassurance to investors by safeguarding their capital against unforeseen events [56].

The following provides a sample funding and investment model for constructing a 4,000 MW Small Modular Reactor (SMR) power station. This model integrates various funding sources, financing strategies, and risk mitigation techniques to address the substantial capital requirements and complexities associated with deploying SMR technology. The station is assumed to consist of multiple 100 MW SMR units, phased over several years.

1. Project Overview and Cost Breakdown

- **Total Installed Capacity**: 4,000 MW (40 x 100 MW SMR units)

- **Estimated Capital Cost**: USD 18 - 25 billion (approx. AUD 27 - 37 billion)

- **Project Duration**: 8-10 years (including planning, construction, and commissioning)

- **Location**: Hypothetical site with access to existing infrastructure

2. Funding Sources and Investment Structure

The funding for this project will be secured through a combination of equity investment, government support, debt financing, and green bonds.

Funding Source	Contribution (%)	Estimated Amount (USD)	Description
Equity Investment (Private)	25%	4.5 - 6.25 billion	Equity stake from institutional investors, utilities, and strategic partners.
Government Grants & Subsidies	15%	2.7 - 3.75 billion	Government support through grants, tax credits, and subsidies for clean energy.
Debt Financing (Project Loans)	40%	7.2 - 10 billion	Long-term loans from banks, export credit agencies (ECAs), and multilateral lenders.
Green Bonds & Climate Finance	20%	3.6 - 5 billion	Issuance of green bonds to finance sustainable infrastructure.
Total	100%	18 - 25 billion	

3. Funding Components Explained

A. Equity Investment (Private)

- **Contributors**: Institutional investors, venture capital funds, private equity, and strategic industry partners.

- **Purpose**: Securing a 25% equity stake in the project to demonstrate commitment and align stakeholders' interests.

- **Key Investors**: Potential investors include energy companies like Shell or BP, private equity funds focused on renewable energy, and infrastructure investment firms.

- **Return on Investment (ROI)**: Equity investors expect long-term returns from electricity sales and carbon credits.

B. Government Grants and Subsidies

- **Purpose**: To reduce the high initial capital cost and incentivize clean energy development.

- **Potential Sources**: Grants from federal and state governments, subsidies for using low-carbon technologies, and tax incentives for nuclear power projects.

- **Justification**: Governments are increasingly prioritizing decarbonization and energy security, making SMRs a key part of their clean energy strategies.

C. Debt Financing (Project Loans)

- **Contributors**: Banks, export credit agencies (ECAs), and international financial institutions like the World Bank.

- **Terms**: Long-term loans with a maturity of 20-25 years at competitive interest rates.

- **Risk Mitigation**: Loan guarantees provided by the government to lower borrowing costs and attract lenders.

- **Purpose**: Covering 40% of the project cost, with the debt repaid through electricity sales over time.

D. Green Bonds and Climate Finance

- **Issuance**: Green bonds targeting environmentally-conscious investors.

- **Allocation**: Funds raised will be used for sustainable infrastructure, waste management systems, and environmental protection measures.

- **Attractiveness**: Green bonds provide lower interest rates due to demand from ESG (Environmental, Social, and Governance) investors.

- **Climate Finance**: Additional funding from international climate funds (e.g., Green Climate Fund) to support clean energy projects.

4. Project Implementation Phases & Cash Flow Plan

Project Phase	Duration (Years)	Estimated Cost (USD)	Funding Source Allocation
Planning & Licensing	2	1.8 billion	Equity & Government Grants
Site Preparation & Civil Works	2	2.5 billion	Debt Financing & Green Bonds
Modular Fabrication	3	6 billion	Debt Financing & Equity
On-Site Assembly & Construction	4	5.7 billion	Debt Financing & Green Bonds
Commissioning & Testing	1	2 billion	Equity & Climate Finance
Total	**8-10**	**18 billion**	

5. Risk Mitigation Strategies

- **Insurance**: Construction and operational risk insurance to cover delays, accidents, and equipment failures.

- **Contingency Fund**: A 10% contingency reserve (~USD 1.8 billion) is set aside to manage unforeseen expenses.

- **Regulatory Support**: Engage early with regulatory authorities to streamline licensing and reduce delays.

- **Environmental & Social Governance (ESG)**: Implement robust ESG practices to attract green investors and mitigate reputational risks.

6. Financial Projections & Revenue Streams

- **Revenue Sources**:

  - **Electricity Sales**: Long-term Power Purchase Agreements (PPAs) with utilities and industrial customers.

  - **Carbon Credits**: Selling carbon credits as SMRs contribute to reducing CO_2 emissions.

  - **Waste Heat Utilization**: Revenue from selling waste heat for district heating or industrial applications.

- **Estimated Payback Period**: 12-15 years after commissioning, depending on electricity prices and carbon credit markets.

- **Internal Rate of Return (IRR)**: Target IRR of 8-12% for equity investors over a 25-year lifespan.

7. Community and Social Impact

- **Job Creation**: Approximately 5,000 jobs during construction and 500 permanent jobs for operation and maintenance.

- **Local Economic Development**: Boost to local suppliers, contractors, and infrastructure development.

- **Public Engagement**: Investment in community outreach programs to build support for the project.

This funding model illustrates a strategic mix of equity, debt, government support, and green finance to deliver a sustainable and financially viable SMR power station. The phased approach reduces financial risk while ensuring timely project completion. By leveraging international collaboration, regulatory support, and innovative financing mechanisms, the project can serve as a model for future clean energy investments.

If the construction timeline for the 4,000 MW Small Modular Reactor (SMR) power station is shortened to 4-6 years, the funding and investment model will need to be optimized to accommodate a more accelerated schedule. This will require a higher initial outlay, tighter project management, and a focus on modular construction efficiencies to reduce the overall project duration.

The following provides an updated funding and investment model to reflect the shorter project timeline:

1. Project Overview and Cost Breakdown (Accelerated Timeline)

- **Total Installed Capacity**: 4,000 MW (40 x 100 MW SMR units)

- **Estimated Capital Cost**: USD 19 - 27 billion (approx. AUD 29 - 40 billion)

- **Project Duration**: 4-6 years (including planning, construction, and commissioning)

- **Location**: Hypothetical site with access to existing infrastructure

2. Funding Sources and Investment Structure

The funding strategy is adjusted to meet the shorter timeline, with a focus on securing upfront capital quickly. The distribution of funding sources is as follows:

Funding Source	Contribution (%)	Estimated Amount (USD)	Description
Equity Investment (Private)	30%	5.7 - 8.1 billion	Higher equity stake due to shorter timeline, attracting strategic investors focused on rapid returns.
Government Grants & Subsidies	15%	2.85 - 4.05 billion	Increased government support to reduce upfront capital costs and accelerate construction.
Debt Financing (Project Loans)	35%	6.65 - 9.45 billion	Short-term loans with favorable interest rates to meet accelerated cash flow needs.
Green Bonds & Climate Finance	20%	3.8 - 5.4 billion	Focus on green bonds to attract ESG-conscious investors for clean energy infrastructure.
Total	100%	19 - 27 billion	

3. Funding Components Explained (Accelerated Model)

A. Equity Investment (Private)

- **Contributors**: Institutional investors, private equity, and strategic partners who are experienced in high-speed infrastructure projects.

- **Purpose**: Securing a larger equity stake of 30% to ensure rapid mobilization of funds.

- **Key Investors**: Targeting energy companies like BP, private equity firms specializing in infrastructure, and sovereign wealth funds.

- **Return on Investment (ROI)**: Shorter payback period (8-10 years) due to accelerated timeline, with higher expected returns.

B. Government Grants and Subsidies

- **Purpose**: Increased government grants to support expedited regulatory approvals and construction.

- **Potential Sources**: Federal and state funding for clean energy, accelerated tax credits, and subsidies for nuclear innovation.

- **Justification**: Governments are incentivized to meet climate goals faster, making SMRs a priority for funding.

C. Debt Financing (Project Loans)

- **Contributors**: Banks, export credit agencies, and international financial institutions.

- **Terms**: Shorter loan maturity of 15-20 years with slightly higher interest rates due to accelerated cash flow requirements.

- **Risk Mitigation**: Government loan guarantees to attract lenders for high-speed infrastructure projects.

- **Purpose**: Covering 35% of the project cost with immediate disbursements to keep the fast-track schedule on target.

D. Green Bonds and Climate Finance

- **Issuance**: Targeting green bond markets for investors focused on sustainable, fast-moving projects.

- **Allocation**: Funds allocated to rapid deployment of cooling systems, waste management, and grid integration.

- **Attractiveness**: High demand for green bonds due to the urgency of climate targets, offering a quick influx of funds.

4. Accelerated Project Implementation Phases & Cash Flow Plan

Project Phase	Duration (Years)	Estimated Cost (USD)	Funding Source Allocation
Planning & Licensing	1	2 billion	Equity & Government Grants
Site Preparation & Civil Works	1	3 billion	Debt Financing & Green Bonds
Modular Fabrication	1.5	7 billion	Equity & Debt Financing
On-Site Assembly & Construction	2	5.5 billion	Debt Financing & Green Bonds
Commissioning & Testing	0.5	1.5 billion	Equity & Climate Finance
Total	**4-6**	**19 billion**	

5. Key Adjustments for Shorter Timeline

- **Accelerated Procurement**: Bulk purchasing of materials and components to reduce lead times.

- **Parallel Construction**: Simultaneous construction of multiple units to compress the schedule.

- **Modular Manufacturing Efficiency**: Increased reliance on off-site prefabrication to reduce on-site construction time.

- **Optimized Regulatory Engagement**: Early and continuous engagement with regulators to expedite approvals.

6. Risk Mitigation Strategies for Accelerated Schedule

- **Insurance**: Comprehensive insurance to cover risks associated with accelerated construction.

- **Contingency Fund**: Increased contingency reserve of 15% (~USD 2.85 billion) due to higher risk of schedule compression.

- **Supply Chain Management**: Secure contracts with multiple suppliers to avoid delays.

- **ESG and Community Engagement**: Fast-track community outreach programs to build public support.

7. Financial Projections & Revenue Streams (Accelerated Model)

- **Revenue Sources**:

  - **Electricity Sales**: Immediate revenue generation through phased commissioning, with earlier units brought online within 4 years.

  - **Carbon Credits**: Higher value due to faster deployment, aligning with urgent climate targets.

  - **Waste Heat Utilization**: Early deployment of waste heat recovery systems for additional revenue streams.

- **Estimated Payback Period**: 8-10 years after commissioning due to higher initial cash flows.

- **Internal Rate of Return (IRR)**: Target IRR of 12-15% due to reduced project timeline and early returns.

8. Community and Social Impact (Accelerated Deployment)

- **Job Creation**: Approximately 6,000 jobs during peak construction and 600 permanent jobs for ongoing operations.

- **Local Economic Boost**: Rapid development of infrastructure, benefiting local suppliers and contractors.

- **Public Support**: Enhanced communication strategies to address concerns related to fast-tracking nuclear projects.

The accelerated funding and investment model demonstrates how a 4,000 MW SMR project can be completed within 4-6 years with a focus on rapid mobilization of resources, streamlined regulatory processes, and efficient modular construction techniques. By leveraging a mix of equity, debt, government support, and green finance, the project can achieve financial viability while contributing to long-term energy security and decarbonization goals.

The funding models developed for constructing a 4,000 MW Small Modular Reactor (SMR) power station are based on a set of key assumptions. These assumptions have been applied to both the standard 8-10 year timeline model and the accelerated 4-6 year model to provide a realistic framework for project execution. Understanding these underlying assumptions is vital for assessing the feasibility and adaptability of the models across different countries and regions.

Project Scope and Configuration: The project is designed to build a 4,000 MW power station by assembling 40 individual 100 MW SMR units. This modular approach allows for factory-based manufacturing of SMR units, which are then transported to the site for assembly. By leveraging modular construction techniques, the project aims to optimize efficiency and reduce construction time. The commissioning process is planned to occur in phases, bringing groups of

SMR units online sequentially to start generating revenue earlier. This phased approach is intended to minimize financial risks and improve cash flow during construction.

Economic and Financial Assumptions: The estimated capital costs for the project vary between the two models. The original model (8-10 years) estimates a capital expenditure of USD 15.5 billion to 25.5 billion, while the accelerated model (4-6 years) increases this to USD 19 billion to 27 billion due to higher costs associated with expedited procurement, labour, and construction. Both models rely on a diversified funding strategy, including contributions from equity investors, government grants, debt financing, and green bonds. Debt is assumed to be secured at interest rates of 4-6%, with loan maturities of 15-20 years. Additionally, government guarantees are presumed to lower risk perceptions among lenders. The use of green bonds is predicated on the assumption that investors are motivated by ESG (Environmental, Social, and Governance) criteria, aligning with global climate initiatives.

Timeline and Scheduling Assumptions: For the standard model (8-10 years), traditional project management methods are assumed, including sequential construction phases and lengthier regulatory approvals. This timeline accommodates the complexity of regulatory compliance and site preparation. In contrast, the accelerated model (4-6 years) relies heavily on parallel construction activities and modular manufacturing to shorten the timeline. Key components are prefabricated to minimize on-site assembly time, while proactive regulatory engagement is assumed to streamline approval processes. Early involvement with regulators and the adoption of innovative construction techniques are critical to achieving the shortened schedule.

Regulatory and Compliance Assumptions: Both models assume a proactive approach to regulatory compliance. Early engagement with regulatory authorities is expected to reduce the time required for permits and approvals. The models also anticipate adherence to international nuclear safety standards, such as those set by the International Atomic Energy Agency (IAEA). Site selection criteria include pre-screening for environmental suitability, seismic stability, and access to water resources for cooling if necessary. These measures are intended to ensure compliance while minimizing project delays.

Labor and Workforce Assumptions: A critical assumption is the availability of a skilled labour pool, including nuclear engineers, construction managers, welders, and control room operators. The project is expected to create thousands of temporary jobs during the construction phase and hundreds of permanent positions for ongoing operations. The accelerated model, in particular, relies on a workforce capable of handling the demands of parallel construction and assembly.

Technological and Operational Assumptions: The models are based on the assumption that the chosen SMR technology is mature, with designs that have already undergone significant testing and validation. Cooling systems are assumed to utilize either water bodies or air-cooled systems, depending on the project site. The SMRs are designed to offer load-following capabilities, allowing them to adjust power output to meet fluctuating grid demand, which is crucial for grid stability in areas with a high penetration of renewables.

Portability of the Models to Other Countries

The funding models developed are intended to be flexible and adaptable to different regions; however, their success depends on local factors such as regulatory environments, financial markets, workforce availability, and infrastructure readiness. In developed economies like the United States, Canada, and the United Kingdom, where financial markets are well-developed and nuclear regulatory frameworks are established, both models are highly portable. These countries can leverage existing capabilities, mature supply chains, and skilled labour to expedite SMR projects.

In emerging economies like India, China, and Brazil, adjustments would be required to address challenges related to regulatory frameworks, funding access, and workforce training. Partnerships with international institutions and public-private funding arrangements could be critical to overcoming these barriers. For developing countries, the models would need to be heavily adapted, particularly in terms of regulatory development, financing mechanisms, and technical training programs to build local capacity.

Ultimately, the portability of these models depends on each country's willingness to invest in nuclear technology, their regulatory agility, and their ability to attract international investment. The flexibility in funding structures, combined with a focus on modular construction, makes the models adaptable but requires careful customization to align with each country's specific conditions.

Projected Return on Investment and Break-Even Points

Projected Return on Investment (ROI) for SMR Projects

Investing in Small Modular Reactors (SMRs) presents a unique opportunity for stakeholders in the energy sector, particularly in light of the increasing demand for low-carbon energy solutions. The projected Return on Investment (ROI) for SMR projects typically ranges from 8% to 12% over their operational lifespan, which can extend from 40 to 60 years [56, 276]. This ROI is influenced by various factors, including construction timelines, operational efficiency, fuel costs, and prevailing market electricity prices. For instance, the modular design of SMRs allows for incremental scaling, enabling additional modules to be added as demand grows without necessitating substantial upfront capital investments [3, 231].

The phased commissioning strategy, where SMRs are brought online in stages, is a critical factor that enhances the financial viability of these projects. This approach allows for early revenue generation, thereby reducing financial risk and improving overall ROI [138, 276]. Moreover, the use of modular construction techniques significantly shortens construction times, with studies indicating that fully modularized SMRs can be constructed in approximately 3.5 years, compared to 5.1 years for larger reactors [138]. This reduction in construction time not only accelerates cash flow but also mitigates the risks associated with long-term capital investments.

Despite the advantages, the initial capital expenditure for SMR projects remains a significant hurdle, estimated between USD 15.5 billion and 27 billion [15]. To enhance financial performance, securing favourable financing options such as low-interest loans and green bonds is essential. Additionally, the economic landscape for SMRs is further bolstered by their potential to operate competitively in liberalized electricity markets, where uncertainties in electricity prices can create substantial option value for investors [277].

Furthermore, the operational efficiency of SMRs, particularly in terms of fuel costs and maintenance, plays a pivotal role in determining their overall economic attractiveness. The modular nature of these reactors allows for simplified designs that can lead to lower operational costs compared to traditional large reactors [30, 245]. As the market for low-carbon energy continues to expand, the strategic deployment of SMRs could serve as a vital component in achieving energy security and sustainability goals.

Mapped Example of Return on Investment (ROI) for a 4,000 MW SMR Power Station Project

The project in this example involves the construction of a **4,000 MW power station** by deploying **40 individual 100 MW SMR units**. This is structured as a phased project where SMR units are commissioned in stages to maximize early revenue generation and minimize financial risks. Below is a detailed breakdown of the projected ROI based on the provided context.

1. Project Overview and Initial Investment

- **Total Capacity**: 4,000 MW (comprising 40 x 100 MW SMR units)

- **Construction Timeline**: Accelerated timeline of 4-6 years, utilizing modular construction techniques.

- **Capital Expenditure**: Estimated between USD 19 billion to 27 billion due to the need for expedited processes.

- **Operational Lifespan**: 40 to 60 years.

- **Phased Commissioning**: Groups of 10 reactors will be brought online sequentially, every 12-18 months, to generate early revenue.

2. Phased Construction and Commissioning Plan

- **Year 0-2**:
    - **Planning, Regulatory Approvals, and Site Preparation**: Engaging early with regulators and securing environmental permits.

- **Factory-Based Prefabrication**: Commence off-site fabrication of reactor components.

- **Year 3**:

  - **On-Site Assembly**: Start assembling the first set of 10 SMR units.

  - **Phased Commissioning**: By the end of year 3, the first 10 units (1,000 MW) are connected to the grid.

  - **Revenue Generation**: The first set starts generating revenue, reducing financial burden and supporting further construction.

- **Year 4-6**:

  - **Sequential Commissioning**: Every 12-18 months, an additional set of 10 units is brought online until the full capacity of 4,000 MW is reached.

  - **Full Plant Operational**: By the end of year 6, the power station is fully operational.

3. Projected Financial Metrics

a. Initial Capital Outlay

- **Capital Investment**: USD 19 billion (lower estimate) to USD 27 billion (upper estimate).

- **Funding Sources**:

  - **Equity Investors**: 30% (~USD 5.7 to 8.1 billion).

  - **Debt Financing**: 50% (~USD 9.5 to 13.5 billion) with an interest rate of 4-6%.

  - **Government Grants and Subsidies**: 10% (~USD 1.9 to 2.7 billion).

  - **Green Bonds**: 10% (~USD 1.9 to 2.7 billion), leveraging ESG criteria.

b. Revenue Projections and ROI

- **Electricity Price**: Assumed average of USD 70 per MWh based on current market rates and contracts.

- **Annual Revenue** (at full capacity):

  - 4,000 MW x 8,000 operating hours/year x USD 70 per MWh = USD 2.24 billion/year.

- **ROI Calculation**:

  - **Total Revenue over 40 years**: USD 89.6 billion.

- o **Operational Costs**: Estimated at 20% of annual revenue, equating to USD 448 million/year.

- o **Net Revenue**: USD 1.792 billion/year after operational costs.

- o **Break-Even Point**: Estimated between 6-8 years after the first set of units begins generating revenue.

- o **ROI**: Projected between 8% to 12% annually over the plant's operational life.

4. Key Benefits Driving ROI

a. Early Revenue from Phased Commissioning

By commissioning the reactors in groups, the project starts generating revenue within 3 years of commencement, thus reducing financial pressure. This phased approach allows for quicker returns, as each set of SMRs begins contributing to cash flow while the remaining units are still under construction.

b. Lower Operating and Maintenance Costs

The modular and factory-built design of SMRs ensures high efficiency and reliability, with lower operational costs compared to traditional reactors. The capacity factor is expected to exceed 90%, ensuring consistent power output and minimal downtime.

c. Load-Following Capability

The SMRs' ability to adjust output to match demand enhances their value in balancing grid stability, especially in markets with high penetration of renewables like solar and wind. This flexibility can attract premium pricing during peak demand periods, further improving revenue potential.

5. Sensitivity Analysis

To ensure robustness, the model considers several risk factors:

- **Electricity Price Volatility**: The financial model assumes a conservative average price of USD 70 per MWh. However, changes in market prices could impact the ROI.

- **Regulatory Delays**: Expedited regulatory engagement is critical. Delays could increase costs and push back revenue timelines.

- **Construction and Labour Costs**: Fluctuations in raw material prices and labour shortages could affect capital expenditure.

6. Portability of the Funding Model

This funding and investment model is adaptable to various countries with strong regulatory frameworks and skilled labour availability. The model is particularly suitable for countries like the United States, Canada, the United Kingdom, and South Korea, where nuclear expertise and infrastructure are well-developed. Additionally, it can be modified to suit emerging markets such as India and Brazil by adjusting the financing structure to include more government support and international development funding.

As an example of practical context, implementing the funding and construction model outlined above for a 4,000 MW Small Modular Reactor (SMR) power station in Australia would involve adapting it to fit the country's specific regulatory, financial, and infrastructural conditions. Australia currently lacks a comprehensive regulatory framework for nuclear energy, largely due to past political resistance and public apprehension. However, with a growing focus on achieving climate targets and ensuring long-term energy security, there is increasing interest in revisiting nuclear options. Establishing a clear regulatory pathway would be essential to the project's success. This would involve engaging with federal and state governments to align with international nuclear safety standards, such as those set by the International Atomic Energy Agency (IAEA). Australia would need to expedite the regulatory processes by creating a dedicated nuclear regulatory body similar to those in the U.S. or UK. Early engagement with regulators and streamlining licensing procedures would be crucial to meet the project's accelerated 4-6 year construction timeline.

Funding the estimated USD 19 to 27 billion (approximately AUD 29 to 41 billion) required for the SMR project would involve leveraging Australia's well-developed financial sector, particularly through green bonds and other ESG (Environmental, Social, and Governance) investments. Given Australia's commitment to decarbonization, SMR projects could attract substantial investment through green bonds, especially if supported by entities like the Clean Energy Finance Corporation (CEFC). Additionally, federal and state governments could contribute grants and subsidies, covering around 10% of the project's costs, to incentivize nuclear energy adoption. This could be further bolstered by public-private partnerships, with organizations such as the Australian Renewable Energy Agency (ARENA) expanding their focus to include advanced nuclear technologies. Collaborating with private equity investors would also help bridge funding gaps and diversify financial support.

A significant challenge for Australia would be the availability of a skilled labor force, given its limited experience with nuclear energy. However, the country possesses a strong base of expertise in related industries like mining, oil and gas, and power generation. By partnering with international nuclear technology firms and universities, Australia could quickly upskill engineers, construction managers, and safety inspectors. Programs through institutions like the University of New South Wales (UNSW) and TAFE (Technical and Further Education) could help accelerate workforce readiness. Moreover, the construction and operation of the SMR station could create thousands of jobs, especially in rural areas, aligning with Australia's economic goals of boosting regional development.

Australia's electricity grid, which currently faces challenges due to the intermittency of renewables like wind and solar, could benefit from the stable, baseload power supply provided by SMRs. The load-following capability of SMRs would enable them to adjust output to complement renewable sources, thereby enhancing grid stability. This flexibility would be particularly beneficial in states like South Australia, where renewable penetration is high. The phased commissioning strategy proposed in the model, where groups of SMRs are brought online sequentially, would allow the project to start generating revenue early, potentially as soon as three years after commencement, thereby supporting cash flow and reducing financial risks.

While the overall capital expenditure is significant, the modular and phased construction approach would help mitigate financial risks. However, the project would require adjustments for Australia's local cost structures, regulatory environment, and supply chain limitations. The projected costs of USD 19 to 27 billion (AUD 29 to 41 billion) might increase due to the need to establish new supply chains for nuclear components and adapt existing infrastructure. Stabilizing electricity prices in Australia, which can fluctuate due to market dynamics, would be another advantage of SMRs. With an assumed average electricity price of USD 70 per MWh (AUD 105 per MWh), SMRs could help stabilize prices while providing a consistent power supply.

The phased construction strategy and modular assembly techniques would enhance the project's financial viability, with early revenue generation offsetting capital expenses. By bringing units online in stages, the project would generate revenue starting as early as year three, supporting ongoing construction and reducing financial pressure. The projected returns on investment range from 8% to 12% annually over the operational life of 40 to 60 years. Additionally, the deployment of SMRs aligns with Australia's climate targets, helping to diversify the energy mix, reduce reliance on imported fuels, and achieve decarbonization goals. This would enable Australia to position itself as a leader in low-carbon energy technologies, particularly in the Asia-Pacific region.

Break-Even Points and Financial Viability

The break-even point for Small Modular Reactors (SMRs) is a critical metric in assessing their financial viability. Research indicates that the break-even point for SMR power stations is typically reached between 8 to 12 years of operation, influenced by factors such as construction time, financing costs, and electricity prices [13, 19]. In a standard construction model of 8-10 years, break-even is generally achieved within 10 years of commencing operations, primarily due to the lower operational costs associated with SMRs compared to traditional large reactors [10, 13]. Furthermore, projects that utilize an accelerated construction timeline of 4-6 years can reach break-even even faster, potentially within 6-8 years post-operation, owing to reduced construction time and earlier revenue generation [13, 19].

Several key factors influence the break-even point for SMRs. Operational efficiency and capacity factor are paramount; SMRs are designed to operate at high capacity factors, often exceeding 90%, ensuring consistent power generation and revenue (Wang, 2024). Additionally, the modular design of SMRs contributes to lower maintenance and fuel costs, enhancing their long-term profitability [19]. The flexibility of SMRs to adjust output and provide load-following capabilities is particularly valuable in grids with a high penetration of intermittent renewable energy sources, thereby increasing potential revenue from peak power pricing [13, 19].

Moreover, the ability to offset costs through waste heat recovery and cogeneration can significantly enhance the financial viability of SMRs. Utilizing waste heat for applications such as district heating or desalination can create additional revenue streams, further accelerating the break-even point [19]. The reliability of SMRs, combined with the intermittent nature of renewable sources that often require backup solutions, positions them as a competitive option in the energy market [13, 19]. Long-term Power Purchase Agreements (PPAs) with industries or municipalities can also secure stable revenue, improving financial predictability and expediting the return on investment [278].

Financial Risks and Mitigation Strategies

Building Small Modular Reactor (SMR) power plants involves significant financial risks due to their high initial capital costs, extended project timelines, and the stringent regulatory requirements in the nuclear sector. One of the primary financial challenges is the substantial capital expenditure required for construction, which can range from USD 15.5 billion to 27 billion (approximately AUD 23.5 billion to 38.7 billion) for a 4,000 MW project. These projects are vulnerable to cost overruns and delays, which can significantly impact their financial viability. To mitigate these risks, adopting phased commissioning allows the generation of early revenue as reactors come online in stages, which helps offset ongoing costs. Additionally, leveraging modular construction techniques and prefabrication can streamline the build process, reducing on-site construction time and associated labor costs. Allocating a contingency budget, typically around 10-15% of total CapEx, can also cushion against unexpected expenses, ensuring smoother cash flow management.

Regulatory and compliance risks are another major concern, as the nuclear industry is heavily regulated. Securing necessary approvals can be a lengthy and costly process, with any delays potentially pushing project timelines and inflating costs. Proactively engaging with regulatory authorities from the planning stages can help expedite the approval process. This may include early pre-licensing activities and comprehensive safety assessments to build confidence with regulators. Establishing dedicated regulatory teams focused on compliance can further streamline these processes. Legal protections and robust insurance coverage are also essential to safeguard against unforeseen legal challenges, particularly in regions where public perception of nuclear projects may lead to political scrutiny.

Financing and interest rate fluctuations can pose additional risks, as SMR projects often require a mix of debt and equity financing. Rising interest rates can increase the cost of debt, thereby affecting the overall budget and profitability. To address this, securing fixed-rate financing agreements can provide more predictable cost management. The use of green bonds and Environmental, Social, and Governance (ESG)-focused funds can attract sustainable investment, potentially lowering borrowing costs. Government-backed guarantees can enhance the creditworthiness of the project, making it more attractive to lenders and reducing the cost of capital.

The ongoing operational and maintenance costs, although lower for SMRs compared to traditional reactors, still pose financial risks related to skilled labour, fuel procurement, and waste management. Entering into long-term maintenance contracts with suppliers can stabilize these expenses, while predictive maintenance technologies can optimize plant performance and reduce unplanned downtime. Investing in workforce training is also crucial, as a skilled and efficient team can mitigate the need for costly external contractors and reduce overall operational expenses.

Market fluctuations in electricity prices can affect revenue projections, particularly in liberalized markets where prices are volatile. Long-term Power Purchase Agreements (PPAs) with utilities or industrial consumers can provide predictable revenue streams, insulating the project from market risks. The ability of SMRs to adjust output to match grid demand also allows them to capitalize on peak pricing periods, enhancing revenue potential. Additionally, the flexibility of SMRs to support cogeneration, such as district heating or desalination, diversifies income sources beyond just electricity sales, thereby reducing financial risk.

Construction delays and supply chain disruptions are other critical risks, particularly in the nuclear industry where specialized components are required. Establishing robust supplier contracts and long-term agreements can secure a steady supply of materials, reducing the risk of delays. Sourcing components locally where possible can mitigate the impact of geopolitical risks and reduce transportation costs. The use of modular prefabrication also minimizes on-site construction challenges, allowing for faster project execution.

Political and public perception risks are inherent in nuclear projects, often leading to additional regulatory hurdles or even project cancellations. Engaging local communities early and maintaining transparent communication with stakeholders can build trust and reduce opposition. Public education campaigns highlighting the safety features, low emissions, and economic benefits of SMRs can improve public acceptance. Corporate social responsibility initiatives, such as job creation and infrastructure development in project areas, can further enhance community support.

Decommissioning and long-term liability risks are significant, as the end-of-life dismantling of SMRs involves managing radioactive waste safely. Establishing dedicated decommissioning funds during the operational phase can ensure that resources are available when needed. Investing in automated decommissioning technologies can lower costs and improve safety during the dismantling process. Adhering to international best practices for decommissioning is essential to minimize legal liabilities and protect against environmental risks.

Long-Term Economic Impact of SMR Power Stations

One of the primary economic advantages of SMRs is their lower capital requirements compared to conventional nuclear reactors. Research indicates that the 'nth-of-a-kind' cost of SMRs is expected to be significantly lower, with estimates around $100 per MWh, making them economically viable even in small island settings [19]. This cost efficiency is further supported by the observation that smaller, co-sited reactors can mitigate the common pitfalls of large-scale projects, such as budget overruns and extended timelines, which often plague megaprojects [175]. The ability to construct SMRs in a modular fashion allows for factory fabrication, which can lead to reduced construction times—potentially up to two years faster than large reactors [232].

Moreover, the deployment of SMRs can enhance energy security, particularly in developing countries where energy needs are growing but financial resources are limited. SMRs can be deployed in a flexible manner, allowing for incremental investments that align with the economic capabilities of these regions [3]. Their smaller size and modular nature make them suitable for areas with limited grid capacity, thus facilitating energy access in remote or underserved communities [279]. This flexibility not only addresses immediate energy needs but also promotes long-term economic development through job creation and infrastructure investment [13].

The socio-economic benefits of SMR deployment extend beyond direct energy production. The construction and operation of SMRs can stimulate local economies through job creation in both the short and long term. The establishment of new employment opportunities is particularly significant in regions that may not have previously had access to nuclear technology [13]. Additionally, the licensing process for SMRs is evolving, with various countries actively supporting the development of this technology to gain a competitive edge in the global energy market [67]. This strategic positioning can lead to increased foreign investment and technological exchange, further bolstering economic growth.

The deployment of Small Modular Reactors (SMRs) is increasingly being seen as a cost-effective alternative to traditional nuclear reactors, especially for countries seeking to diversify their energy portfolios and enhance energy security. The economic advantages of SMRs, particularly their lower capital requirements and modular construction capabilities, make them highly attractive on a global scale. In comparison to conventional large-scale reactors, which often suffer from budget overruns and prolonged construction timelines, SMRs offer a more agile solution. Research indicates that, once SMR technology reaches maturity, the "nth-of-a-kind" cost could drop significantly, with estimates suggesting a levelized cost of electricity (LCOE) of around $100 per MWh. This cost-effectiveness makes SMRs economically feasible for a range of applications, including small island nations and remote communities, where energy infrastructure is less developed.

Countries like the United States, Canada, and the United Kingdom are leveraging the modular construction approach of SMRs to achieve faster deployment. For instance, the factory-based prefabrication of components can reduce construction timelines by up to two years compared to traditional nuclear projects. This advantage is particularly beneficial for enhancing competitiveness in the global energy market. In the U.S., companies like NuScale Power are at the forefront of SMR technology, supported by government grants and a strong regulatory framework that facilitates quicker approvals. Similarly, the UK's focus on investing in Rolls-Royce's SMR program positions the country as a leader in the next generation of nuclear technology. By streamlining the licensing process, these countries are strategically positioning themselves to attract foreign investment and capitalize on technological exports.

For developing nations with growing energy needs but limited financial resources, SMRs present an appealing solution to enhance energy security. Countries like India and Brazil can benefit from the flexibility of SMR deployment, allowing for incremental investments that align with their economic capabilities. The smaller size and modular nature of SMRs enable them to be deployed in regions with limited grid infrastructure, effectively addressing energy access in remote and underserved areas. This not only provides immediate relief to energy shortages but also supports long-term economic development through infrastructure investments and job creation. In India, where energy demand is soaring, the deployment of SMRs could be crucial in reducing dependence on coal and lowering greenhouse gas emissions. Brazil, with its vast remote regions, could similarly use SMRs to provide reliable power, thereby enhancing energy independence.

In Africa, where energy access remains a critical challenge, countries like South Africa are exploring SMRs to diversify their energy mix and reduce their reliance on coal. The economic feasibility of SMRs in these contexts is enhanced by the technology's scalability, which allows governments to invest gradually while expanding capacity as demand grows. By enabling localized energy production, SMRs can also stimulate economic activity in rural regions, driving social development and reducing urban migration pressures.

The socio-economic benefits of SMR deployment go beyond direct energy production, contributing to broader economic growth and technological advancement. The construction and operation of SMRs can stimulate local economies by creating both short-term construction jobs and long-term operational roles. This is particularly impactful in regions that lack advanced industrial sectors. For instance, Australia's potential shift towards nuclear energy, driven by SMRs, could create thousands of skilled jobs, particularly in rural areas where employment opportunities are scarce. By building the necessary regulatory and technical expertise, Australia could also position itself as a regional leader in nuclear technology within the Asia-Pacific, further strengthening its energy independence.

Moreover, countries actively supporting SMR development are poised to gain a competitive edge in the global energy market. For example, South Korea, already a leader in conventional nuclear power, is exploring SMR technology to enhance its export potential. The evolving licensing processes in these countries are designed to streamline approvals, reducing time-to-market and fostering technological innovation. This strategic positioning can attract foreign investment, promote technological exchange, and accelerate economic growth. Additionally, countries that invest in SMRs are likely to see increased collaboration with international partners, contributing to global efforts in achieving energy security and decarbonization goals.

Chapter 8
Environmental Impact and Sustainability of SMRs

Carbon Footprint and Emissions Reductions

Small Modular Reactors (SMRs) are increasingly recognized as a viable solution for reducing carbon emissions in the global energy landscape. Their design allows for lower capital costs, enhanced safety features, and the ability to provide reliable baseload power, making them an attractive alternative to fossil fuels and a complement to renewable energy sources such as wind and solar [65, 280, 281]. The operational phase of SMRs is particularly noteworthy, as they produce zero carbon emissions due to their reliance on nuclear fission rather than fossil fuel combustion. This characteristic positions SMRs favourably in the context of achieving net-zero carbon targets, as they can significantly reduce greenhouse gas (GHG) emissions compared to conventional energy sources [101, 102].

The carbon footprint associated with SMRs primarily arises from their construction, fuel cycle, and waste management processes. However, lifecycle assessments indicate that the total lifecycle emissions for nuclear power, including SMRs, are substantially lower than those of coal and natural gas. For instance, studies have estimated the lifecycle GHG emissions of SMRs to be around 9.1 gCO_2/kWh, which is comparable to wind energy at approximately 11 gCO_2/kWh and significantly lower than natural gas at around 450 gCO_2/kWh and coal at approximately 820 gCO_2/kWh [13, 280]. This data underscores the potential of SMRs to contribute to decarbonization efforts effectively.

Moreover, the integration of SMRs into the energy grid can enhance overall system reliability and stability, particularly as the share of intermittent renewable energy sources increases. By providing a steady supply of electricity, SMRs can help mitigate the challenges posed by the variability of renewables, thus supporting a more resilient energy infrastructure [65, 282]. The modular nature of SMRs also allows for flexible deployment, enabling them to be situated in remote locations or clustered to form larger power plants, further enhancing their utility in diverse energy contexts [281].

The efficiency of construction and modular design has gained significant attention in recent years, particularly in the context of sustainable development and reduced carbon emissions. One of the most compelling advantages of modular construction is its ability to streamline the building process, which is particularly relevant in the construction of Small Modular Reactors (SMRs). The modular approach allows for factory-based prefabrication, which significantly shortens construction timelines and minimizes the carbon footprint associated with traditional on-site construction methods. Studies have shown that modular construction can reduce on-site activities to as little as 10-15% of total construction efforts, leading to delivery time reductions of up to 40% compared to conventional methods [283, 284]. This efficiency not only accelerates project timelines but also facilitates quicker emissions reductions by enabling the transition from fossil fuel-based power sources to cleaner energy alternatives like SMRs, which can be operational within 4 to 6 years.

The inherent design of SMRs contributes to their operational efficiency. Advanced reactor designs, such as high-temperature gas-cooled reactors (HTGRs) and molten salt reactors (MSRs), can achieve thermal efficiencies exceeding 45% [285]. This higher efficiency is a direct result of their ability to operate at elevated temperatures, which enhances the conversion of heat to electricity. The modular design of these reactors allows for the use of alternative coolants, such as helium or molten salts, which further optimizes their thermal performance compared to traditional water-cooled reactors. This capability not only maximizes electricity generation per unit of fuel but also contributes to a significant reduction in the overall carbon footprint associated with energy production.

In addition to the environmental benefits, modular design enhances flexibility and adaptability in construction projects. By breaking down complex structures into manageable modules, construction processes can be more easily adjusted to meet specific project requirements or constraints [286, 287]. This adaptability is particularly beneficial in urban settings where space and resources may be limited, allowing for innovative solutions that maintain high standards of efficiency and sustainability. Furthermore, the integration of smart manufacturing technologies with modular design principles can lead to substantial reductions in energy consumption and emissions during the production phase, reinforcing the overall sustainability of modular construction practices [286].

The decarbonization of hard-to-abate sectors is a significant challenge in the global effort to reduce greenhouse gas (GHG) emissions. Small Modular Reactors (SMRs) present a promising solution to this issue, particularly in applications beyond electricity generation, such as district heating, desalination, and high-temperature industrial processes. The integration of SMRs into these sectors can effectively replace fossil fuel usage, thereby mitigating emissions from industries that contribute substantially to global GHG emissions.

SMRs can be utilized for district heating, which is critical in regions where traditional heating methods rely heavily on fossil fuels. A study by Teräsvirta et al. [227] demonstrates the potential of SMRs to replace combined heat and power (CHP) systems in Nordic district heating networks, indicating significant reductions in conventional air pollutants, including particulate matter and nitrogen oxides. Furthermore, the operational resilience of SMR-integrated microgrids can enhance the heating needs of consumers, as discussed by Poudel et al. [288], who emphasize the flexibility and reliability of SMR-DER (Distributed Energy Resources) systems. This adaptability is crucial for ensuring a stable and low-carbon energy supply in urban settings.

In addition to district heating, SMRs are well-suited for industrial applications that require high-temperature heat, such as steel and cement manufacturing. Vanatta [289] highlights that SMRs can achieve outlet temperatures up to 850 °C, which can address a significant portion of emissions from large industrial facilities. The ability of SMRs to provide high-temperature process heat positions them as a viable alternative to fossil fuels in energy-intensive industries. Moreover, the coupling of SMRs with hydrogen production processes offers a pathway to generate low-carbon hydrogen, essential for decarbonizing sectors like heavy transport and chemicals. Research by Kim et al. [290] indicates that SMRs can be integrated into hydrogen production systems, where the heat generated can significantly reduce production costs.

Desalination is another area where SMRs can contribute to decarbonization efforts. Ghazaie et al. [38] conducted a thermodynamic analysis of SMR coupling with various desalination technologies, demonstrating the feasibility of using SMRs to power both thermal and membrane desalination processes. This integration not only addresses water scarcity issues but also aligns with global sustainability goals by providing a low-carbon alternative to conventional desalination methods that typically rely on fossil fuels.

Small Modular Reactors (SMRs) present a promising solution for achieving long-term emissions reduction while ensuring grid stability, particularly as the integration of intermittent renewable energy sources like solar and wind continues to grow. The inherent variability of these renewable sources necessitates a reliable and flexible power supply to maintain grid stability. SMRs, characterized by their ability to operate flexibly and adjust output according to demand, can effectively complement renewable energy systems. This flexibility is crucial as it allows SMRs to serve as a consistent, low-carbon baseload power source, thereby reducing reliance on natural gas peaker plants, which are significant sources of CO_2 emissions when ramped up to meet peak demand [59, 281].

Countries such as Canada, the United Kingdom, and the United States are actively exploring the deployment of SMRs as part of their strategies to phase out aging coal plants and meet ambitious climate targets. For instance, Canada's initiative to introduce SMRs in provinces like Saskatchewan and New Brunswick is aimed at decreasing coal dependency and achieving net-zero emissions by 2050 [59]. Similarly, the UK's investment in SMR technology aligns with its mid-century net-zero emissions goal while also addressing energy security concerns [59]. The ability of SMRs to provide reliable power can significantly enhance the resilience of the electric grid, especially in regions transitioning away from fossil fuels [291].

Moreover, the modular design of SMRs allows for easier integration into existing energy systems, providing a pathway to enhance grid reliability amidst the increasing share of renewable energy. Research indicates that SMRs can be designed to improve frequency stability and overall grid performance, making them a sustainable solution to the challenges posed by the growing share of intermittent energy sources [59, 292]. The operational characteristics of SMRs, including their capacity for load-following and cogeneration, further support their role in stabilizing the grid while contributing to emissions reduction [248].

Waste Management and Fuel Recycling Options

SMRs generate three primary categories of radioactive waste: low-level waste (LLW), intermediate-level waste (ILW), and high-level waste (HLW). The classification is based on the radioactivity levels, with HLW posing the greatest risk due to its high radioactivity and long half-life [274]. Notably, SMRs are engineered to produce lower volumes of waste compared to traditional reactors, primarily due to their enhanced fuel efficiency and innovative designs [274, 293]. The majority of waste from SMRs consists of LLW and ILW, which are generally easier to manage and can be disposed of in near-surface facilities designed to isolate them from the environment [294, 295]. In contrast, HLW, which includes spent nuclear fuel, requires more complex management strategies due to its hazardous nature [293, 294].

The management of spent nuclear fuel, classified as HLW, is critical due to its long-term radiotoxicity. Upon removal from the reactor, spent fuel is initially stored in wet storage pools, where it is cooled and shielded by water [293]. After several years, the fuel can be transferred to dry storage in robust casks designed to contain radiation and prevent environmental contamination [293]. SMRs, with their smaller core sizes and lower fuel burn-up rates, produce less spent fuel per unit of electricity generated, which is a significant advantage in waste management [274, 293]. Furthermore, some advanced SMR designs utilize innovative fuel types, such as high-assay low-enriched uranium (HALEU) or thorium, which can further reduce the volume of long-lived radioactive waste [274, 293].

Fuel recycling is a pivotal strategy for minimizing the environmental impact of nuclear waste. Advanced SMR designs can incorporate fuel recycling technologies that extract usable isotopes from spent fuel, significantly reducing the volume and radiotoxicity of waste [274, 293]. Countries like France and Japan have established successful fuel recycling programs for traditional reactors, which could be adapted for SMRs [293]. By reprocessing spent fuel, up to 97% of its material can be reused, leaving only about 3% as waste requiring deep geological disposal [274, 293]. This not only conserves resources but also diminishes the long-term storage burden associated with HLW [293].

Next-generation SMRs, such as molten salt reactors (MSRs) and high-temperature gas-cooled reactors (HTGRs), are designed to optimize fuel usage and minimize waste generation. MSRs, for instance, utilize liquid fuel that can be continuously reprocessed during operation, thereby reducing the production of long-lived waste products [274, 293]. HTGRs can also utilize a broader range of fuel types, including thorium, which further decreases the generation of long-lived radioactive waste [274, 293]. Fast reactors, another type of advanced SMR, can consume long-lived isotopes found in HLW, effectively reducing the radiotoxicity of waste and minimizing the need for deep geological repositories [293].

Despite advancements in recycling and waste minimization, some radioactive waste will require long-term isolation. Deep geological disposal is currently considered the most secure method for managing HLW, involving the placement of waste in stable geological formations [294, 295]. Countries such as Finland and Sweden are at the forefront of developing deep geological repositories designed to safely contain waste for thousands of years [294, 295]. The modular nature of SMRs could facilitate centralized waste processing and disposal for multiple units, achieving economies of scale and reducing the environmental impact of transportation and storage [294, 295].

The integration of advanced waste management and fuel recycling technologies with SMR deployment presents a sustainable approach to nuclear power. By minimizing waste volumes, recycling valuable isotopes, and employing innovative reactor designs, SMRs can effectively address the environmental concerns associated with traditional nuclear power plants. As nations strive to meet climate goals and transition to low-carbon energy systems, the adoption of SMRs, coupled with enhanced waste management capabilities, could play a crucial role in ensuring energy security and environmental protection.

SMRs in the Context of Sustainable Energy Goals

SMRs have the potential to significantly contribute to decarbonization efforts by providing a low-emission alternative to fossil fuel-based power generation. Unlike traditional coal or gas plants, SMRs operate with virtually no greenhouse gas emissions during their operational phase, making them a key player in achieving the net-zero targets outlined in the Paris Agreement [11]. Countries like the United States, Canada, and the United Kingdom are incorporating SMRs into their long-term energy strategies, recognizing their ability to replace aging coal-fired plants and reduce air pollution [11]. Furthermore, SMRs can decarbonize hard-to-abate sectors, such as heavy industry, which require high-temperature heat that renewable sources alone cannot efficiently provide [11].

The integration of renewable energy sources, such as wind and solar, presents challenges due to their inherent intermittency. SMRs can provide stable baseload power, essential for maintaining grid stability, particularly in regions with high renewable penetration [269]. Their load-following capabilities allow SMRs to adjust output in response to demand fluctuations, complementing intermittent renewable sources and reducing reliance on fossil fuel-based peaking plants [269]. This flexibility enhances the resilience of power grids, ensuring a stable electricity supply as the share of renewables increases [269].

The modular construction of SMRs allows for factory-based manufacturing, significantly reducing on-site construction time and costs compared to traditional nuclear power plants [28]. This approach minimizes environmental disruption and makes SMRs suitable for deployment in remote or underserved areas where large-scale infrastructure projects may not be feasible [28]. Additionally, SMRs require a smaller land footprint, making them more compatible with densely populated regions or areas with land use constraints [28].

Safety concerns have historically hindered the expansion of nuclear energy. However, SMRs incorporate advanced safety features, such as passive cooling systems that do not rely on external power sources, thereby reducing the risk of accidents [16]. The smaller cores and lower operating pressures of SMRs further enhance their safety profile [16]. By improving safety measures, SMRs can foster greater public acceptance of nuclear energy, which is crucial for their successful deployment [296]. Countries like Canada and the UK are actively engaging the public to build trust in SMR technology as a safe and reliable energy source [296].

SMRs are designed to be more fuel-efficient than traditional reactors, often utilizing advanced fuel types that produce less waste [11]. Some designs achieve higher fuel burnup rates, extracting more energy from the same amount of nuclear fuel, which reduces the volume of spent fuel that needs to be managed [11]. This improved fuel efficiency aligns with sustainability principles by optimizing resource use and minimizing environmental impact [11]. Furthermore, SMRs can leverage existing waste disposal facilities or be integrated into a closed fuel cycle, where spent fuel is reprocessed and reused, as seen in countries like France [11].

The economic viability of SMRs is enhanced by their modular construction and scalability, which can lead to significant cost reductions over time [11]. The deployment of SMRs is expected to create jobs in construction, manufacturing, and ongoing operations, particularly in regions transitioning from traditional industries like coal mining [11]. For instance, Canada's investment in SMR technology is projected to generate thousands of high-skilled jobs, supporting economic diversification while transitioning to a low-carbon economy [11].

SMRs offer versatile applications beyond electricity generation, including providing process heat for industries with high carbon footprints, such as steel and cement production [42]. They can also be utilized for desalination, addressing water scarcity in regions with limited freshwater resources [42]. Countries in the Middle East are exploring SMRs for desalination, allowing them to meet growing water needs without increasing carbon emissions [42]. This versatility positions SMRs as valuable assets for achieving broader sustainability goals, including water security and industrial decarbonization [42].

Investing in SMR technology can enhance a country's global competitiveness in the clean energy sector. By leading in SMR deployment, nations can position themselves at the forefront of the nuclear technology market, creating opportunities for exports and international collaboration [11]. Additionally, SMRs can contribute to national energy independence by reducing reliance on imported fossil fuels, providing a stable, domestically-produced energy supply that is less vulnerable to geopolitical tensions [11].

SMRs and Integration with Renewable Energy

The integration of Small Modular Reactors (SMRs) with renewable energy sources, such as wind and solar, presents a promising strategy for achieving a balanced and low-carbon energy system. SMRs are characterized by their ability to provide stable baseload power, which is essential for complementing the intermittent nature of renewable energy generation. Unlike traditional large-scale nuclear reactors, SMRs offer flexibility in operation, allowing them to be scaled according to demand, thereby addressing the variability associated with renewable sources. For instance, during periods when solar generation is low, such as at night, or when wind generation is insufficient, SMRs can maintain a consistent power supply, ensuring grid reliability [221, 267].

One of the significant advantages of integrating SMRs with renewable energy is the enhancement of grid stability. As the share of renewables increases globally, grid operators face challenges in managing the fluctuations in electricity supply. SMRs can provide load-following capabilities, adjusting their output in real-time to match grid demand. This flexibility is crucial in regions with high renewable penetration, where maintaining frequency and voltage stability is often problematic. For example, during peak solar output, SMRs can reduce their output, and conversely, they can quickly ramp up when solar generation diminishes [269, 281]. This capability not only helps in balancing supply and demand but also mitigates the risks associated with the variability of renewable energy sources [221].

From an economic perspective, the integration of SMRs with renewables can lead to a reduction in the overall cost of energy production over time. Although the initial capital investment for SMRs is substantial, their operational efficiency and longevity contribute to lower long-term costs. Moreover, this integration diminishes the reliance on expensive energy storage solutions that are typically necessary to manage the intermittency of renewable sources. Instead of depending solely on battery storage, SMRs can act as a reliable backup, thus reducing the need for extensive storage infrastructure and making the energy transition more economically viable [13, 175]. This hybrid approach not only lowers costs but also enhances the feasibility of decarbonizing the energy sector [297].

Furthermore, the deployment of SMRs alongside renewable energy sources can significantly benefit remote and off-grid communities. In areas where access to stable electricity is limited, the combination of SMRs and renewables can provide a robust energy solution. SMRs can operate independently, delivering continuous power even in regions with inconsistent weather patterns, which is particularly advantageous in large, sparsely populated countries like Canada and Australia [65, 298]. This integration fosters energy self-sufficiency while simultaneously reducing environmental impacts, aligning with global climate goals [299].

The environmental benefits of integrating SMRs with renewable energy are substantial. While renewable sources are inherently zero-emission, their intermittent nature can necessitate backup generation from fossil fuels, which contributes to greenhouse gas emissions. SMRs, operating without greenhouse gas emissions during their lifecycle, can effectively replace fossil fuel-based power plants that are currently used to balance renewable output. This transition not only reduces carbon emissions but also supports global efforts to mitigate climate change, as outlined in agreements such as the Paris Agreement [300, 301].

Implementation Case Example

The following considers the construction costs, practicality, Infrastructure requirements, ongoing costs and environmental impact over a 50 year period for a hypothetical country with the following characteristics:

- Population 30 million

- Average hourly usage 6570 KW per hour per household

- the average energy used per day by a household with three people is about 18.71 kilowatt-hours (kWh).

- A household with four people typically uses around 21.355 kWh/day; if there are five or more people, it's about 25.43 kWh/day.

- On average, small businesses use 3,600–3,650 kWh of electricity per quarter.

- Land mass *7 688 287km²*

- Summer runs from December to March (average temperature 29 °C), autumn from March to May, winter runs from June to August (average 13 °C) and spring from September to November

if the country were to supply its electricity using each of the following options:

- Gas alone

- Renewables alone (wind and solar) alone

- SMR alone

- Combination of gas and renewables

- Combination of SMR and renewables

- Combination of gas, renewables and SMR

To analyse the most suitable energy strategy for a country with the given demographic, geographic, and energy usage characteristics, we need to compare various options for electricity generation. Each option has distinct construction costs, practicality, infrastructure requirements, ongoing operational costs, and environmental impacts. Let's delve into each scenario over a 50-year period.

Assumptions

Before we begin, let's outline some key assumptions that will influence the analysis:

1. **Population and Household Energy Consumption**: The country has a population of 30 million, with an average household consumption of approximately 18.71 to 25.43 kWh/day depending on household size. We assume a mix of small businesses with quarterly usage of around 3,600–3,650 kWh.

2. **Electricity Demand**: Based on the above data, the estimated national electricity demand is calculated, accounting for residential, commercial, and industrial consumption, as well as seasonal variations. We will assume an annual growth rate of 1.5% to account for population and economic growth.

3. **Land Area**: The country has a vast land area (7.69 million km^2), which provides ample space for infrastructure development such as solar farms, wind farms, or SMR facilities.

4. **Climate Considerations**: The climate features hot summers and mild winters, influencing both energy consumption (for cooling and heating) and the efficiency of renewable sources like solar power.

5. **Timeframe**: The analysis is conducted over a 50-year period to account for construction, operational lifespan, maintenance, and decommissioning.

Exploring the different energy options:

1. Gas Alone

Construction Costs: Gas power plants have relatively low initial capital costs compared to nuclear or renewable installations. Building a gas-fired power plant costs approximately USD 800 to 1,200 per kW, leading to an estimated cost of USD 8-12 billion for a capacity of 10 GW to meet the country's energy needs.

Practicality: Gas plants can be constructed relatively quickly (3-4 years) and are highly flexible in terms of scaling up capacity. However, they depend heavily on the availability of natural gas, either domestically or through imports.

Infrastructure Requirements: Gas infrastructure includes pipelines, storage facilities, and power plants. If the country does not have significant natural gas reserves, it would need to invest in LNG import terminals.

Ongoing Costs: Ongoing costs are primarily driven by fuel prices, which are volatile and could increase due to geopolitical factors. The operational lifespan of a gas plant is around 30 years, after which significant refurbishment or replacement would be required.

Environmental Impact: Gas is a cleaner fossil fuel compared to coal, but it still emits greenhouse gases, particularly methane and CO_2. Over a 50-year period, a gas-only approach would contribute to climate change and may not align with global decarbonization targets.

2. Renewables Alone (Wind and Solar)

Construction Costs: Solar power plants have an average cost of USD 1,000 to 1,500 per kW, while wind farms range from USD 1,200 to 2,000 per kW. Given the country's landmass and favourable climate for solar energy, constructing a capacity of 10 GW could cost approximately USD 12-18 billion.

Practicality: Renewables are intermittent sources, meaning they require storage solutions (e.g., batteries) or grid management to ensure reliability. The country's large land area and sunny climate are ideal for large-scale solar farms. However, wind energy may be limited to certain coastal areas.

Infrastructure Requirements: Large-scale solar and wind farms require significant land, grid enhancements, and investments in battery storage systems to store excess energy for use during low production periods.

Ongoing Costs: Renewables have low ongoing operational costs once constructed, with minimal fuel costs. However, the need for periodic maintenance, particularly for wind turbines, and the replacement of solar panels after about 25-30 years would incur expenses.

Environmental Impact: Renewables have a near-zero carbon footprint during operation, making them the most environmentally friendly option. However, the production and disposal of solar panels and batteries have environmental impacts that need to be managed.

3. SMR Alone

Construction Costs: Building a Small Modular Reactor (SMR) plant is estimated to cost USD 5,000 to 6,000 per kW, resulting in a total expenditure of USD 20-24 billion for 4,000 MW capacity. The costs are high due to extensive regulatory compliance and the use of advanced safety technologies.

Practicality: SMRs offer consistent, baseload power with load-following capabilities, making them ideal for grid stability. However, Australia would need to establish a robust regulatory framework for nuclear energy.

Infrastructure Requirements: SMRs require secure sites, cooling systems, waste management facilities, and grid connectivity. Additional investments in specialized labour training and nuclear safety are needed.

Ongoing Costs: The operational lifespan of SMRs can exceed 60 years, with relatively low ongoing fuel costs and maintenance expenses compared to traditional reactors. The main long-term cost involves waste management and decommissioning.

Environmental Impact: SMRs have minimal greenhouse gas emissions during operation. However, the challenge lies in managing radioactive waste and ensuring safe disposal, which remains a concern for public acceptance.

4. Combination of Gas and Renewables

Construction Costs: The hybrid approach involves setting up gas plants and renewable sources, with an estimated cost of USD 10-14 billion. Gas provides baseload power, while renewables cover peak demand periods.

Practicality: This option is flexible and scalable, allowing for a balanced approach to energy production. It can leverage existing gas infrastructure while gradually increasing the share of renewables.

Infrastructure Requirements: Requires both gas pipelines and renewable energy installations, along with grid enhancements to manage variable renewable inputs.

Ongoing Costs: Moderate operational costs, with gas plants providing reliability during periods of low renewable output. Fuel costs for gas remain a variable factor.

Environmental Impact: Reduces emissions compared to gas alone but still contributes to greenhouse gases due to natural gas combustion. However, it offers a feasible pathway to transition towards more renewables over time.

5. Combination of SMR and Renewables

Construction Costs: This approach involves a mix of SMR units and renewable installations, with total costs estimated at USD 25-30 billion. The higher upfront cost is offset by long-term benefits in energy security and emissions reduction.

Practicality: SMRs provide stable baseload power, while renewables supplement during peak times. This combination ensures grid stability and lowers reliance on fossil fuels.

Infrastructure Requirements: Requires investment in nuclear infrastructure and renewable energy systems. The country would need to build expertise in nuclear safety and waste management.

Ongoing Costs: Lower operational costs due to the efficiency of SMRs and the minimal ongoing costs of renewables. The long lifespan of SMRs (up to 60 years) reduces replacement costs.

Environmental Impact: Significantly lower emissions compared to gas-based options, with effective waste management systems to handle nuclear byproducts.

6. Combination of Gas, Renewables, and SMR

Construction Costs: This diversified approach would cost around USD 30-35 billion, providing a balance between reliability, sustainability, and flexibility.

Practicality: This model offers the highest resilience and flexibility by combining the strengths of all three sources. Gas provides backup during peak demand, renewables offer low-cost energy, and SMRs ensure consistent baseload power.

Infrastructure Requirements: The country would need comprehensive infrastructure, including gas pipelines, renewable installations, and SMR facilities. Grid management systems would be required to optimize energy flow from diverse sources.

Ongoing Costs: This approach has moderate ongoing costs due to the combination of low-cost renewables, stable SMR operation, and flexible gas support.

Environmental Impact: By integrating renewables and SMRs, this option significantly reduces carbon emissions while ensuring grid reliability. Gas would be used sparingly, primarily as a backup, minimizing its environmental footprint.

Summarized in a table format:

Criteria	Gas Alone	Renewables Alone (Wind & Solar)	SMR Alone	Combination of Gas & Renewables	Combination of SMR & Renewables	Combination of Gas, Renewables & SMR
Construction Costs	USD 8-12 billion (AUD 12-18 billion)	USD 12-18 billion (AUD 18-27 billion)	USD 20-24 billion (AUD 30-36 billion)	USD 10-14 billion (AUD 15-21 billion)	USD 25-30 billion (AUD 37.5-45 billion)	USD 30-35 billion (AUD 45-52.5 billion)
Practicality	Quick to deploy but dependent on fuel imports	Ideal for sunny and windy regions; intermittent	Stable baseload power but requires nuclear expertise	Flexible and scalable; uses existing gas infra	Stable baseload power with peak support from renewables	Most flexible and resilient; balanced supply from all sources
Infrastructure Requirements	Pipelines, storage, power plants	Extensive land for solar/wind farms, batteries	Nuclear safety facilities, cooling systems	Pipelines, renewable installations, storage	Nuclear facilities, solar/wind farms, grid upgrades	Comprehensive infrastructure for all energy types
Ongoing Costs	High fuel costs; moderate maintenance	Low, but requires battery replacements	Low fuel costs; high decommissioning expenses	Moderate due to mixed fuel use	Low due to long SMR lifespan and renewables' efficiency	Moderate; relies on balance of gas for peak and SMRs for baseload

Criteria	Gas Alone	Renewables Alone (Wind & Solar)	SMR Alone	Combination of Gas & Renewables	Combination of SMR & Renewables	Combination of Gas, Renewables & SMR
Environmental Impact	Moderate emissions (methane, CO_2)	Near-zero operational emissions but has disposal impacts	Low emissions, challenges with radioactive waste	Reduced emissions vs gas; still uses fossil fuels	Significantly lower emissions; waste management needed	Lowest emissions; fossil fuels used sparingly
Operational Lifespan	30 years	25-30 years (solar panels, wind turbines)	60+ years	30-60 years depending on mix	60+ years due to SMR's longevity	60+ years with diversified sources
Energy Security	High reliance on imported gas	Vulnerable to weather variability	High stability, low import dependence	Moderate; relies on both imports and renewables	High; stable baseload with renewables supplement	Very high; diversified energy mix reduces risks
Flexibility & Load-Following	High flexibility, but emissions-intensive	Limited without storage; needs battery backup	High flexibility with load-following capabilities	Moderate; gas covers renewables' intermittency	Excellent; SMRs handle baseload, renewables cover peaks	Best flexibility; combination covers baseload and peaks
Job Creation & Economic Impact	Short-term construction jobs; limited long-term	High during construction; lower after completion	High due to specialized nuclear industry	High; combines construction and maintenance jobs	Significant due to dual industries	Very high; combines benefits of all sectors
Carbon Emissions Over 50 Years	High cumulative emissions	Low cumulative emissions	Near-zero emissions during operation	Moderate emissions	Low emissions, primarily from renewables & SMRs	Lowest cumulative emissions
Decommissioning Costs	Moderate, especially for older plants	Moderate; replacement of panels, turbines	High due to nuclear waste management	Moderate to high, depending on gas component	High but manageable with advanced planning	High but spread across multiple technologies

From an economic, environmental, and practical standpoint, a combination of SMR and renewables appears to be the most balanced approach for a country with the given characteristics. This hybrid solution leverages the stable, low-emission power of SMRs while utilizing the flexibility and cost-effectiveness of renewables. It provides a sustainable pathway for achieving long-term energy security and emissions reduction while addressing the challenges of intermittency and scalability. The initial capital investment is substantial, but the long-term benefits in terms of reduced emissions, energy independence, and economic resilience outweigh the costs, particularly in a world moving towards net-zero emissions by mid-century.

References

1. Dong, Z., *Saturated Adaptive Output-Feedback Power-Level Control for Modular High Temperature Gas-Cooled Reactors.* Energies, 2014. **7**(11): p. 7620-7639.
2. Vujic, J., et al., *Small Modular Reactors: Simpler, Safer, Cheaper?* Energy, 2012. **45**(1): p. 288-295.
3. Kessides, I.N. and B.B. Кузнецов, *Small Modular Reactors for Enhancing Energy Security in Developing Countries.* Sustainability, 2012. **4**(8): p. 1806-1832.
4. Brown, N.R., A. Worrall, and M. Todosow, *Impact of Thermal Spectrum Small Modular Reactors on Performance of Once-Through Nuclear Fuel Cycles With Low-Enriched Uranium.* Annals of Nuclear Energy, 2017. **101**: p. 166-173.
5. Memmott, M., et al., *The Primary Reactor Coolant System Concept of the Integral, Inherently-Safe Light Water Reactor.* Annals of Nuclear Energy, 2017. **100**: p. 53-67.
6. Sovacool, B.K. and M.V. Ramana, *Back to the Future.* Science Technology & Human Values, 2014. **40**(1): p. 96-125.
7. Boarin, S. and M.E. Ricotti, *An Evaluation of SMR Economic Attractiveness.* Science and Technology of Nuclear Installations, 2014. **2014**: p. 1-8.
8. Zhang, X.Y., et al., *Perspective on Site Selection of Small Modular Reactors.* Journal of Environmental Informatics Letters, 2020.
9. Dong, Z., *MLP Compensated Pd Power-Level Control for Modulator High Temperature Gas-Cooled Reactors.* 2014.
10. Wang, Y., et al., *Small Modular Reactors: An Overview of Modeling, Control, Simulation, and Applications.* Ieee Access, 2024. **12**: p. 39628-39650.
11. Xiong, W.B., et al., *Features and Application Analysis of Advanced Small Nuclear Power Reactors.* Advanced Materials Research, 2014. **986-987**: p. 315-321.
12. Guner, T., O.S. Bursi, and S. Erlicher, *Optimization and Performance of Metafoundations for Seismic Isolation of Small Modular Reactors.* Computer-Aided Civil and Infrastructure Engineering, 2022. **38**(12): p. 1558-1582.
13. Locatelli, G., C.M. Bingham, and M. Mancini, *Small Modular Reactors: A Comprehensive Overview of Their Economics and Strategic Aspects.* Progress in Nuclear Energy, 2014. **73**: p. 75-85.
14. Saleh, W., et al., *Advancing Small Modular Reactor Technology Assessment in the Czech Republic, Egypt, and Poland.* Science and Technology of Nuclear Installations, 2023. **2023**: p. 1-16.
15. Zohuri, B., *Navigating the Regulatory Challenges and Economic Viability of Small Modular Nuclear Reactors: A Key to Future Nuclear Power Generation in the USA and Beyond.* Journal of Material Sciences & Manfacturing Research, 2023: p. 1-4.
16. Yin, S., et al., *Accident Process and Core Thermal Response During a Station Blackout Initiated Study for Small Modular Reactor.* Frontiers in Energy Research, 2018. **6**.
17. Kanyolo, T.N., H.C. Oyando, and C.-k. Chang, *Acceleration Analysis of Canned Motors for SMR Coolant Pumps.* Energies, 2023. **16**(15): p. 5733.
18. Maio, F.D., L. Bani, and E. Zio, *The Contribution of Small Modular Reactors to the Resilience of Power Supply.* Journal of Nuclear Engineering, 2022. **3**(2): p. 152-162.

19.   Hong, S. and B.W. Brook, *Economic Feasibility of Energy Supply by Small Modular Nuclear Reactors on Small Islands: Case Studies of Jeju, Tasmania and Tenerife.* Energies, 2018. **11**(10): p. 2587.

20.   Dong, Z., *A Differential-Algebraic Model for the Once-Through Steam Generator of MHTGR-Based Multimodular Nuclear Plants.* Mathematical Problems in Engineering, 2015. **2015**: p. 1-12.

21.   Dong, Z., *Model-Free Coordinated Control for MHTGR-Based Nuclear Steam Supply Systems.* Energies, 2016. **9**(1): p. 37.

22.   Olumayegun, O., M. Wang, and G. Kelsall, *Thermodynamic Analysis and Preliminary Design of Closed Brayton Cycle Using Nitrogen as Working Fluid and Coupled to Small Modular Sodium-Cooled Fast Reactor (SM-SFR).* Applied Energy, 2017. **191**: p. 436-453.

23.   Zhang, J., *Lead–<scp>B</Scp>ismuth Eutectic (<scp>LBE</Scp>): A Coolant Candidate for Gen. <scp>IV</Scp> Advanced Nuclear Reactor Concepts.* Advanced Engineering Materials, 2013. **16**(4): p. 349-356.

24.   Wang, X., et al., *An Experimental Study on Transient Characteristics of a Nuclear Reactor Coolant Pump in Coast-Down Process Under Power Failure Condition.* Frontiers in Energy Research, 2020. **8**.

25.   Crawford, D.C., et al., *First Principles Analysis of Heat Exchanger Concepts and Designs for a Closed CO_2 Brayton Cycle With Regeneration for a Lunar Fission to Surface Power System.* Proceedings of the Institution of Mechanical Engineers Part G Journal of Aerospace Engineering, 2011. **225**(2): p. 194-203.

26.   Keating, R., et al., *ASME Boiler and Pressure Vessel Code Roadmap for Compact Heat Exchangers in High Temperature Reactors.* Journal of Nuclear Engineering and Radiation Science, 2020. **6**(4).

27.   Forsberg, C.W., *Separating Nuclear Reactors From the Power Block With Heat Storage to Improve Economics With Dispatchable Heat and Electricity.* Nuclear Technology, 2021. **208**(4): p. 688-710.

28.   Fetterman, R.J., et al., *An Overview of the Westinghouse Small Modular Reactor.* 2011.

29.   Ingersoll, D.T., *An Overview of the Safety Case for Small Modular Reactors.* 2011: p. 369-373.

30.   Pannier, C. and R. Škoda, *Comparison of Small Modular Reactor and Large Nuclear Reactor Fuel Cost.* Energy and Power Engineering, 2014. **06**(05): p. 82-94.

31.   Greenwood, M.S., et al., *Dynamic System Models for Informing Licensing and Safeguards Investigations of Molten Salt Reactors.* 2018.

32.   Yu, Y., N. Ma, and S. Wang, *Effect of Air Temperature on Passive Containment Cooling System Reliability in AP1000.* Proceedings of the Institution of Mechanical Engineers Part O Journal of Risk and Reliability, 2014. **229**(4): p. 310-318.

33.   Santiko Tri Sulaksono, N., et al., *CFD Analysis of Natural Convection on the Outer Surface of the Containment of APWR Model.* Journal of Advanced Research in Fluid Mechanics and Thermal Sciences, 2024. **115**(1): p. 19-29.

34.   Juarsa, M., et al., *Estimation of Natural Circulation Flow Based on Temperature in the FASSIP-02 Large-Scale Test Loop Facility.* Iop Conference Series Earth and Environmental Science, 2018. **105**: p. 012091.

35.   D'Auria, F.S., N. Aksan, and H. Glaeser, *Physical Phenomena in Nuclear Thermal Hydraulics and Current Status.* Tecnica Italiana-Italian Journal of Engineering Science, 2021. **65**(1): p. 1-11.

36. Holan, J., P. Bílý, and R. Štefan, *Study of the Underground Placement of a Reinforced Concrete Containment Building*. Advances in Science and Technology, 2021. **108**: p. 35-44.

37. Fernández-Arias, P., D. Vergara, and Á. Antón-Sancho, *Bibliometric Review and Technical Summary of PWR Small Modular Reactors*. Energies, 2023. **16**(13): p. 5168.

38. Ghazaie, S.H., et al., *Comparative Analysis of Hybrid Desalination Technologies Powered by SMR*. Energies, 2020. **13**(19): p. 5006.

39. Shin, Y.-H., et al., *Advanced Passive Design of Small Modular Reactor Cooled by Heavy Liquid Metal Natural Circulation*. Progress in Nuclear Energy, 2015. **83**: p. 433-442.

40. Dong, Z., et al., *Dynamic Modeling and Control Characteristics of the Two-Modular HTR-PM Nuclear Plant*. Science and Technology of Nuclear Installations, 2017. **2017**: p. 1-19.

41. Refaey, M.K.S.H.A., *Predication of the SMR Critical Core Performance Under Zero Power*. International Journal of Systems Engineering, 2021. **5**(1): p. 18.

42. Ingersoll, D.T., et al., *Integration of NuScale SMR With Desalination Technologies*. 2014.

43. Setiawan, M.A., et al., *An Approach for Integration of User Requirement and Anthropometry Data in the Process Design of Reactor Main Control Room*. Jurnal Teknologi Reaktor Nuklir Tri Dasa Mega, 2023. **25**(3): p. 107.

44. Toshinsky, G.I. and V.V. Petrochenko, *Modular Lead-Bismuth Fast Reactors in Nuclear Power*. Sustainability, 2012. **4**(9): p. 2293-2316.

45. Hollrah, B., L. Zou, and R. Hu, *Preliminary Primary System Thermal Fluids Analysis of a Horizontal Compact HTGR*. 2022.

46. Zohuri, B., *Nuclear Micro Power Reactor: The New Generation of Innovative Small Reactors*. 2021.

47. Zhao, P., et al., *CFD Analysis of the Primary Cooling System for the Small Modular Natural Circulation Lead Cooled Fast Reactor SNRLFR-100*. Science and Technology of Nuclear Installations, 2016. **2016**: p. 1-12.

48. Chen, H., et al., *Preliminary Design of a Medium-Power Modular Lead-Cooled Fast Reactor With the Application of Optimization Methods*. International Journal of Energy Research, 2018. **42**(11): p. 3643-3657.

49. Aydoğan, F., et al., *Quantitative and Qualitative Comparison of Light Water and Advanced Small Modular Reactors*. Journal of Nuclear Engineering and Radiation Science, 2015. **1**(4).

50. Arocena, P., D.S. Saal, and T. Coelli, *Vertical and Horizontal Scope Economies in the Regulated <scp>U</Scp>.<scp>S</Scp>. Electric Power Industry*. Journal of Industrial Economics, 2012. **60**(3): p. 434-467.

51. Chabinsky, H., *Navigating Nuclear: Microreactors, SMRs, and Traditional Plants*. 2023, Last Energy.

52. Earth Science Australia, *SMR – Small Modular Nuclear Reactors in Australia*. 2024, Earth Science Australia,.

53. World Nuclear Association, *Outline History of Nuclear Energy*. 2024, World Nuclear Association.

54. World Nuclear Association, *Nuclear Power in the World Today*. 2024, World Nuclear Association.

55. Marcus, G.H., *Nuclear Power After Fukushima*. Mechanical Engineering, 2011. **133**(12): p. 27-29.

56.   Андрианов, А.А., Т.А. Osipova, and O.N. Andrianova, *Comparative Analysis of the Investment Attractiveness of Nuclear Power Plant Concepts Based on Small and Medium Sized Reactor Modules and a Large Nuclear Reactor.* Nuclear Energy and Technology, 2020. **6**(3): p. 167-173.

57.   Chalkiadakis, N., et al., *A New Path Towards Sustainable Energy Transition: Techno-Economic Feasibility of a Complete Hybrid Small Modular Reactor/Hydrogen (SMR/H2) Energy System.* Energies, 2023. **16**(17): p. 6257.

58.   Asiamah, G.L. and C.-k. Chang, *Microgrid-Based Small Modular Reactor for a High-Renewable-Energy Penetration Grid in Ghana.* Energies, 2024. **17**(5): p. 1136.

59.   Boudot, C., et al., *Small Modular Reactor-Based Solutions to Enhance Grid Reliability: Impact of Modularization of Large Power Plants on Frequency Stability.* Epj Nuclear Sciences & Technologies, 2022. **8**: p. 16.

60.   Ramana, M.V., *Small Modular and Advanced Nuclear Reactors: A Reality Check.* Ieee Access, 2021. **9**: p. 42090-42099.

61.   Darby, L., A. Hansson, and C.A. Tisdell, *Small-Scale Nuclear Energy.* Case Studies in the Environment, 2020. **4**(1).

62.   Locatelli, G., et al., *Load Following With Small Modular Reactors (SMR): A Real Options Analysis.* Energy, 2015. **80**: p. 41-54.

63.   Peng, M.J., et al., *Preliminary Design and Study of a Small Modular Chlorine Salt Fast Reactor Cooled by Supercritical Carbon Dioxide.* Energies, 2023. **16**(13): p. 4862.

64.   Ghimire, L. and E. Waller, *Small Modular Reactors: Opportunities and Challenges as Emerging Nuclear Technologies for Power Production.* Journal of Nuclear Engineering and Radiation Science, 2023. **9**(4).

65.   Zarębski, P. and D. Katarzyński, *Small Modular Reactors (SMRs) as a Solution for Renewable Energy Gaps: Spatial Analysis for Polish Strategy.* Energies, 2023. **16**(18): p. 6491.

66.   Agar, A.S., et al., *Expected Accuracy Range of Cost Estimates for Small Modular Reactors at the Early Concept Design Stage.* 2018.

67.   Sainati, T., G. Locatelli, and N. Brookes, *Small Modular Reactors: Licensing Constraints and the Way Forward.* Energy, 2015. **82**: p. 1092-1095.

68.   World Nuclear Association, *Small Nuclear Power Reactors.* 2024, World Nuclear Association.

69.   Almalki, R., J.M. Piwowar, and J. Siemer, *Geographical Considerations in Site Selection for Small Modular Reactors in Saskatchewan.* Geosciences, 2019. **9**(9): p. 402.

70.   Middleton, B. and C.M. Mendez, *Integrating Safety, Operations, Security, and Safeguards (ISOSS) Into the Design of Small Modular Reactors : A Handbook.* 2013.

71.   Cha, A. and R. Tscherning, *Exploring a Coal-to-Nuclear Transition: Repurposing of Legacy Coal Assets to Locate Small Modular Reactors in Alberta.* 2024. **17**(1).

72.   Rahman, M.W., M.Z. Abedin, and M.S. Chowdhury, *Efficiency analysis of nuclear power plants: A comprehensive review.* World Journal of Advanced Research and Reviews, 2023. **19**(2): p. 527-540.

73.   Andrianov, A.A., et al., *Comparative analysis of the investment attractiveness of nuclear power plant concepts based on small and medium sized reactor modules and a large nuclear reactor.* Nuclear Energy and Technology, 2020. **6**(3): p. 167-173.

74.   Rowan, D., *Issues and challenges in assessing ecological and human health risk from the siting of SMRs in Canada.* CNL Nuclear Review (Online), 2020. **9**(1): p. 93-97.

75. Lee, S.-C., et al., *Comparative studies of the structural and transport properties of molten salt FLiNaK using the machine-learned neural network and reparametrized classical forcefields.* The Journal of Physical Chemistry B, 2021. **125**(37): p. 10562-10570.

76. Glass, J., A. Burgess, and T. Okugawa, *Investigation of thermodynamic factors influencing Thorium reactor efficiencies.* PAM Review Energy Science & Technology, 2015. **2**: p. 14-31.

77. Felmy, H.M., et al., *On-line monitoring of gas-phase molecular iodine using Raman and fluorescence spectroscopy paired with chemometric analysis.* Environmental Science & Technology, 2021. **55**(6): p. 3898-3908.

78. Serp, J., et al., *The molten salt reactor (MSR) in generation IV: overview and perspectives.* Progress in Nuclear Energy, 2014. **77**: p. 308-319.

79. Bradshaw, M., *Rolls-Royce SMR to Begin Regulatory Assessment.* 2022, MA Business London.

80. Agbalalah, E.H., R. Solorzano, and J. Rosenzweig, *Human Resources Challenges in The Nuclear Industry (Knowledge Retention and Transfer (Krt)) In the United Kingdom (UK) As Reference.* European Journal of Human Resource, 2020. **4**(1): p. 63-76.

81. Doyle, J., *Acclimatizing nuclear? Climate change, nuclear power and the reframing of risk in the UK news media.* International Communication Gazette, 2011. **73**(1-2): p. 107-125.

82. Lebedev, V. and A. Deev, *Heat Storage as a Way to Increase Energy Efficiency and Flexibility of NPP in Isolated Power System.* Applied Sciences, 2023. **13**(24): p. 13130.

83. Zverev, D.L., et al., *RITM-200: New-Generation Reactor for a New Nuclear Icebreaker.* Atomic Energy, 2013. **113**(6): p. 404-409.

84. Rogalev, N., et al., *An Overview of Small Nuclear Power Plants for Clean Energy Production: Comparative Analysis of Distributed Generation Technologies and Future Perspectives.* Energies, 2023. **16**(13): p. 4899.

85. Guo, Y. and L. He, *Communicating in Diverse Local Cultures: Analyzing Chinese Government Communication Programs Around Nuclear Power Projects.* Review of Policy Research, 2023. **41**(4): p. 654-678.

86. King, A. and M.V. Ramana, *The China Syndrome? Nuclear Power Growth and Safety After Fukushima.* Asian Perspective, 2015. **39**(4): p. 607-636.

87. Liu, B., J. Liu, and L. Shen, *Low-Temperature Nuclear Heating Reactors: Characteristics and Application of Licensing Law in China.* Frontiers in Energy Research, 2023. **10**.

88. Park, J.B. and C. Dahoon, *'Eco-Nuclear' Energy Transformation? Authoritarian Environmentalism and Regulatory Policy in China.* Journal of Asian and African Studies, 2022. **59**(4): p. 1263-1286.

89. Taylor, T.H., D.N. Ford, and K.F. Reinschmidt, *Impact of Public Policy and Societal Risk Perception on U.S. Civilian Nuclear Power Plant Construction.* Journal of Construction Engineering and Management, 2012. **138**(8): p. 972-981.

90. Woo, J., et al., *Public Attitudes Toward the Construction of New Power Plants in South Korea.* Energy & Environment, 2017. **28**(4): p. 499-517.

91. Kim, H., *Economic and Environmental Implications of the Recent Energy Transition on South Korea's Electricity Sector.* Energy & Environment, 2018. **29**(5): p. 752-769.

92. Richardson, L., *Protesting Policy and Practice in South Korea's Nuclear Energy Industry.* 2017: p. 133-154.

93.   Hong, S. and B.W. Brook, *At the Crossroads: An Uncertain Future Facing the Electricity-generation Sector in South Korea*. Asia & the Pacific Policy Studies, 2018. **5**(3): p. 522-532.

94.   Lee, K.-H., et al., *Recent Advances in Ocean Nuclear Power Plants*. Energies, 2015. **8**(10): p. 11470-11492.

95.   Park, E., *Social Acceptance of Renewable Energy Technologies in the Post-Fukushima Era*. Frontiers in Psychology, 2021. **11**.

96.   Foroughimehr, N., et al., *Design and Implementation of a Specialised Millimetre-Wave Exposure System for Investigating the Radiation Effects of 5G and Future Technologies*. Sensors, 2024. **24**(5): p. 1516.

97.   Burke, P.J., *On the Way Out: Government Revenues From Fossil Fuels in Australia*. Australian Journal of Agricultural and Resource Economics, 2023. **67**(1): p. 1-17.

98.   Lü, B., et al., *A Zero-Carbon, Reliable and Affordable Energy Future in Australia*. Energy, 2021. **220**: p. 119678.

99.   Lapanporo, B.P., Z. Su'ud, and A.P.A. Mustari, *Neutronic Design of Small Modular Long-life Pressurized Water Reactor Using Thorium Carbide Fuel at a Power Level of 300–500 MWth*. Eastern-European Journal of Enterprise Technologies, 2024. **1**(8 (127)): p. 18-27.

100.   Fernández-Arias, P., D. Vergara, and J.A. Orosa, *A Global Review of PWR Nuclear Power Plants*. Applied Sciences, 2020. **10**(13): p. 4434.

101.   D'yakov, A.F., *Small Modular Nuclear Reactors: Development Prospects*. World Economy and International Relations, 2023. **67**(6): p. 47-60.

102.   Zhi-tao, L. and J. Fan, *Technology Readiness Assessment of Small Modular Reactor (SMR) Designs*. Progress in Nuclear Energy, 2014. **70**: p. 20-28.

103.   Dunlap, C., *Rethinking Nuclear Cooperation in Argentina's and Brazil's Competition for Prestige, 1972–1980*. Latin American Research Review, 2021. **56**(2): p. 385-399.

104.   Hurtado, D., *Semi-Periphery and Capital-Intensive Advanced Technologies: The Construction of Argentina as a Nuclear Proliferation Country*. Journal of Science Communication, 2015. **14**(02): p. A05.

105.   Colombo, S., C. Guglielminotti, and M.N. Vera, *El Desarrollo Nuclear De Argentina Y El Régimen De No Proliferación*. Perfiles Latinoamericanos, 2017. **25**(49): p. 119-139.

106.   Leotlela, M., *Establishing an Effective Nuclear Regulatory Regime: A Case Study of South Africa*. Science Technology & Public Policy, 2021. **5**(1): p. 1.

107.   Cano, J. and K.E.S. Millán-Rivas, *Nuclear Energy in Latin America in the Face of Economic and Environmental Challenges*. International Journal of Scientific Research and Management, 2019. **7**(04).

108.   Frano, R.L., *Benefits of Seismic Isolation for Nuclear Structures Subjected to Severe Earthquake*. Science and Technology of Nuclear Installations, 2018. **2018**: p. 1-11.

109.   Ismail, M., J. Rodellar, and F. Pozo, *Passive and Hybrid Mitigation of Potential Near-Fault Inner Pounding of a Self-Braking Seismic Isolator*. Soil Dynamics and Earthquake Engineering, 2015. **69**: p. 233-250.

110.   Zhong, W., et al., *Investigations of the Effects of a Passive Bumper on the Seismic Response of Base-Isolated Buildings: Experimental Study and Parameter Optimization*. Journal of Vibration and Control, 2022. **29**(11-12): p. 2842-2853.

111.   Bhardwaj, S.R., A.H. Varma, and S.R. Malushte, *Minimum Requirements and Section Detailing Provisions for Steel-Plate Composite (SC) Walls in Safety-Related Nuclear Facilities*. 2017. **54**(2): p. 89-108.

112.   McIntyre, M.L., S.A. Murphy, and C.A.T. Sirsly, *Do Firms Seek Social License to Operate When Stakeholders Are Poor? Evidence From Africa.* Corporate Governance, 2015. **15**(3): p. 306-314.

113.   Hoedl, S., *A Social License for Nuclear Technologies.* 2018: p. 19-44.

114.   Barich, A., et al., *Social License to Operate in Geothermal Energy.* Energies, 2021. **15**(1): p. 139.

115.   Hendra, H. and I.K. Rachmawati, *Pengaruh Stakeholder Engagement, Social Mapping Dan Penerapan Corporate Social Responsibility Terhadap Ketercapaian Social License to Operate).* Jurnal Ilmiah Global Education, 2023. **4**(3): p. 1963-1978.

116.   Martin, O.S., et al., *The Winds of Change: The Role of Community Engagement and Benefit-Sharing in Wind Farm Developments.* Iop Conference Series Earth and Environmental Science, 2022. **1073**(1): p. 012006.

117.   Hall, N., et al., *Social Licence to Operate: Understanding How a Concept Has Been Translated Into Practice in Energy Industries.* Journal of Cleaner Production, 2015. **86**: p. 301-310.

118.   Bice, S., M. Brueckner, and C. Pforr, *Putting Social License to Operate on the Map: A Social, Actuarial and Political Risk and Licensing Model (SAP Model).* Resources Policy, 2017. **53**: p. 46-55.

119.   Reed, M., et al., *A Theory of Participation: What Makes Stakeholder and Public Engagement in Environmental Management Work?* Restoration Ecology, 2017. **26**(S1).

120.   Stephens, S. and B. Robinson, *The Social License to Operate in the Onshore Wind Energy Industry: A Comparative Case Study of Scotland and South Africa.* Energy Policy, 2021. **148**: p. 111981.

121.   Fristikawati, Y., *Legal Analysis Regarding Nuclear Power Plant and Its Relation to the Protection of Environment and Society.* International Journal of Research in Business and Social Science (2147-4478), 2022. **11**(1): p. 290-297.

122.   Omotehinse, A.O. and G.d. Tomi, *A Social License to Operate: Pre-Mining Effects and Activities Perspective.* Rem - International Engineering Journal, 2019. **72**(3): p. 523-527.

123.   Zeliang, C., et al., *Integral PWR-Type Small Modular Reactor Developmental Status, Design Characteristics and Passive Features: A Review.* Energies, 2020. **13**(11): p. 2898.

124.   Siegel, J., et al., *An Expert Elicitation of the Proliferation Resistance of Using Small Modular Reactors (SMR) for the Expansion of Civilian Nuclear Systems.* Risk Analysis, 2017. **38**(2): p. 242-254.

125.   Yahya, M.S. and Y.H. Kim, *An Innovative Core Design for a Soluble-Boron-Free Small Pressurized Water Reactor.* International Journal of Energy Research, 2017. **42**(1): p. 73-81.

126.   Shahmirzaei, A., G.R. Ansarifar, and A. Koraniany, *Assessment of Gadolinium Concentration Effects on the NuScale Reactor Parameters and Optimizing the Fuel Composition via Machine Learning Method.* International Journal of Energy Research, 2022. **46**(7): p. 8838-8871.

127.   Zhu, G., et al., *Low Enriched Uranium and Thorium Fuel Utilization Under Once-through and Offline Reprocessing Scenarios in Small Modular Molten Salt Reactor.* International Journal of Energy Research, 2019. **43**(11): p. 5775-5787.

128.   Son, S.M., et al., *Radionuclide Transport in a Long-term Operation Supercritical CO_2-cooled Direct-cycle Small Nuclear Reactor.* International Journal of Energy Research, 2020. **44**(5): p. 3905-3921.

129.   Sabharwall, P., et al., *Small Modular Molten Salt Reactor (SM-MSR).* 2011.

130. Albadawi, G., S. Wang, and X. Cao, *Numerical and Experimental Study of Thermal Stratification Outside a Small SMR Containment Vessel*. Sustainability, 2018. **10**(7): p. 2332.

131. Gaikwad, A.J., et al., *Ensuring Safety of New, Advanced Small Modular Reactors for Fundamental Safety and With an Optimal Main Heat Transport Systems Configuration*. Kerntechnik, 2023. **88**(4): p. 475-490.

132. Lee, J.I., *Review of Small Modular Reactors: Challenges in Safety and Economy to Success*. Korean Journal of Chemical Engineering, 2024: p. 1-20.

133. Edwards, H., A. Locke, and A. Jackson, *Modular construction of nuclear power stations*. 2016, Ingenia.

134. Stewart, W.R. and K. Shirvan, *Capital Cost Estimation for Advanced Nuclear Power Plants*. Renewable and Sustainable Energy Reviews, 2022. **155**: p. 111880.

135. Wrigley, P., et al., *Module Layout Optimization Using a Genetic Algorithm in Light Water Modular Nuclear Reactor Power Plants*. Nuclear Engineering and Design, 2019. **341**: p. 100-111.

136. Upadhyaya, B.R., et al., *Monitoring Pump Parameters in Small Modular Reactors Using Electric Motor Signatures*. Journal of Nuclear Engineering and Radiation Science, 2016. **3**(1).

137. Lloyd, C. and A.R.M. Roulstone, *A Methodology to Determine SMR Build Schedule and the Impact of Modularisation*. 2018.

138. Stewart, W.R., J. Gregory, and K. Shirvan, *Impact of Modularization and Site Staffing on Construction Schedule of Small and Large Water Reactors*. Nuclear Engineering and Design, 2022. **397**: p. 111922.

139. Baliza, A.R., A.Z. Mesquita, and Y. Morghi, *Review of Regulatory Requirements for Implementing an Aging Management System for the IPR-R1 Triga Reactor of CDTN*. Brazilian Journal of Radiation Sciences, 2022. **10**(3A).

140. Alhudaidi, M., D. Ilić, and M. Gnjatovic, *Strategic Defence Implications of Hazardous Material Transport*. 2019: p. 147.

141. Liu, J. and W. Dai, *Overview of Nuclear Waste Treatment and Management*. E3s Web of Conferences, 2019. **118**: p. 04037.

142. Weichselbraun, A., *From Accountants to Detectives: How Nuclear Safeguards Inspectors Make Knowledge at the International Atomic Energy Agency*. Polar Political and Legal Anthropology Review, 2020. **43**(1): p. 120-135.

143. Backovsky, D., *Atoms for Climate: Institutional Development, Critical Junctures, and the IAEA's Positioning and Communication on Nuclear Energy and Climate Change*. 2023.

144. Barahona, A., *Radiation Risk in Cold War Mexico: Local and Global Networks*. NTM Zeitschrift Für Geschichte Der Wissenschaften Technik Und Medizin, 2022. **30**(2): p. 245-270.

145. Salminen, E., J. Iżewska, and P. Andreo, *IAEA's Role in the Global Management of Cancer-Focus on Upgrading Radiotherapy Services*. Acta Oncologica, 2005. **44**(8): p. 816-824.

146. Gutiérrez-Villamil, C., et al., *Impact of International Atomic Energy Agency Support to the Development of Nuclear Cardiology in Low-and-Middle-Income Countries: Case of Latin America and the Caribbean*. Journal of Nuclear Cardiology, 2019. **26**(6): p. 2048-2054.

147. Zaharieva, N., et al., *Comparative Evaluation of Different Options for Energy System Development in Small Countries.* Iop Conference Series Earth and Environmental Science, 2023. **1128**(1): p. 012021.

148. Steur, R., F. Depisch, and J. Kupitz, *The Status of the IAEA International Project on Innovative Nuclear Reactors and Fuel Cycles (INPRO) and the Ongoing Activities of the Phase 1B of INPRO.* 2004.

149. Baldus, J., H. Müller, and C. Wunderlich, *The Global Nuclear Order and the Crisis of the Nuclear Non-Proliferation Regime: Taking Stock and Moving Forward.* Zeitschrift Für Friedens- Und Konfliktforschung, 2021. **10**(2): p. 195-218.

150. Coppen, T., *Relevant Provisions of the Nuclear Non-Proliferation Treaty.* 2017: p. 123-190.

151. Bernstein, A., et al., *A Prototype Experiment for Cooperative Monitoring of Nuclear Reactors With Cubic Meter Scale Antineutrino Detectors.* 2005.

152. Ritchie, N. and K. Egeland, *The Diplomacy of Resistance: Power, Hegemony and Nuclear Disarmament.* Global Change Peace & Security, 2018. **30**(2): p. 121-141.

153. Clarke, M., *Nuclear Disarmament and the 2010 NPT Review Conference.* Global Policy, 2010. **1**(1): p. 101-107.

154. Bess, J.D., et al., *Intrinsic Value of the International Benchmark Projects, ICSBEP and IRPhEP, for Advanced Reactor Development.* Frontiers in Energy Research, 2023. **11**.

155. Fleming, M., et al., *New Features and Improvements in the NEA Nuclear Data Tool Suite.* Epj Web of Conferences, 2020. **239**: p. 19003.

156. Fiorito, L., J. Dyrda, and M. Fleming, *JEFF-3.3 Covariance Application to ICSBEP Using SANDY and NDAST.* Epj Web of Conferences, 2019. **211**: p. 07003.

157. Ford, M.J. and A. Abdulla, *New Methods for Evaluating Energy Infrastructure Development Risks.* Risk Analysis, 2021. **43**(3): p. 624-640.

158. Yin, M. and K. Zou, *The Implementation of the Precautionary Principle in Nuclear Safety Regulation: Challenges and Prospects.* Sustainability, 2021. **13**(24): p. 14033.

159. Shykinov, N., R.P. Rulko, and D. Mroz, *Importance of Advanced Planning of Manufacturing for Nuclear Industry.* Management and Production Engineering Review, 2016. **7**(2): p. 42-49.

160. Brooks, T., *After Fukushima Daiichi: New Global Institutions for Improved Nuclear Power Policy.* Ethics Policy & Environment, 2012. **15**(1): p. 63-69.

161. Visschers, V. and M. Siegrist, *How a Nuclear Power Plant Accident Influences Acceptance of Nuclear Power: Results of a Longitudinal Study Before and After the Fukushima Disaster.* Risk Analysis, 2012. **33**(2): p. 333-347.

162. Ho, S.S., et al., *Science Literacy or Value Predisposition? A Meta-Analysis of Factors Predicting Public Perceptions of Benefits, Risks, and Acceptance of Nuclear Energy.* Environmental Communication, 2018. **13**(4): p. 457-471.

163. Shykinov, N., R. Rulko, and D. Mroz, *Importance of advanced planning of manufacturing for nuclear industry.* Management and Production Engineering Review, 2016. **7**(2): p. 42--49.

164. Zhang, X., et al., *Perspective on site selection of small modular reactors.* J. Environ. Inform. Lett, 2020. **3**: p. 40-49.

165. Burger, J., J. Clarke, and M. Gochfeld, *Information needs for siting new, and evaluating current, nuclear facilities: ecology, fate and transport, and human health.* Environmental monitoring and assessment, 2011. **172**: p. 121-134.

166.   Grimston, M., W.J. Nuttall, and G. Vaughan, *The siting of UK nuclear reactors*. Journal of Radiological Protection, 2014. **34**(2): p. R1.

167.   Basri, N., et al., *Regulatory requirements for nuclear power plant site selection in Malaysia—a review*. Journal of Radiological Protection, 2016. **36**(4): p. R96.

168.   O'Hara, J., J. Higgins, and S. Fleger. *Control Room Design Review Guidance: Technical Criteria Updates*. in *Proceedings of the Human Factors and Ergonomics Society Annual Meeting*. 2014. SAGE Publications Sage CA: Los Angeles, CA.

169.   Gattie, D. and M. Hewitt, *National Security as a Value-Added Proposition for Advanced Nuclear Reactors: A U.S. Focus*. Energies, 2023. **16**(17): p. 6162.

170.   Hosseini, S.A., et al., *Small Modular Reactors Licensing Process Based on BEPU Approach: Status and Perspective*. Sustainability, 2023. **15**(8): p. 6636.

171.   Sakurahara, T., et al., *Global Importance Measure Methodology for Integrated Probabilistic Risk Assessment*. Proceedings of the Institution of Mechanical Engineers Part O Journal of Risk and Reliability, 2019. **234**(2): p. 377-396.

172.   Aras, E. and M.A. Diaconeasa, *A Critical Look at the Need for Performing Multi-Hazard Probabilistic Risk Assessment for Nuclear Power Plants*. Eng—advances in Engineering, 2021. **2**(4): p. 454-467.

173.   Li, X. and J. Gong, *Fragility and Leakage Risk Assessment of Nuclear Containment Structure Under Loss-of-Coolant Accident Conditions Considering Liner Corrosion*. Applied Sciences, 2024. **14**(6): p. 2407.

174.   Borsoi, S.S., et al., *Risk-Based Design of Electric Power Systems for Non-Conventional Nuclear Facilities at Shutdown Modes*. Brazilian Journal of Radiation Sciences, 2022. **10**(3A).

175.   Asuega, A., B.J. Limb, and J.C. Quinn, *Techno-Economic Analysis of Advanced Small Modular Nuclear Reactors*. Applied Energy, 2023. **334**: p. 120669.

176.   Värri, K. and S. Syri, *The Possible Role of Modular Nuclear Reactors in District Heating: Case Helsinki Region*. Energies, 2019. **12**(11): p. 2195.

177.   Popov, O.O., et al., *Perspectives of Nuclear Energy Development in Ukraine on the Global Trends Basis*. Iop Conference Series Earth and Environmental Science, 2023. **1254**(1): p. 012108.

178.   Geysmans, R., et al., *Broadening and Strengthening Stakeholder Engagement in Emergency Preparedness, Response and Recovery*. Radioprotection, 2020. **55**: p. S219-S225.

179.   Zhu, Y., *Industry Stakeholders Perspectives on Assessing the Effect of Government Policy on Renewable Energy Investment in China*. International Journal of Energy Economics and Policy, 2023. **13**(4): p. 563-573.

180.   Ivanova, Z., *Self-Government Elements in Town-Planning of the Russian Federation: Public Hearings Experience*. Matec Web of Conferences, 2017. **106**: p. 08014.

181.   Perko, T., M. Martell, and C. Turcanu, *Transparency and Stakeholder Engagement in Nuclear or Radiological Emergency Management*. Radioprotection, 2020. **55**: p. S243-S248.

182.   Waris, M., et al., *Analyzing the Constructs of Stakeholder Engagement Towards Renewable Energy Projects Success in Malaysia: A PLS Approach*. Kne Social Sciences, 2019.

183.   Lindner, R., C. Jaca, and J. Hernantes, *A Good Practice for Integrating Stakeholders Through Standardization—The Case of the Smart Mature Resilience Project*. Sustainability, 2021. **13**(16): p. 9000.

184. Corradini, M.L., et al., *Light Water Reactor Sustainability Program Reactor Safety Technologies Pathway Technical Program Plan*. 2016.

185. Shin, D.-W., Y. Shin, and G.H. Kim, *Comparison of Risk Assessment for a Nuclear Power Plant Construction Project Based on Analytic Hierarchy Process and Fuzzy Analytic Hierarchy Process*. Journal of Building Construction and Planning Research, 2016. **04**(03): p. 157-171.

186. Chen, S.-K., Y.W. Ti, and K.-Y. Tsai, *Nuclear Power Plant Construction Scheduling Problem With Time Restrictions: A Particle Swarm Optimization Approach*. Science and Technology of Nuclear Installations, 2016. **2016**: p. 1-9.

187. Kim, W.-J., D. Ryu, and Y. Jung, *Application of Linear Scheduling Method (LSM) for Nuclear Power Plant (NPP) Construction*. Nuclear Engineering and Design, 2014. **270**: p. 65-75.

188. Zio, E. and N. Pedroni, *Monte Carlo Simulation-Based Sensitivity Analysis of the Model of a Thermal–hydraulic Passive System*. Reliability Engineering & System Safety, 2012. **107**: p. 90-106.

189. Vijayan, P.K., et al., *Safety Features in Nuclear Power Plants to Eliminate the Need of Emergency Planning in Public Domain*. Sadhana, 2013. **38**(5): p. 925-943.

190. Leng, L., *Fault Tree Reliability Analysis for Passive Medium Pressure Safety Injection System in Nuclear Power Plant*. Energy and Power Engineering, 2013. **05**(04): p. 264-268.

191. Cole, D.G., *Advanced I&C for Fault-Tolerant Supervisory Control of Small Modular Reactors*. 2018.

192. Sun, L., et al., *Conceptual Design and Analysis of a Passive Residual Heat Removal System for a 10 MW Molten Salt Reactor Experiment*. Progress in Nuclear Energy, 2014. **70**: p. 149-158.

193. Leyer, S. and M. Wich, *The Integral Test Facility Karlstein*. Science and Technology of Nuclear Installations, 2012. **2012**: p. 1-12.

194. Dai, S., J. Chun-nan, and Y. Chen, *Passive Cooldown Performance of Integral Pressurized Water Reactor*. Energy and Power Engineering, 2013. **05**(04): p. 505-509.

195. Salem, M.E., et al., *Proposed Modeling of Natural Convection Cooling Heat Pipe*. Journal of Nuclear Technology in Applied Science, 2022. **10**(1): p. 1-10.

196. Elshahat, A., et al., *Simulation of the Westinghouse AP1000 Response to SBLOCA Using RELAP/SCDAPSIM*. International Journal of Nuclear Energy, 2014. **2014**: p. 1-9.

197. Lin, C.J., T.-L. Hsieh, and C.W. Yang, *Human Factors in Control Room Modernisation*. Measurement and Control, 2015. **48**(3): p. 92-96.

198. Surbakti, T.S.T. and P. Purwadi, *Analysis of Neutronic Safety Parameters of the Multi-Purpose Reactor–Gerrit Augustinus Siwabessy (RSG-GAS) Research Reactor at Serpong*. Jurnal Penelitian Fisika Dan Aplikasinya (Jpfa), 2019. **9**(1): p. 78.

199. Mwanga, E., *Tanzanian Environmental Impact Assessment Laws and Practice for Projects in World Heritage Sites*. The Journal of Environment & Development, 2022. **31**(1): p. 88-107.

200. Karimi, H., S. Neamat, and S. Galali, *Application of Mathematical Matrices for Environmental Impact Assessment, a Case Study of Thermal Power Plant*. Journal of Applied Science and Technology Trends, 2020. **1**(1): p. 13-16.

201. Thamir, A.D., et al., *Environmental Impact Assessment of Halfaya Oilfield Project*. Engineering and Technology Journal, 2019. **37**(2C): p. 258-267.

202. Merk, B., et al., *On a Long Term Strategy for the Success of Nuclear Power*. 2017.

203.   Hyatt, N.C. and M.I. Ojovan, *Special Issue: Materials for Nuclear Waste Immobilization.* Materials, 2019. **12**(21): p. 3611.

204.   Solomon, B.D., M. Andrén, and U. Strandberg, *Three Decades of Social Science Research on High-Level Nuclear Waste: Achievements and Future Challenges.* Risk Hazards & Crisis in Public Policy, 2010. **1**(4): p. 13-47.

205.   Davy, M.K., M.K. Levy, and H. Agripa, *A Brief Overview of Radiation Waste Management and Nuclear Safety.* Physics & Astronomy International Journal, 2023. **7**(2): p. 150-151.

206.   Du, W., W. You, and Z. Xu, *Review and Prospect of Legal Development in Commercial Nuclear Energy.* Energies, 2022. **15**(12): p. 4310.

207.   Li, C., et al., *Potential Impacts of Fukushima Nuclear Leakage on China's Carbon Neutrality—an Investigation on Nuclear Power Avoidance and Regional Heterogeneity.* Frontiers in Environmental Science, 2021. **9**.

208.   Chen, Y., et al., *Prospects in China for Nuclear Development Up to 2050.* Progress in Nuclear Energy, 2018. **103**: p. 81-90.

209.   Wu, Y. and K. Chen, *Nuclear Power's Potential for Carbon Emission Reduction in Chinese Mainland Based on Real Data and Scenario Analysis.* American Journal of Software Engineering and Applications, 2016. **5**(6): p. 46.

210.   Portugal-Pereira, J., et al., *Better Late Than Never, but Never Late Is Better: Risk Assessment of Nuclear Power Construction Projects.* Energy Policy, 2018. **120**: p. 158-166.

211.   Lee, S., J. Lim, and C.-G. Yi, *The Improvement of the Regional Regulatory Governance System for Radiation Risk Management: Spatial Analysis on Radiation Hazards in South Korea.* Sustainability, 2022. **14**(2): p. 966.

212.   Rabbi, M.F. and M. Sabharwal, *Dynamics of Nuclear Energy Policies in India: A Case Study on the Emergence of Nuclear Safety Regulatory Authority.* Indian Journal of Public Administration, 2018. **64**(4): p. 664-685.

213.   Abbasi, S.A., et al., *Is Nuclear Power Generation a Viable Alternative to the Energy Needs of Pakistan? Swot-Rii Analysis.* International Journal of Energy Economics and Policy, 2021. **11**(4): p. 529-536.

214.   Peng, H., et al., *A Multi-Criteria Decision Support Framework for Inland Nuclear Power Plant Site Selection Under Z-Information: A Case Study in Hunan Province of China.* Mathematics, 2020. **8**(2): p. 252.

215.   Fernández-Arias, P., D. Vergara, and Á. Antón-Sancho, *Global Review of International Nuclear Waste Management.* Energies, 2023. **16**(17): p. 6215.

216.   Pioro, I., et al., *Current Status of Reactors Deployment and Small Modular Reactors Development in the World.* Journal of Nuclear Engineering and Radiation Science, 2020. **6**(4).

217.   Jyothi, R.K., L.G.T.C.d. Melo, and H.-S. Yoon, *An Overview of Thorium as a Prospective Natural Resource for Future Energy.* Frontiers in Energy Research, 2023. **11**.

218.   Wang, J. and S. Kim, *Comparative Analysis of Public Attitudes Toward Nuclear Power Energy Across 27 European Countries by Applying the Multilevel Model.* Sustainability, 2018. **10**(5): p. 1518.

219.   Badora, A., K. Kud, and M. Woźniak, *Nuclear Energy Perception and Ecological Attitudes.* Energies, 2021. **14**(14): p. 4322.

220.   Association, W.N., *Design Maturity and Regulatory Expectations for Small Modular Reactors.* London, UK, 2021.

221. El-Emam, R.S. and M.H. Subki, *Small Modular Reactors For <scp>nuclear-renewable</Scp> Synergies: Prospects and Impediments*. International Journal of Energy Research, 2021. **45**(11): p. 16995-17004.

222. Hidayatullah, H., S.S.S. Susyadi, and M.H. Subki, *Design and Technology Development for Small Modular Reactors – Safety Expectations, Prospects and Impediments of Their Deployment*. Progress in Nuclear Energy, 2015. **79**: p. 127-135.

223. Mignacca, B., G. Locatelli, and T. Sainati, *Deeds Not Words: Barriers and Remedies for Small Modular Nuclear Reactors*. Energy, 2020. **206**: p. 118137.

224. Bell, C.D., *Approach to UK SMR Component Design*. 2018.

225. Barattino, W.J., et al., *The Business Case for SMRs on DOD Installations*. 2011.

226. Wang, L., et al., *Intercalated Architecture of MA2Z4 Family Layered Van Der Waals Materials With Emerging Topological, Magnetic and Superconducting Properties*. Nature Communications, 2021. **12**(1).

227. Teräsvirta, A., S. Syri, and P. Hiltunen, *Small Nuclear Reactor—Nordic District Heating Case Study*. Energies, 2020. **13**(15): p. 3782.

228. Frick, K., et al., *Thermal Energy Storage Configurations for Small Modular Reactor Load Shedding*. Nuclear Technology, 2018. **202**(1): p. 53-70.

229. Stewart, W.R., et al., *Evaluating Labor Needs for Fleet-Scale Deployments of Large vs. Small Modular Light Water Reactors*. 2024.

230. Momin, R., *Nuclear Reactor Safety: A Two-Layered Paradigm With (Zr3Si2) Neutron Reflector and Bio Shield in BWRX-300 SMR*. 2024.

231. Jain, S., F. Roelofs, and C.W. Oosterlee, *Valuing Modular Nuclear Power Plants in Finite Time Decision Horizon*. Energy Economics, 2013. **36**: p. 625-636.

232. Abdulla, A., I.L. Azevedo, and M.G. Morgan, *Expert Assessments of the Cost of Light Water Small Modular Reactors*. Proceedings of the National Academy of Sciences, 2013. **110**(24): p. 9686-9691.

233. Agar, A.S., et al., *Stakeholder Perspectives on the Cost Requirements of Small Modular Reactors*. Progress in Nuclear Energy, 2019. **112**: p. 51-62.

234. Webbe-Wood, D. and W.J. Nuttall, *Calculations of Net Present Value for a Small Modular Fusion Power Plant*. Proceedings of the Institution of Civil Engineers - Energy, 2023. **176**(4): p. 187-196.

235. Pham, B.T., G.L. Hawkes, and J.J. Einerson, *Improving thermal model prediction through statistical analysis of irradiation and post-irradiation data from AGR experiments*. Nuclear Engineering and Design, 2014. **271**: p. 209-216.

236. Sakon, A., et al. *Reactor Noise Analysis for a Graphite-moderated and-reflected core in KUCA*. in *EPJ Web of Conferences*. 2021. EDP Sciences.

237. Herlina, H., S.T. Wicaksono, and A.K. Rivai, *OXIDATION BEHAVIOR AND MICRO STRUCTURE ANALYSIS OF NUCLEAR GRAPHITE IG-110 AT 520 °C UNDER AIR ENVIRONMENT*. MATERIALS RESEARCH COMMUNICATIONS, 2021. **2**(1): p. 7-12.

238. Zhou, X.-w., et al., *Nuclear graphite for high temperature gas-cooled reactors*. New Carbon Materials, 2017. **32**(3): p. 193-204.

239. Bakhri, S., *Investigation of rod control system reliability of pwr reactors*. KnE Energy, 2016: p. 94-105.

240. Luthfi, W., et al., *Measured and calculated integral reactivity of control rods in RSG-GAS first core*. Jurnal Teknologi Reaktor Nuklir Tri Dasa Mega, 2022. **24**(1): p. 37-44.

241. Hinks, J., et al. *Transmission electron microscopy study of graphite under in situ ion irradiation*. in *Journal of Physics: Conference Series*. 2012. IOP Publishing.

242.   Szondy, D., *Nuclear SMR welding breakthrough: A year's work now takes a day*. 2024, New Atlas.

243.   Meyer, R.M., et al., *Research Gaps and Technology Needs in Development of PHM for Passive AdvSMR Components*. 2014.

244.   Procházková, D., *Generic Model for Management of Safety of Technical Installations Powered by Small Modular Reactors*. Design Construction Maintenance, 2023. **3**: p. 7-12.

245.   Carelli, M.D., et al., *Economic Features of Integral, Modular, Small-to-Medium Size Reactors*. Progress in Nuclear Energy, 2010. **52**(4): p. 403-414.

246.   Vautier, J.-F., et al., *Benefits of Systems Thinking for a Human and Organizational Factors Approach to Safety Management*. Environment Systems & Decisions, 2018. **38**(3): p. 353-366.

247.   Bhowmik, P., et al., *Advances in Integral and Separate Effects Experiments for Water-Cooled Small Modular Reactors*. 2023: p. 1698-1713.

248.   Chang, C.-k. and H.C. Oyando, *Review of the Requirements for Load Following of Small Modular Reactors*. Energies, 2022. **15**(17): p. 6327.

249.   Foro Nuclear, *Characteristics and advantages of Small Modular Reactors*. 2020, Foro Nuclear,.

250.   Quintero, J., *Megaproject Management*. 2020.

251.   Khan, S.U.-D., Z. Almutairi, and M. Alanazi, *Techno-economic assessment of fuel cycle facility of system integrated modular advanced reactor (SMART)*. Sustainability, 2021. **13**(21): p. 11815.

252.   Korpysa, J., M. Halicki, and A. Lopatka, *Entrepreneurial management of project supply chain–a model approach*. Problems and Perspectives in Management, 2020. **18**(3): p. 211.

253.   Bankvall, L., et al., *Interdependence in supply chains and projects in construction*. Supply chain management: an international journal, 2010. **15**(5): p. 385-393.

254.   Khan, S.U.D., et al., *Safety analysis of pool-type double containment of system-integrated modular advanced reactor: a case study for Saudi Arabia*. International Journal of Energy Research, 2021. **45**(8): p. 12047-12058.

255.   Skiba, R., *Water industry cyber security human resources and training needs*. 2020.

256.   Favari, E. and F. Cantoni, *Megaproject Management: A Multidisciplinary Approach to Embrace Complexity and Sustainability*. 2020: Springer Nature.

257.   Betancourt, M.C., et al., *Computational model for thermohydraulic analysis of an integral pressurized water reactor with mixed oxide fuel (Th, Pu) O2*. Brazilian Journal of Radiation Sciences, 2022. **10**(3A (Suppl.)).

258.   Tükez, E.T. and A. Kaya, *SCADA System for Next-Generation Smart Factory Environments*. Icontech International Journal, 2022. **6**(1): p. 48-52.

259.   Teixeira, M.A., et al., *Flow-based Intrusion Detection Algorithm for Supervisory Control and Data Acquisition Systems: A Real-time Approach*. Iet Cyber-Physical Systems Theory & Applications, 2021. **6**(3): p. 178-191.

260.   Xia, G., M. Peng, and X. Du, *Analysis of Load-Following Characteristics for an Integrated Pressurized Water Reactor*. International Journal of Energy Research, 2013. **38**(3): p. 380-390.

261.   Zhang, Y., Y. Xiang, and L. Wang, *Reliability Analysis of Power Grids With Cyber Vulnerability in SCADA System*. 2014.

262. Upadhyay, D. and S. Sampalli, *SCADA (Supervisory Control and Data Acquisition) Systems: Vulnerability Assessment and Security Recommendations.* Computers & Security, 2020. **89**: p. 101666.

263. Alimi, O.A., et al., *A Review of Research Works on Supervised Learning Algorithms for SCADA Intrusion Detection and Classification.* Sustainability, 2021. **13**(17): p. 9597.

264. Katkar, S.R. and S.V. Umredkar, *Intelligent Scada for Load Control.* International Journal of Power System Operation and Energy Management, 2013: p. 160-164.

265. Shbib, R., S. Zhou, and K.I.H. Alkadhimi, *SCADA System Security, Complexity, and Security Proof.* 2013: p. 405-410.

266. Liu, M., M. Yuan, and G. Li, *Design Private Cloud of Oil and Gas SCADA System.* Icst Transactions on Scalable Information Systems, 2014. **1**(3): p. e5.

267. Rahman, J. and J. Zhang, *Steady-State Modeling of Small Modular Reactors for Multi-Timescale Power System Operations With Temporally Coupled Sub-Models.* Ieee Transactions on Power Systems, 2024: p. 1-13.

268. Joshi, K., B. Poudel, and R. Gokaraju, *Investigating Small Modular Reactor's Design Limits for Its Flexible Operation With Photovoltaic Generation in Microcommunities.* Journal of Nuclear Engineering and Radiation Science, 2021. **7**(3).

269. Shrestha, R., I. Al-Anbagi, and D. Wagner, *Siting of Small Modular Reactors With Renewable Power Generation Support.* Iet Renewable Power Generation, 2022.

270. Islam, M.R. and H.A. Gabbar, *Study of Small Modular Reactors in Modern Microgrids.* International Transactions on Electrical Energy Systems, 2014. **25**(9): p. 1943-1951.

271. Yu, Y., et al., *Optimal Configuration of Power-to-Heat Equipment Considering Peak-Shaving Ancillary Service Market.* Energies, 2023. **16**(19): p. 6860.

272. Chen, H., et al., *Optimizing Energy Storage Participation in Emerging Power Markets.* 2015: p. 1-6.

273. World Nuclear Association, *What is nuclear waste, and what do we do with it?* 2024, World Nuclear Association.

274. Krall, L., A. Macfarlane, and R.C. Ewing, *Nuclear Waste From Small Modular Reactors.* Proceedings of the National Academy of Sciences, 2022. **119**(23).

275. Boldon, L. and P. Sabharwall, *Small Modular Reactor: First-of-a-Kind (FOAK) and NTH-of-a-Kind (NOAK) Economic Analysis.* 2014.

276. Vegel, B. and J.C. Quinn, *Economic Evaluation of Small Modular Nuclear Reactors and the Complications of Regulatory Fee Structures.* Energy Policy, 2017. **104**: p. 395-403.

277. Berthélémy, M. and L.E. Rangel, *Nuclear Reactors' Construction Costs: The Role of Lead-Time, Standardization and Technological Progress.* Energy Policy, 2015. **82**: p. 118-130.

278. Chen, Y., *The Impact of Interest Rate Uncertainty on Renewable Energy Investments: Compared With Conventional Energy Investments.* Applied and Computational Engineering, 2024. **60**(1): p. 76-82.

279. Black, G., et al., *Carbon Free Energy Development and the Role of Small Modular Reactors: A Review and Decision Framework for Deployment in Developing Countries.* Renewable and Sustainable Energy Reviews, 2015. **43**: p. 83-94.

280. Carless, T.S., W.M. Griffin, and P.S. Fischbeck, *The Environmental Competitiveness of Small Modular Reactors: A Life Cycle Study.* Energy, 2016. **114**: p. 84-99.

281. Frick, K., J.M. Doster, and S.M. Bragg-Sitton, *Design and Operation of a Sensible Heat Peaking Unit for Small Modular Reactors.* Nuclear Technology, 2018. **205**(3): p. 415-441.

282. Banaszkiewicz, M. and M. Skwarło, *Numerical Investigations of Transient Thermal Loading of Steam Turbines for SMR Plants*. Archives of Thermodynamics, 2024: p. 197-220.

283. Jang, H.-B., Y. Ahn, and S. Roh, *Comparison of the Embodied Carbon Emissions and Direct Construction Costs for Modular and Conventional Residential Buildings in South Korea*. Buildings, 2022. **12**(1): p. 51.

284. Baú, G. and R.J.J. Oviedo-Haito, *Mapping Activities and Inputs of Modular Construction With Steel 3D Modules in Brazil*. Iop Conference Series Earth and Environmental Science, 2022. **1101**(4): p. 042002.

285. Liu, X., et al., *Analysis of the Brayton Cycle Coupled With a Small Fluoride Salt-Cooled High-Temperature Reactor*. Frontiers in Energy Research, 2023. **10**.

286. Xiaoyi, D., *Study on Sustainable Development Strategies for Enterprises: Application of Modular Design and Smart Manufacturing in Low-Carbon Production*. 2024. **2**(1): p. 111-114.

287. Davies, P., et al., *Boundary Negotiations: A Paradox Theoretical Approach for Efficient and Flexible Modular Systems*. International Journal of Operations & Production Management, 2021. **41**(5): p. 574-597.

288. Poudel, B., et al., *Operational Resilience of Nuclear-Renewable Integrated-Energy Microgrids*. Energies, 2022. **15**(3): p. 789.

289. Vanatta, M., W. Stewart, and M. Craig, *The Role of Policy and Module Manufacturing Learning in Industrial Decarbonization by Small Modular Reactors*. 2024.

290. Kim, J., M. Rweyemamu, and B. Purevsuren, *Machine Learning-Based Approach for Hydrogen Economic Evaluation of Small Modular Reactors*. Science and Technology of Nuclear Installations, 2022. **2022**: p. 1-9.

291. Greene, S.R., *How Nuclear Power Can Transform Electric Grid and Critical Infrastructure Resilience*. Journal of Critical Infrastructure Policy, 2020. **1**(2): p. 37-72.

292. Boudot, C., et al., *Multiphysics Analysis of Power Transients Based on Power System and Nuclear Dynamics Software Chaining*. 2023.

293. Kim, T., et al., *Nuclear Waste Attributes of SMRs Scheduled for Near-Term Deployment*. 2022.

294. Hosan, M.I., *Radioactive Waste Classification, Management and Environment*. Engineering International, 2017. **5**(2): p. 53-62.

295. Khelurkar, N., S.M.R. Shah, and H. Jeswani, *A Review of Radioactive Waste Management*. 2015: p. 1-6.

296. Hlaváček, M., et al., *Nuclear Reactor at Home? Public Acceptance of Small Nuclear Reactors in the Neighborhood*. Frontiers in Energy Research, 2023. **11**.

297. Hermawan, E. and U. Sudjadi, *Integrated Nuclear-Renewable Energy System for Industrialization in West Nusa Tenggara Province, Indonesia: Economic, Potential Site, and Policy Recommendation*. International Journal of Energy Economics and Policy, 2022. **12**(4): p. 146-159.

298. Imani, L., A. Setiawan, and M.K. Ridwan, *Demand and Electricity Energy Mix in Indonesia 2030 With Small Modular Reactor Nuclear Power Plant and Renewable Energy Scenario*. Iop Conference Series Earth and Environmental Science, 2021. **927**(1): p. 012025.

299. Bogdanov, D., et al., *Radical Transformation Pathway Towards Sustainable Electricity via Evolutionary Steps*. Nature Communications, 2019. **10**(1).

300. Clarke, D. and P.D. Ezhilchelvan, *Assessing the Attack Resilience Capabilities of a Fortified Primary-Backup System.* 2010: p. 182-187.

301. Al-Bassam, A.M., J.A. Conner, and V.I. Manousiouthakis, *Natural-Gas-Derived Hydrogen in the Presence of Carbon Fuel Taxes and Concentrated Solar Power.* Acs Sustainable Chemistry & Engineering, 2018. **6**(3): p. 3029-3038.

Index

Air Cooling Systems, 73, 91, 103, 365

Australia, 20, 21, 31, 70, 114, 131, 132, 260, 261, 262, 263, 270, 294, 295, 296, 297, 298, 350, 351, 352, 354, 403, 404, 408, 415, 418, 425, 428

Automation, 172, 196, 208, 209, 279, 300, 330, 331, 341, 343, 365

Baseload Power, 22, 23, 24, 25, 28, 29, 73, 179, 201, 205, 317, 353, 354, 371, 387, 388, 403, 409, 411, 413, 415, 418, 419, 420, 421

Baseload Power Supply, 23, 24, 25, 28, 29, 73, 179, 201, 353, 354, 371, 387, 388, 403, 409, 411, 413, 415, 418, 419, 420

Battery Storage, 316, 364, 418, 420

Bodies, Regulatory, 228, 233, 333

Bonds, Green, 392

Canada, 30, 31, 63, 70, 71, 76, 77, 79, 87, 88, 114, 123, 124, 135, 223, 224, 246, 249, 256, 261, 263, 269, 270, 285, 286, 287, 288, 300, 306, 309, 313, 335, 339, 340, 348, 349, 350, 352, 373, 389, 390, 399, 402, 407, 411, 413, 414, 415, 426

Capital Costs, 1, 18, 23, 25, 29, 72, 74, 76, 78, 85, 86, 87, 89, 97, 104, 106, 107, 109, 127, 129, 140, 148, 161, 164, 166, 171, 191, 192, 193, 194, 195, 197, 198, 199, 200, 201, 202, 203, 205, 206, 207, 208, 209, 220, 223, 229, 245, 269, 301, 305, 306, 309, 310, 311,312, 313, 314, 315, 316, 317, 318, 320, 322, 329, 331, 332, 341, 351, 352, 354, 355, 356, 362, 373, 383, 384, 385, 386, 387, 388, 389, 390, 392, 394, 398, 399,

400, 402, 403, 404, 405, 406, 409, 411, 414, 415, 416, 417, 418, 419, 420, 422

China, 31, 70, 71, 77, 78, 81, 83, 88, 127, 128, 136, 137, 162, 205, 218, 225, 226, 256, 257, 258, 259, 260, 269, 270, 276, 277, 278, 279, 303, 304, 305, 337, 338, 349, 399, 427, 432, 434

Cogeneration, 296, 405, 406, 412

Community Engagement, 24, 44, 110, 111, 112, 113, 117, 122, 134, 141, 146, 149, 152, 153, 154, 155, 156, 157, 158, 159, 160, 161, 163, 178, 201, 204, 207, 217, 230, 231, 232, 248, 252, 256, 262, 311, 333, 355, 358, 386, 387, 393, 396, 406

Compliance Standards, 24, 34, 45, 74, 76, 98, 99, 100, 103, 110, 117, 126, 127, 128, 133, 146, 150, 156, 159, 162, 163, 176, 197, 214, 216, 218, 219, 223, 224, 227, 230, 231, 232, 233, 240, 241, 242, 243, 244, 245, 246, 248, 251, 252, 253, 254, 255, 256, 257, 258, 259, 262, 263, 265, 266, 267, 268, 269, 270, 272, 276, 279, 280, 282, 283, 285, 286, 288, 289, 290, 292, 294, 295, 296, 306, 309, 311, 319, 324, 325, 333, 334, 335, 337, 338, 340, 341, 342, 343, 348, 350, 354, 355, 358, 367, 368, 375, 380, 384, 385, 386, 387, 398, 405, 418

Construction Costs, 1, 2, 3, 4, 5, 16, 22, 23, 30, 40, 43, 44, 52, 62, 64, 70, 71, 72, 73, 74, 76, 77, 79, 85, 86, 87, 88, 89, 93, 97, 98, 100, 101, 103, 104,

105, 106, 109, 110, 111, 113, 117,
120, 121, 122, 123, 124, 125, 127,
128, 129, 132, 134, 135, 136, 140,
141, 142, 143, 144, 145, 146, 147,
150, 151, 153, 159, 161, 162, 164,
166, 167, 168, 169, 171, 174, 175,
178, 181, 188, 191, 192, 193, 194,
195, 196, 197, 198, 199, 200, 201,
202, 203, 204, 205, 206, 207, 208,
209, 210, 211, 212, 219, 220, 222,
223, 224, 225, 226, 227, 228, 229,
230, 231, 232, 234, 239, 241, 242,
243, 245, 246, 248, 249, 252, 253,
255, 256, 258, 259, 260, 261, 262,
263, 266, 267, 268, 269, 270, 271,
272, 273, 274, 275, 276, 277, 278,
279, 280, 281, 282, 283, 284, 285,
286, 287, 288, 289, 290, 291, 292,
293, 294, 295, 296, 297, 298, 299,
300, 301, 302, 303, 304, 305, 306,
309, 310, 311, 312, 313, 314, 316,
317, 318, 319, 320, 321, 322, 323,
325, 326, 327, 328, 329, 330, 331,
332, 333, 334, 335, 336, 337, 338,
339, 341, 342, 348, 349, 350, 351,
353, 354, 355, 356, 357, 358, 362,
367, 368, 384, 386, 387, 388, 390,
393, 394, 395, 396, 397, 398, 399,
400, 401, 402, 403, 404, 405, 406,
407, 408, 409, 410, 414, 416, 417,
421, 436
Construction Risks, 164, 202
Construction, Modular, 85, 193, 194,
197, 269, 352, 430
Cooling Systems, 4, 6, 7, 13, 14, 15, 16,
17, 24, 27, 28, 29, 38, 40, 43, 45, 48,
52, 73, 74, 75, 77, 79, 81, 82, 90, 91,
93, 95, 96, 98, 101, 102, 103, 107,
108, 114, 115, 118, 120, 141, 142,
145, 146, 159, 160, 162, 165, 169,
170, 171, 172, 173, 175, 176, 177,
178, 179, 180, 181, 182, 183, 184,

185, 186, 187, 188, 189, 199, 202,
205, 206, 208, 211, 212, 228, 231,
233, 240, 241, 244, 247, 248, 292,
307, 308, 312, 329, 330, 331, 332,
333, 335, 339, 341, 342, 357, 359,
360, 364, 365, 366, 367, 372, 374,
376, 377, 381, 383, 384, 395, 398,
417, 418, 420
Cooling systems, Passive, 27
Cooling Towers, 91, 95, 107, 108, 142,
170, 176, 179, 185, 199, 208, 307,
329, 331, 341, 360, 365
Cost Reduction Strategies, 77, 193,
194, 197, 200, 209, 314, 414
Cybersecurity Measures, 91, 95, 113,
135, 172, 173, 211, 240, 282, 288,
290, 291, 294, 297, 356, 365, 366,
368, 373, 374
Decarbonization, 24, 25, 73, 76, 206,
209, 234, 263, 298, 321, 388, 389,
392, 397, 403, 404, 408, 409, 410,
411, 413, 414, 417
Decommissioning, 22, 34, 36, 38, 78,
121, 221, 234, 241, 246, 258, 263,
264, 267, 269, 274, 276, 277, 279,
280, 285, 286, 288, 289, 291, 294,
353, 379, 386, 387, 406, 417, 418,
420
Deployment, SMR, 87, 114, 115, 116,
117, 125, 160, 161, 163, 206, 223,
249, 254, 256, 262, 263, 305, 326,
407, 408, 413, 414
Desalination, 2, 16, 73, 75, 76, 78, 82,
86, 87, 88, 106, 115, 166, 206, 209,
363, 364, 405, 406, 410, 411, 414
Economic Impact, 85, 106, 125, 174,
193, 194, 208, 414
Electricity Market, 2, 87, 131, 133, 195,
201, 203, 206, 209, 222, 260, 306,
314, 352, 389, 399, 400, 401, 402,
404, 405, 406, 407, 408, 414

Electricity Supply, 1, 2, 4, 5, 6, 7, 8, 10,
    11, 12, 16, 17, 23, 26, 27, 28, 30, 31,
    32, 33, 34, 35, 36, 37, 38, 39, 40, 41,
    42, 44, 45, 46, 52, 65, 68, 69, 70, 71,
    72, 73, 74, 75, 76, 78, 80, 81, 82, 86,
    87, 88, 92, 93, 95, 98, 99, 103, 104,
    106, 114, 130, 167, 170, 171, 172,
    176, 179, 191, 203, 205, 209, 210,
    211, 302, 305, 307, 314, 317, 332,
    333, 339, 353, 361, 363, 364, 365,
    368, 369, 370, 371, 372, 373, 375,
    376, 388, 390, 391, 392, 393, 399,
    400, 403, 404, 406, 407, 409, 410,
    412, 413, 414, 415, 416, 417
Emissions Reduction, 2, 45, 46, 54, 69,
    71, 73, 78, 79, 100, 101, 117, 119,
    121, 122, 124, 125, 130, 131, 132,
    145, 160, 161, 179, 201, 203, 204,
    209, 226, 231, 253, 256, 260, 272,
    277, 284, 287, 311, 327, 337, 372,
    378, 386, 388, 390, 393, 406, 408,
    409, 410, 411, 412, 413,414, 415,
    418, 419, 420, 421, 422
Energy Efficiency, 5, 6, 8, 16, 26, 27, 29,
    30, 31, 32, 34, 36, 37, 38, 39, 40, 41,
    45, 48, 50, 51, 52, 68, 69, 76, 80, 81,
    82, 83, 84, 85, 87, 93, 97, 98, 99, 102,
    103, 105, 107, 108, 115, 118, 126,
    127, 137, 145, 146, 164, 165, 167,
    168, 169, 171, 172, 174, 175, 176,
    177, 179, 191, 192, 193, 194, 195,
    196, 197, 198, 199, 200, 202, 203,
    204, 205, 206, 207, 209, 212, 216,
    223, 227, 228, 230, 233, 242, 265,
    267, 276, 279, 283, 290, 296, 300,
    301, 304, 306, 311, 314, 317, 318,
    319, 321, 322, 323, 325, 326, 327,
    328, 330, 331, 332, 333, 337, 339,
    340, 349, 357, 360, 361, 362, 363,
    364, 365, 366, 370, 372, 373, 375,
    384, 385, 386, 388, 397, 399, 400,
    402, 404, 407, 410, 412, 414, 415,
    417, 419, 420
Energy Storage, 12, 13, 36, 75, 84, 91,
    92, 95, 96, 98, 108, 121, 136, 143,
    167, 173, 174, 176, 181, 196, 206,
    212, 217, 241, 270, 285, 291, 307,
    315, 316, 317, 328, 330, 363, 364,
    367, 368, 376, 377, 379, 380, 385,
    386, 412, 413, 415, 417, 418, 420,
    421
Enrichment, Fuel, 80
Environmental Impact, 4, 29, 44, 69,
    73, 75, 77, 78, 102, 103, 106, 109,
    111, 113, 117, 121, 123, 125, 130,
    134, 135, 157, 160, 162, 174, 175,
    176, 178, 179, 185, 204, 228, 231,
    234, 242, 244, 247, 251, 252, 254,
    262, 263, 266, 267, 268, 269, 272,
    275, 277, 279, 281, 284, 285, 287,
    290, 293, 295, 298, 306, 311, 312,
    337, 338, 340, 342, 355, 358, 360,
    364, 373, 376, 380, 383, 386, 412,
    413, 414, 415, 416, 418
Export Opportunities, 407, 414
Finland, 69, 116, 224, 251, 252, 274,
    376, 379, 413
France, 30, 69, 70, 71, 83, 125, 126,
    136, 206, 224, 251, 264, 270, 274,
    288, 289, 290, 291, 348, 349, 376,
    412, 414
Fuel Recycling, 6, 30, 53, 75, 82, 88,
    173, 175, 294, 376, 412, 413
Generators, Helical coil steam, 361
Government Subsidies, 391, 395, 403
Green Bonds, 389, 391, 394, 395, 398,
    400, 403, 405
Grid Flexibility, 1, 2, 4, 5, 8, 16, 17, 18,
    21, 24, 25, 28, 30, 40, 72, 73, 74, 75,
    78, 79, 80, 84, 85, 86, 89, 99, 100,
    103, 105, 107, 131, 137, 146, 147,
    164, 167, 177, 179, 181, 193, 198,
    200, 209, 220, 223, 243, 247, 271,

282, 306, 314, 322, 326, 354, 358, 363, 371, 372, 384, 388, 399, 402, 403, 404, 406, 407, 408, 410, 411, 413, 415, 419, 420, 421, 422
Grid Integration, 309, 395
Grid Stability, 2, 4, 8, 16, 17, 18, 22, 23, 24, 43, 44, 52, 71, 72, 73, 74, 75, 78, 82, 84, 86, 87, 88, 89, 90, 91, 92, 93, 96, 97, 98, 99, 103, 104, 114, 116, 117, 131, 132, 148, 151, 159, 160, 161, 162, 166, 167, 171, 172, 179, 181, 184, 198, 201, 206, 208, 226, 302,305, 308, 309, 310, 315, 317, 332, 333, 339, 343, 352, 358, 362, 364, 365, 368, 369, 370, 371, 372, 373, 374, 383, 384, 387, 388, 389, 395, 398, 401, 402, 403, 406, 407, 408, 409, 411, 412, 413, 415, 418, 419, 420
Heat exchangers, 210, 319, 337
Heat Management, 2, 5, 6, 7, 8, 9, 10, 12, 13, 16, 17, 25, 26, 27, 30, 31, 34, 36, 37, 38, 39, 40, 41, 43, 46, 48, 50, 51, 53, 61, 65, 73, 76, 78, 80, 81, 82, 83, 84, 86, 87, 88, 91, 92, 95, 98, 102, 103, 106, 107, 114, 115, 116, 165, 168, 169, 170, 171, 173, 176, 179, 180, 181, 182, 183, 184, 185, 186, 187, 188, 190, 191, 192, 197, 199, 206, 209, 210, 211, 229, 239, 289, 296, 307, 318, 319, 320, 321, 322, 323, 334, 335, 336, 337, 341, 351, 353, 357, 358, 360, 361, 362, 363, 364, 365, 374, 376, 381, 393, 397, 405, 410, 411, 413, 414
Hydrogen Production, 16, 31, 39, 40, 41, 44, 50, 51, 53, 65, 73, 75, 80, 81, 82, 88, 130, 166, 168, 206, 209, 411
India, 30, 43, 44, 70, 71, 79, 80, 180, 218, 253, 254, 255, 256, 264, 270, 279, 280, 281, 282, 349, 399, 403, 408, 434

Industrial Applications, 14, 28, 34, 38, 51, 52, 69, 70, 71, 72, 78, 86, 87, 89, 115, 125, 129, 161, 193, 200, 202, 203, 206, 216, 217, 225, 226, 229, 244, 251, 256, 257, 260, 272, 279, 292, 320, 321, 322, 325, 336, 342, 348, 349, 350, 351, 375, 376, 389, 391, 405, 406, 413, 421, 431, 436
Infrastructure Requirements, 1, 2, 4, 5, 6, 8, 17, 22, 23, 24, 25, 29, 36, 44, 72, 74, 75, 76, 77, 90, 91, 92, 93, 96, 98, 101, 102, 103, 104, 105, 107, 108, 110, 111, 112, 113, 114, 115, 116, 121, 124, 127, 130, 139, 140, 141, 142, 146, 147, 148, 150, 151, 155, 158, 161, 171, 172, 173, 176, 178, 179, 181, 193, 198, 201, 204, 208, 226, 242, 248, 249, 256, 258, 260, 263, 277, 278, 280, 281, 284, 291, 293, 294, 296, 297, 299, 304, 305, 309, 311, 312, 313, 315, 316, 320, 326, 327, 328, 330, 333, 339, 343, 348, 349, 350, 352, 353, 357, 358, 359, 362, 364, 366, 367, 368, 369, 370, 373, 375, 380, 381, 383, 384, 387, 388, 389, 390, 391, 392, 393, 394, 395, 397, 399, 403, 404, 406, 407, 408, 409, 414, 415, 416, 417, 419, 420
Integration with Renewables, 1, 22, 23, 24, 120, 263, 317, 353, 371, 372, 387, 388, 398, 402, 403, 409, 413, 415, 416, 419, 420, 421, 422
Integration, Grid, 352
International Collaboration, 44, 45, 46, 60, 70, 77, 78, 79, 80, 106, 119, 120, 127, 129, 131, 133, 134, 136, 138, 155, 159, 160, 162, 166, 175, 209, 213, 214, 216, 217, 218, 219, 220, 221, 222, 223, 226, 228, 231, 234, 239, 241, 242, 249, 255, 256, 257, 260, 261, 262, 267, 269, 270, 276,

277, 279, 280, 282, 283, 286, 288,
294, 296, 298, 310, 319, 325, 333,
334, 335, 338, 340, 349, 350, 351,
352, 353, 355, 381, 388, 390, 392,
393, 395, 398, 399, 403, 406, 408,
414, 436

Investment Strategies, 1, 2, 18, 22, 24,
25, 40, 74, 75, 89, 104, 125, 126, 137,
148, 158, 203, 207, 232, 352, 353,
354, 383, 384, 386, 387, 389, 390,
391, 394, 397, 399, 402, 403, 404,
405, 407, 408, 411, 414, 415, 419,
422, 426

Japan, 20, 21, 30, 49, 50, 53, 63, 69, 70,
71, 81, 83, 88, 129, 130, 137, 222,
226, 349, 366, 376, 412

Job Creation, 111, 114, 117, 155, 158,
197, 344, 406, 407, 408

Licensing Requirements, 5, 24, 75, 76,
77, 100, 108, 122, 123, 124, 125,
127, 134, 135, 141, 166, 207, 209,
219, 220, 221, 223, 224, 226, 227,
228, 229, 230, 231, 233, 234, 242,
243, 244, 246, 247, 249, 250, 251,
252, 253, 254, 256, 257, 258, 260,
261, 262, 266, 267, 268, 269, 270,
274, 277, 282, 285, 286, 289, 292,
295, 298, 300, 306, 309, 310, 311,
313, 333, 351, 354, 393, 403, 405,
407, 408

Load-Following Capability, 17, 25, 73,
81, 87, 194, 332, 337, 339, 343, 371,
372, 387, 398, 403, 404, 412, 413,
415, 418, 421

Logistics Management, 104, 105, 121,
148, 151, 198, 199, 298, 307, 311,
313, 318, 325, 326, 327, 354, 384

Maintenance Costs, 15, 16, 17, 21, 23,
24, 28, 38, 39, 46, 49, 84, 88, 90, 92,
93, 96, 98, 105, 107, 111, 117, 146,
165, 166, 173, 176, 206, 208, 209,
245, 253, 272, 273, 294, 317, 323,

340, 344, 349, 351, 353, 357, 358,
362, 365, 370, 373, 374, 375, 377,
385, 386, 393, 400, 404, 406, 417,
418, 420, 421

Management, Supply chain, 436

Modular Construction, 1, 4, 5, 7, 8, 15,
16, 21, 22, 25, 27, 28, 39, 40, 53, 72,
73, 74, 76, 77, 78, 79, 81, 82, 85, 86,
87, 89, 90, 92, 93, 95, 96, 97, 98, 99,
101, 104, 105, 109, 117, 121, 124,
127, 128, 129, 132, 134, 135, 141,
162, 164, 166, 167, 168, 169, 173,
174, 175, 177, 178, 180, 181, 191,
192, 193, 194, 195, 196, 197, 198,
199, 200, 201, 202, 203, 204, 205,
206, 207, 208, 209, 211, 212, 219,
220, 221, 223, 224, 225, 226, 227,
228, 231, 232, 234, 245, 249, 254,
256, 260, 264, 266, 267, 269, 292,
298, 299, 300, 304, 305, 306, 307,
309, 310, 311, 313, 314, 317, 318,
319, 320, 321, 322, 323, 325, 326,
327, 328, 329, 330, 331, 337, 339,
349, 351, 353, 354, 356, 357, 358,
361, 362, 367, 369, 372, 384, 385,
386, 388, 394, 397, 398, 399, 400,
402, 404, 405, 406, 407, 408, 410,
412, 413, 414, 431, 436

Modular Design, 1, 4, 16, 23, 27, 28, 29,
40, 41, 74, 76, 77, 79, 93, 97, 99, 100,
171, 207, 212, 361

Monitoring Sensors, 112, 139, 141, 337,
338, 340, 365, 374

Natural Gas, 2, 26, 30, 31, 34, 35, 36,
37, 50, 51, 53, 59, 64, 68, 73, 77, 80,
81, 82, 91, 93, 95, 96, 98, 117, 119,
130, 132, 169, 171, 180, 181, 205,
206, 208, 209, 253, 303, 304, 319,
350, 355, 363, 365, 378, 388, 403,
408, 409, 410, 411, 412, 413, 415,
416, 417, 418, 419, 420, 421, 427,
435

Natural Gas Pipelines, 364, 367, 417, 419, 420

Nuclear energy, 72, 256, 387

Nuclear Energy, 1, 2, 4, 5, 6, 7, 8, 9, 10, 11, 12, 13, 14, 16, 17, 21, 22, 23, 24, 25, 26, 27, 28, 29, 30, 31, 33, 34, 35, 36, 37, 38, 39, 40, 41, 43, 44, 45, 46, 48, 50, 51, 52, 53, 54, 55, 56, 57, 58, 60, 61, 62, 63, 64, 65, 66, 68, 69, 70, 71, 72, 73, 74, 75, 76, 77, 78, 79, 80, 81, 82, 83, 84, 85, 86, 88, 89, 90, 93, 95, 97, 98, 99, 100, 102, 104, 105, 106, 108, 110, 112, 113, 114, 115, 116, 117, 119, 120, 121, 122, 123, 124, 125, 126, 127, 128, 129, 130, 131, 132, 133, 134, 135, 136, 137, 138, 146, 147, 158, 160,161, 164, 165, 166, 167, 168, 169, 171, 172, 173, 174, 175, 178, 179, 180, 181, 183, 184, 185, 186, 187, 188, 189, 190, 191, 192, 193, 194, 195, 196, 197, 198, 199, 200, 201, 202, 203, 204, 205, 206, 207, 208, 209, 210, 213, 214, 215, 216, 217, 218, 219, 220, 221, 222, 223, 224, 225, 226, 227, 228, 229, 230, 231, 232, 233, 234, 239, 240, 241, 242, 243, 246, 249, 250, 251, 252, 253, 254, 255, 256, 257, 258, 259, 260, 261, 262, 263, 264, 265, 266, 269, 270, 271, 272, 273, 274, 275, 276, 277, 278, 279, 280, 281, 282, 283, 284, 285, 286, 287, 288, 289, 290, 291, 292, 293, 294, 295, 296, 297, 298, 299, 300, 302, 303, 304, 305, 306, 308, 309, 310, 311, 313, 317, 318, 319, 320, 321, 322, 323, 324, 325, 326, 333, 334, 336, 339, 340, 343, 344, 348, 349, 350, 351, 352, 353, 354, 355, 358, 359, 364, 366, 367, 368, 371, 372, 373, 375, 376, 380, 384, 385, 386, 389, 390, 391, 395, 397, 398, 399, 402, 403, 404, 405, 406, 407, 408, 409, 412, 413, 414, 415, 417, 418, 419, 420, 421, 426, 427, 430, 431, 432, 435, 437

Nuclear Fuel, 5, 7, 12, 13, 15, 24, 25, 30, 31, 32, 34, 35, 36, 37, 38, 39, 40, 41, 44, 46, 48, 51, 52, 53, 58, 61, 66, 68, 70, 71, 73, 75, 76, 77, 78, 79, 80, 81, 82, 83, 84, 85, 87, 88, 89, 91, 92, 96, 98, 100, 104, 105, 115, 117, 121, 128, 130, 131, 165, 166, 167, 168, 171, 173, 174, 175, 176, 178, 179, 181, 182, 183, 184, 185, 186, 187, 189, 190, 192, 196, 206, 208, 210, 212, 217, 219, 221, 233, 241, 250, 253, 264, 270, 273, 281, 299, 308, 358, 367, 368, 372, 373, 375, 376, 377, 379, 380, 384, 385, 386, 388, 399, 400, 404, 406, 409, 410, 412, 413, 414, 415, 417, 418, 420, 436

Nuclear Fuel Cycle, 31, 40, 48, 53, 75, 80, 82, 84, 165, 167, 174, 179, 206, 219, 253, 409, 414, 436

Nuclear Reactor, 1, 5, 6, 7, 10, 12, 13, 14, 15, 16, 17, 19, 23, 24, 25, 26, 27, 28, 29, 30, 31, 32, 33, 34, 35, 36, 37, 38, 39, 40, 41, 43, 44, 45, 46, 48, 49, 50, 52, 56, 61, 62, 63, 64, 65, 66, 68, 69, 70, 71, 72, 78, 79, 80, 81, 82, 83, 84, 85, 86, 87, 88, 90, 91, 92, 93, 95, 96, 97, 98, 99, 100, 101, 102, 104, 107, 108, 112, 115, 116, 118, 119, 122, 123, 124, 125, 127, 128, 131, 132, 133, 135, 137, 138, 140, 141, 142, 144, 160, 162, 165, 167, 168, 169, 170, 171, 172, 173, 174, 175, 176, 177, 178, 180, 181, 182, 183, 184, 185, 186, 187, 188, 189, 190, 191, 192, 193, 194, 197, 198, 199, 200, 201, 202, 204, 205, 206, 207, 208, 209, 210, 211, 214, 216, 220, 221, 222, 226, 227, 228, 229, 230,

233, 239, 242, 243, 244, 247, 248,
251, 252, 253, 255, 257, 259, 262,
263, 264, 265, 266, 267, 268, 269,
271, 272, 276, 278, 279, 283, 284,
287, 291, 292, 293, 294, 296, 297,
298, 299, 300, 302, 303, 304, 305,
306, 307, 308, 310, 311, 312, 317,
318, 319, 320, 321, 322, 323, 324,
325, 326, 327, 328, 329, 330, 331,
332, 333, 334, 335, 336, 337, 338,
339, 340, 341, 342, 348, 349, 350,
351, 353, 355, 357, 358, 360, 361,
362, 363, 364, 365, 366, 368, 370,
371, 372, 373, 374, 375, 376, 377,
381, 386, 389, 401, 410, 412, 413,
426, 427, 436

Nuclear Waste, 13, 22, 30, 34, 46, 48,
50, 51, 53, 73, 75, 80, 82, 83, 84, 88,
90, 91, 92, 96, 98, 99, 101, 108, 110,
121, 122, 136, 154, 156, 157, 158,
165, 171, 173, 174, 175, 176, 195,
197, 206, 212, 214, 217, 221, 224,
231, 241, 246, 250, 263, 264, 268,
270, 273, 274, 277, 278, 281, 284,
285, 287, 288, 290, 291, 293, 294,
297, 301, 311, 353, 363, 364, 367,
368, 375, 376, 377, 378, 379, 380,
384, 385, 386, 392, 393, 395, 397,
405, 406, 409, 412, 413, 414, 418,
419, 421, 437

Operational Efficiency, 5, 10, 12, 14,
16, 17, 18, 27, 34, 36, 38, 39, 46, 48,
66, 71, 73, 74, 82, 83, 84, 85, 86, 89,
90, 91, 92, 93, 97, 98, 99, 100, 105,
106, 107, 111, 117, 120, 122, 125,
127, 130, 134, 135, 146, 147, 152,
167, 172, 177, 179, 185, 194, 202,
205, 207, 210, 213, 216, 220, 223,
227, 228, 231, 233, 234, 239, 242,
243, 245, 246, 248, 250, 253, 255,
258, 259, 263, 267, 269, 270, 272,
273, 274, 275, 276, 277, 279, 280,

282, 284, 285, 286, 287, 288, 289,
290, 292, 294, 295, 296, 297, 299,
305, 306, 309, 324, 330, 331, 333,
336, 339, 341, 342, 344, 350, 353,
359, 364, 365, 366, 367, 368, 372,
374, 375, 381, 386, 387, 393, 403,
404, 407, 408, 412, 415, 418, 420,
421

Partnerships, Public-private, 351, 389

Passive Cooling Systems, 6, 15, 22, 27,
43, 79, 82, 105, 106, 107, 121, 170,
172, 183, 184, 186, 188, 189, 190,
191, 202, 206, 207, 211, 221, 226,
245, 255, 259, 293, 334, 339, 357,
358, 360, 364, 372, 374, 375, 381,
385, 386, 414

Perception, Public, 230, 264

Phased Commissioning, 69, 71, 211,
245, 252, 255, 258, 259, 269, 277,
280, 292, 299, 300, 301, 302, 303,
304, 305, 306, 309, 310, 311, 328,
329, 330, 331, 332, 339, 341, 343,
348, 349, 355, 390, 393, 394, 397,
399, 402, 404, 405

Pollution Control, 102, 120, 121, 179,
185, 278, 290, 338, 360, 368, 413

Power Generation, 2, 4, 5, 6, 7, 8, 11,
12, 14, 16, 17, 21, 23, 24, 25, 28, 30,
31, 32, 34, 35, 36, 38, 39, 40, 41, 45,
51, 52, 65, 66, 69, 70, 71, 72, 73, 75,
76, 79, 80, 81, 82, 85, 86, 87, 88, 92,
96, 97, 109, 115, 117, 122, 124, 125,
126, 132, 160, 164, 166, 167,
171,172, 174, 178, 180, 190, 197,
201, 203, 204, 205, 206, 208, 209,
210, 222, 230, 233, 261, 263, 283,
304, 309, 311, 314, 315, 327, 332,
350, 357, 362, 363, 364, 367, 370,
371, 372, 373, 375, 376, 380, 384,
385, 388, 397, 399, 400, 403, 404,
405, 407, 410, 412, 413, 414, 415,
416, 427, 428

Private Investment, 1, 2, 18, 22, 24, 25, 40, 74, 75, 89, 104, 125, 126, 137, 148, 158, 203, 207, 232, 352, 353, 354, 383, 384, 386, 387, 389, 390, 391, 394, 397, 399, 402, 403, 404, 405, 407, 408, 411, 414, 415, 419, 422, 426

Process, Licensing, 242

Project Financing, 28, 74, 201, 203, 207, 306, 311, 313, 384, 389, 390, 391, 394, 398, 399, 400, 403, 404, 405

Public Consultation, 101, 135, 154, 159, 231, 248, 261

Public Education, 111, 155, 158, 351, 386, 387, 406

Public Perception, 51, 99, 110, 131, 158, 163, 260, 261, 264, 405, 406

Radiation Protection, 8, 13, 14, 15, 26, 27, 34, 44, 45, 54, 59, 61, 85, 90, 91, 92, 95, 96, 97, 98, 99, 100, 101, 106, 110, 121, 126, 154, 157, 162, 165, 168, 169, 172, 173, 174, 176, 181, 189, 190, 192, 202, 211, 214, 224, 239, 240, 241, 248, 250, 252, 254, 255, 257, 259, 261, 262, 263, 264, 273, 274, 277, 280, 281, 283, 285, 288, 289, 293, 295, 319, 323, 325, 333, 335, 338, 339, 342, 343, 367, 374, 376, 377, 378, 380, 412

Radioactive Isotopes, 12, 46, 54, 60, 64, 174, 192, 208, 367, 412, 413

Reactor Containment, 6, 7, 8, 13, 14, 15, 16, 22, 25, 26, 27, 30, 34, 36, 38, 42, 43, 48, 49, 51, 52, 79, 80, 84, 90, 93, 96, 97, 99, 101, 102, 105, 106, 107, 112, 121, 142, 146, 168, 169, 175, 176, 178, 181, 183, 186, 187, 188, 189, 190, 192, 199, 202, 208, 211, 228, 229, 239, 244, 247, 248, 255, 259, 278, 281, 283, 292, 293, 296, 307, 318, 319, 320, 328, 331, 333, 335, 337, 338, 341, 358, 360, 361, 366, 367, 374, 376, 377, 378, 380, 381, 436

Reactor design, 37, 87

Reactors, Molten salt, 124, 130

Recycling, Fuel, 412

Regulatory Approvals, 24, 46, 75, 89, 100, 101, 108, 110, 111, 113, 123, 126, 127, 129, 135, 136, 145, 147, 150, 156, 159, 166, 200, 206, 208, 209, 221, 228, 229, 233, 243, 244, 246, 248, 249, 251, 255, 257, 258, 259, 261, 266, 267, 268, 288, 292, 295, 300, 303, 305, 307, 308, 309, 311, 325, 333, 354, 390, 398, 405

Regulatory Compliance, 52, 98, 106, 109, 110, 113, 125, 126, 128, 129, 134, 136, 137, 159, 166, 209, 213, 214, 218, 219, 222, 223, 224, 225, 226, 230, 240, 242, 243, 244, 246, 249, 250, 252, 253, 256, 257, 258, 263, 264, 265, 270, 271, 272, 273, 274, 275, 276, 277, 278, 279, 280, 281, 282, 283, 284, 285, 286, 287, 288, 289, 290, 291, 292, 294, 295, 297, 300, 319, 333, 340, 341, 342, 355, 360, 368

Research and Development, 2, 30, 31, 34, 51, 55, 56, 57, 58, 59, 60, 61, 62, 63, 64, 65, 66, 70, 71, 77, 83, 85, 124, 126, 130, 160, 194, 196, 197, 213, 250, 253, 261, 295, 303, 304, 305, 350, 351, 359

Risk Management, 232, 282, 286

Russia, 30, 43, 44, 49, 56, 57, 58, 70, 71, 78, 82, 83, 88, 126, 127, 136, 161, 206, 224, 225, 256, 270, 292, 293, 294, 302, 305, 337, 349, 370, 373, 376

Safety Features, 4, 5, 6, 14, 16, 22, 23, 27, 30, 40, 43, 52, 54, 69, 72, 74, 76, 79, 81, 83, 87, 89, 91, 99, 100, 105,

106, 107, 110, 112, 121, 124, 127, 128, 129, 130, 131, 132, 134, 135, 137, 138, 141, 154, 167, 169, 170, 173, 175, 180, 181, 182, 186, 189, 191, 192, 200, 201, 202, 205, 207, 209, 219, 220, 221, 223, 225, 227, 233, 244, 245, 247, 251, 254, 255, 256, 259, 262, 292, 293, 311, 324, 335, 341, 359, 372, 375, 381, 385, 386, 406, 409, 414

Safety Systems, 1, 2, 4, 5, 6, 7, 13, 14, 15, 16, 17, 22, 23, 24, 25, 27, 29, 30, 31, 32, 34, 38, 39, 40, 41, 43, 44, 45, 46, 48, 51, 52, 53, 54, 69, 72, 73, 74, 75, 76, 78, 79, 80, 81, 82, 83, 84, 87, 89, 90, 91, 92, 93, 95, 97, 99, 100, 101, 102, 105, 106, 107, 108, 110, 111, 112, 113, 114, 115, 117, 118, 119, 120, 121, 122, 123, 124, 125, 126, 127, 128, 129, 130, 131, 132, 133, 134, 135, 136, 137, 138, 140, 141, 142, 145, 146, 148, 149, 150, 151, 152, 153, 154, 155, 156, 157, 158, 160, 162, 164, 165, 166, 167, 168, 169, 170, 172, 173, 175, 176, 177, 178, 180, 181, 182, 183, 184, 185, 186, 187, 188, 189, 190, 191, 192, 193, 194, 195, 196, 197, 198, 199, 200, 201, 202, 204, 205, 206, 207, 208, 209, 210, 211, 214, 216, 217, 218, 219, 220, 221, 222, 223, 224, 225, 226, 227, 228, 229, 230, 231, 232, 233, 234, 239, 240, 241, 242, 243, 244, 245, 246, 247, 248, 249, 250, 251, 252, 253, 254, 255, 256, 257, 258, 259, 260, 261, 262, 263, 264, 265, 266, 267, 268, 269, 270, 271, 272, 273, 274, 275, 276, 277, 278, 279, 280, 281, 282, 283, 284, 285, 286, 287, 288, 289, 290, 291, 292, 293, 294, 295, 296, 297, 298, 299, 300, 301, 303, 304, 306,

308, 309, 310, 311, 318, 319, 320, 321, 322, 323, 324, 325, 328, 329, 330, 331, 332, 333, 334, 335, 336, 337, 338, 339, 340, 341, 342, 343, 344, 348, 349, 350, 351, 355, 356, 357, 358, 359, 360, 361, 364, 365, 366, 367, 368, 370, 371, 372, 373, 374, 375, 380, 381, 382, 383, 384, 385, 386, 387, 388, 390, 398, 403, 405, 406, 409, 414, 418, 419, 420

Scalability, 8, 16, 21, 25, 27, 28, 29, 40, 72, 74, 99, 105, 127, 129, 134, 167, 168, 177, 197, 200, 201, 254, 322, 362, 383, 388, 408, 414, 422

Seawater Desalination, 2, 16, 73, 75, 76, 78, 82, 86, 87, 88, 106, 115, 166, 206, 209, 363, 364, 405, 406, 410, 411, 414

Skilled Labor, 195, 403, 405

Small Modular Reactors, 2, 5, 6, 7, 8, 14, 15, 16, 17, 18, 20, 21, 22, 23, 25, 26, 27, 28, 31, 32, 34, 39, 46, 50, 51, 73, 75, 77, 78, 79, 86, 87, 89, 90, 91, 92, 93, 95, 96, 97, 98, 99, 100, 101, 102, 103, 104, 105, 106, 107, 108, 109, 110, 111, 112, 113, 114, 115, 116, 117, 121, 122, 123, 124, 125, 126, 127, 128, 132, 133, 134, 135, 136, 137, 138, 139, 141, 142, 144, 145, 146, 147, 150, 153, 154, 155, 157, 158, 159, 160, 161, 162, 163, 164, 166, 167, 168, 169, 171, 172, 173, 174, 175, 176, 177, 178, 180, 181, 182, 183, 184, 185, 186, 189, 190, 191, 192, 193, 194, 196, 197, 198, 199, 200, 201, 203, 204, 205, 206, 207, 208, 209, 210, 211, 220, 221, 222, 223, 224, 225, 226, 227, 228, 229, 230, 233, 234, 241, 242, 244, 245, 246, 247, 248, 249, 250, 251, 252, 253, 254, 255, 256, 258, 259, 260, 262, 263, 265, 267, 268,

269, 270, 272, 273, 274, 275, 276,
277, 278, 279, 280, 281, 282, 283,
284, 285, 286, 287, 288, 290, 291,
292, 293, 294, 295, 296, 297, 298,
299, 300, 301, 302, 303, 305, 306,
309, 310, 311, 312, 314, 315, 316,
317, 318, 319, 320, 321, 322, 323,
324, 325, 326, 327, 330, 332, 333,
334, 335, 336, 337, 338, 339, 340,
343, 344, 348, 349, 350, 351, 352,
353, 354, 355, 357, 359, 360, 361,
362, 363, 364, 365, 366, 367, 368,
369, 370, 371, 372, 373, 374, 375,
377, 379, 380, 381, 383, 384, 385,
386, 387, 388, 389, 390, 393, 394,
397, 398, 399, 400, 401, 403, 404,
405, 406, 407, 408, 411, 412, 413,
414, 416, 417, 418, 419, 420, 421,
422, 423, 425, 426, 427, 428, 429,
430, 435, 436, 438

Small Modular Reactors (SMRs), 2, 4, 5,
6, 14, 16, 19, 21, 23, 25, 28, 31, 53,
72, 73, 74, 75, 102, 108, 110, 117,
122, 125, 126, 129, 131, 132, 134,
138, 159, 167, 168, 169, 172, 173,
177, 178, 180, 182, 183, 184, 186,
188, 189, 191, 192, 193, 195, 197,
199, 200, 201, 202, 203, 205, 206,
210, 219, 222, 223, 227, 228, 231,
232, 233, 234, 239, 240, 242, 243,
246, 249, 253, 256, 260, 265, 288,
292, 295, 300, 302, 304, 314, 317,
319, 320, 322, 323, 324, 325, 326,
327, 330, 359, 360, 361, 363, 367,
369, 370, 371, 373, 375, 380, 383,
387, 399, 404, 407, 409, 410, 411,
415, 423, 424, 425, 426, 429, 430,
432, 433, 434, 435, 436, 437, 438

Stability, 2, 10, 17, 22, 24, 30, 42, 73,
83, 84, 85, 96, 101, 102, 115, 119,
120, 137, 138, 139, 140, 141, 142,
143, 145, 146, 147, 150, 159, 163,

167, 168, 184, 192, 247, 258, 269,
275, 283, 310, 315, 317, 332, 333,
358, 366, 367, 369, 370, 371, 372,
373, 374, 387, 388, 389, 398, 402,
403, 409, 411, 412, 413, 415, 418,
419, 421

Storage Facilities, 75, 91, 96, 212, 368,
380, 417

Supply Chain, 121, 203, 224, 256, 260,
263, 298, 299, 303, 348, 351, 353,
354, 356, 387, 399, 404, 406, 436

Sustainability Goals, 2, 5, 7, 44, 51, 53,
80, 81, 101, 102, 117, 120, 130, 132,
173, 195, 201, 202, 203, 204, 205,
206, 257, 264, 275, 279, 286, 290,
358, 364, 367, 388, 400, 410, 411,
414, 419

Sweden, 115, 413

Synchronization, Grid, 339

Technological Innovation, 14, 28, 75,
78, 79, 125, 126, 133, 179, 192, 205,
219, 246, 249, 250, 260, 321, 348,
351, 395, 408

Thermal Storage, 364

Thorium Fuel, 25, 30, 31, 40, 80, 81, 83,
84, 85, 130, 165, 167, 174, 210, 264,
412, 413

United Kingdom, 31, 35, 36, 37, 38, 68,
70, 77, 82, 87, 114, 124, 125, 135,
136, 161, 194, 222, 224, 251, 252,
261, 270, 274, 282, 283, 284, 285,
300, 309, 313, 320, 321, 348, 349,
350, 351, 352, 389, 399, 402, 403,
407, 411, 413, 414, 427, 432, 434,
435

United States of America, 7, 30, 31, 33,
34, 52, 56, 59, 60, 61, 62, 63, 65, 66,
68, 69, 70, 71, 77, 81, 83, 87, 114,
122, 135, 159, 160, 218, 223, 228,
229, 234, 242, 243, 246, 249, 256,
261, 263, 269, 270, 302, 309, 348,

350, 376, 389, 399, 402, 407, 411,
413, 423
Uranium Fuel, 5, 7, 8, 9, 10, 11, 12, 13,
25, 30, 31, 32, 34, 35, 36, 37, 38, 40,
48, 54, 55, 56, 58, 59, 60, 61, 63, 64,
66, 68, 69, 71, 75, 77, 78, 79, 80, 81,
82, 83, 84, 85, 115, 121, 165, 167,
168, 174, 189, 210, 376, 412
Waste Disposal, 22, 174, 222, 287, 288,
414
Waste management, 353

Waste, High-level, 379, 384
Waste, Intermediate-level, 378
Water Desalination, 2, 16, 73, 75, 76,
78, 82, 86, 87, 88, 106, 115, 166, 206,
209, 363, 364, 405, 406, 410, 411,
414
Workforce Training, 24, 77, 96, 99, 113,
155, 157, 178, 263, 309, 343, 344,
350, 351, 352, 353, 355, 380, 381,
399, 406, 418, 436

Made in United States
Troutdale, OR
12/22/2024

27136658R00252